Seismic Reservoir Modeling

Seismic Reservoir Modeling

Theory, Examples, and Algorithms

Dario Grana
University of Wyoming
Laramie, WY, USA

Tapan Mukerji
Stanford University
Stanford, CA, USA

Philippe Doyen
Independent Consultant
London, UK

WILEY Blackwell

This edition first published 2021

Registered Offices
John Wiley & Sons, Inc., 111 River Street, Hoboken, NJ 07030, USA
John Wiley & Sons Ltd, The Atrium, Southern Gate, Chichester, West Sussex, PO19 8SQ, UK

Editorial Office
9600 Garsington Road, Oxford, OX4 2DQ, UK

For details of our global editorial offices, customer services, and more information about Wiley products visit us at www.wiley.com.

Wiley also publishes its books in a variety of electronic formats and by print-on-demand. Some content that appears in standard print versions of this book may not be available in other formats.

Library of Congress Cataloging-in-Publication Data applied for

ISBN 9781119086185 (Hardback)

Cover Design: Wiley
Cover Images: © Courtesy of Dario Grana

Set in 9.5/12.5pt STIXTwoText by SPi Global, Pondicherry, India

C9781119086185_290321

Printed and bound by CPI Group (UK) Ltd, Croydon CR0 4YY

To our loved families

Contents

Preface

Geophysical data are commonly acquired to understand the structure of the subsurface for a variety of applications, such as energy resource exploration, natural hazards, groundwater, and surface processes. Seismic reservoir modeling is an interdisciplinary field that integrates physics, mathematics, geology, statistics, and computer science. The goal of seismic reservoir characterization studies is to predict three-dimensional models of rock and fluid properties in the subsurface based on the available geophysical data, such as borehole measurements and surface geophysical surveys. Mathematical methods are required to solve the so-called inverse problem in which the properties of interest are estimated from their geophysical response. The solution to this problem is generally non-unique owing to the data errors, their limited resolution, the physics approximations, and the natural variability and heterogeneity of porous rocks. Multiple model realizations of rock and fluid properties, constrained by geophysical measurements and prior geological information, can be generated by combining statistical methods and computer science algorithms.

The main goal of this book is to bring together in one place basic and advanced methodologies of the current state of the art in seismic reservoir characterization. This work finds inspiration in the book *Seismic Reservoir Characterization* by Philippe Doyen. For the rock physics part, it strongly relies on *The Rock Physics Handbook* by Gary Mavko, Tapan Mukerji, and Jack Dvorkin, whereas for the geostatistics part, it relies on *Geostatistical Reservoir Modeling* by Michael Pyrcz and Clayton Deutsch, and *Multiple-Point Geostatistics* by Gregoire Mariethoz and Jef Caers. Unlike other textbooks on seismic reservoir modeling, this book offers a detailed description of the mathematical–physical methods used for quantitative seismic interpretation. Indeed, it focuses on mathematical methods for the estimation of petrophysical properties, such as porosity, mineral volumes, and fluid saturations from geophysical data and attributes. Owing to the non-uniqueness of the solution, we present a set of probabilistic methods that aim to estimate the most likely model as well as the uncertainty associated with the predictions. The model uncertainty can be quantified using probability distributions, confidence intervals, or a set of model realizations.

Chapter 1 reviews the main mathematical and statistical concepts used in the methods proposed in this book. We first present a review of probability and statistics to familiarize the reader with the concept of probability distribution. We then introduce the notion of a mathematical inverse problem, where the model variables are estimated from a set of measurements, based on the physics operator that links the model variables to the measured data. The inverse problem is formulated in a Bayesian setting.

Chapter 2 includes an overview of rock physics models. A rock physics model is defined as one or multiple equations that compute the elastic response (P-wave and S-wave velocity and density) of a porous rock given its petrophysical properties (porosity, mineral volumes, and fluid saturations). Empirical and theoretical models are presented with examples in different geological environments.

Chapters 3 and 4 focus on geostatistical methods for spatial interpolation and simulation of multiple realizations of subsurface properties. Geostatistics is a branch of spatial statistics that aims to analyze and predict properties associated with spatiotemporal phenomena in geosciences. Spatial statistics notions and algorithms are commonly used in geoscience to mimic the spatial and temporal continuity of geological processes. Chapter 3 describes methods for the interpolation and stochastic simulation of continuous properties, such as petrophysical and elastic variables, whereas Chapter 4 extends these algorithms to discrete properties, such as rock types and facies. An example of application of geostatistical methods to elevation and temperature maps in the Yellowstone National Park is presented to illustrate the algorithms and compare the predictions with the exact measurements.

Chapter 5 summarizes the developments in seismic and petrophysical inverse problems of the past two decades. The chapter includes three main topics: Bayesian linearized AVO inversion, Bayesian rock physics inversion, and geostatistical inversion. Bayesian linearized AVO inversion is an elegant and efficient algorithm proposed by Arild Buland and Henning Omre for the prediction of elastic properties from measured seismic data. Bayesian rock physics inversion refers to a set of probabilistic methods for the prediction of petrophysical properties from elastic attributes, based on different statistical assumptions for the distribution of the model variables and different linear or non-linear rock physics models. Geostatistical inversion methods include multiple stochastic algorithms that aim to predict petrophysical properties from measured seismic data, by iteratively perturbing and updating an initial realization or a set of realizations. All the methodologies are illustrated through one-dimensional examples based on borehole data.

Chapter 6 extends the Bayesian inversion methodology to facies classification and to joint inversion of facies and petrophysical properties from seismic data. This chapter discusses traditional Bayesian facies classification methods based on seismic data and seismically derived attributes, and introduces recent advances in stochastic sampling methods for the joint prediction of facies and reservoir properties, integrating Bayesian inverse theory and geostatistical algorithms.

Chapter 7 introduces additional sources of uncertainty associated with data processing, natural heterogeneity, and geological interpretation, and elaborates on the integration of seismic reservoir characterization methods in other domains of reservoir modeling. The chapter discusses the application of seismic and petrophysical inversion methods to time-lapse geophysical data, the use of different geophysical datasets, such as electromagnetic data, and the assimilation of seismic data in fluid flow simulation for updating the reservoir model in monitoring studies. These topics introduce recent research challenges and directions. Probabilistic model predictions are also used in decision-making studies associated with the value of information of geophysical data.

Chapter 8 presents several case studies previously published in peer-reviewed journals. The applications include two hydrocarbon reservoirs in the Norwegian Sea, a carbonate field offshore Brazil, and a CO_2 sequestration study offshore Norway.

The Appendix contains the description of the Seismic Reservoir Modeling (SeReM) MATLAB package including codes for rock physics, geostatistics, inversion, and facies modeling. The Matlab package SeReM and the Python version SeReMpy are available at the following link: https://seismicreservoirmodeling.github.io/SeReM/.

Acknowledgments

We completed our education as PhD students of the Stanford Rock Physics and Borehole Geophysics Project at Stanford University, under the supervision of Professor Amos Nur and Professor Gary Mavko. Amos and Gary continuously promoted innovations in theoretical and experimental rock physics for geophysical studies. We have also collaborated with the Stanford Center for Reservoir Forecasting group founded by Professor André Journel, who pioneered innovations in geostatistics theory and applications. Amos, Gary, and André's knowledge and mentorship have contributed to our professional growth as geoscientists.

We acknowledge our colleagues Jack Dvorkin, Henning Omre, Jo Eidsvik, Jef Caers, Per Avseth, Patrick Connolly, Brian Russell, Subhashis Mallick, Ezequiel González, Mita Sengupta, Lucy MacGregor, Alessandro Amato del Monte, and Leonardo Azevedo, as well as our collaborators in academia and industry for the constructive discussions and their help throughout these years. A special acknowledgment goes to Leandro de Figueiredo and Mingliang Liu for their help with the examples and Ernesto Della Rossa for the meticulous review.

The examples included in this book have been made possible by the availability of data and open source software. We would like to thank Equinor (operator of the Norne field) and its license partners Eni and Petoro for the release of the Norne data and the Center for Integrated Operations at NTNU for cooperation and coordination of the Norne case. We also thank SINTEF for providing the MATLAB Reservoir Simulation Toolbox. We thank the Society of Exploration Geophysicists for the permission to include some examples originally published in *Geophysics*.

We acknowledge the School of Energy Resources of the University of Wyoming and the Nielson Energy Fellowship, the sponsors of the Stanford Center for Earth Resources Forecasting, the Stanford Rock Physics and Borehole Geophysics Project, and the Dean of the Stanford School of Earth, Energy, and Environmental Sciences, Professor Steve Graham, for their continued support.

We want to thank all our co-authors of previous publications, especially the MSc and PhD students that contributed to the recent developments in our research. This book would not have been possible without the contribution of numerous students, colleagues, and friends who have constantly inspired and motivated us. Finally, we thank our families for their love, encouragement, and support.

1

Review of Probability and Statistics

Statistics and probability notions and methods are commonly used in geophysics studies to describe the uncertainty in the data, model variables, and model predictions. Statistics and probability are two branches of mathematics that are often used together in applied science to estimate parameters and predict the most probable outcome of a physical model as well as its uncertainty. Statistical methods aim to build numerical models for variables whose values are uncertain (e.g. seismic velocities or porosity in the subsurface) from measurements of observable data (e.g. measurements of rock properties in core samples and boreholes). Probability is then used to make predictions about unknown events (e.g. porosity value at a new location) based on the statistical models for uncertain variables. In reservoir modeling, for example, we can use statistics to create multiple reservoir models of porosity and water saturation using direct measurements at the well locations and indirect measurements provided by geophysical data, and then apply probability concepts and tools to make predictions about the total volume of hydrocarbon or water in the reservoir. The predictions are generally expressed in the form of a probability distribution or a set of statistical estimators such as the most-likely value and its variability. For example, the total fluid volume can be described by a Gaussian distribution that is completely defined by two parameters, the mean and the variance, that represent the most-likely value and the uncertainty of the property prediction, respectively. Probability and statistics have a vast literature (Papoulis and Pillai 2002), and it is not the intent here to do a comprehensive review. Our goal in this chapter is to review some basic concepts and establish the notation and terminology that will be used in the following chapters.

1.1 Introduction to Probability and Statistics

The basic concept that differentiates statistics and probability from other branches of mathematics is the notion of the random variable. A random variable is a mathematical variable such that the outcome is unknown but the likelihood of each of the possible outcomes is known. For example, the value of the P-wave velocity at a given location in the reservoir might be unknown owing to the lack of direct measurements; however, the available data might suggest that velocity is likely to be between 2 and 6 km/s with an expected value of 4 km/s. We model our lack of knowledge about the P-wave velocity by describing it as a

Seismic Reservoir Modeling: Theory, Examples, and Algorithms, First Edition. Dario Grana, Tapan Mukerji, and Philippe Doyen.

random variable. The expected value is the mean of the random variable and the lower and upper limits of the confidence interval represent the range of its variability. Any decision-making process involving random variables in the subsurface should account for the uncertainty in the predictions, because the predicted value, for example the mean of the random variable, is not necessarily the true value of the property and its accuracy depends on the uncertainty of the measurements, the approximations in the physical models, and the initial assumptions. All these concepts can be formalized using statistics and probability definitions.

In probability and statistics, we view a problem involving random variables as an experiment (i.e. a statistical experiment) where the variable of interest can take different possible values and we aim to predict the outcome of the experiment. We can formulate the main notions of probability using set theory. In Kolmogorov's formulation (Papoulis and Pillai 2002), sets are interpreted as events and probability is a mathematical measure on a class of sets. The sample space S is the collection of all the possible outcomes of the experiment. In reservoir modeling studies, an example of sample space could be the set of all possible reservoir models of porosity generated using a geostatistical method (Chapter 3). An event E is a subset of the sample space. For example, an event E could represent all the reservoir models with average porosity less than 0.20.

If the sample space is large enough, we can use a frequency-based approach to estimate the probability of an event E. In this setting, we can define the probability of an event E as the number of favorable outcomes divided by the total number of outcomes. In other words, the probability of E is the cardinality of E (i.e. the number of elements in the set E) divided by the cardinality of the sample space S (i.e. the number of elements in the set S). In our example, the probability of a reservoir model having an average porosity lower than 0.20 can be computed as the number of models with average porosity less than 0.20, divided by the total number of reservoir models. For instance, if the sample space includes 1000 geostatistical models of porosity and the event E includes 230 models, then the probability $P(E)$ that a reservoir model has average porosity less than 0.20 is $P(E) = 230/1000 = 0.23$.

In general, there are two main interpretations of probability: the frequentist approach and the Bayesian approach. The frequentist approach is based on the concept of randomness and this interpretation is related to experiments dealing with aleatory uncertainty owing to natural variations. Statistical events associated with tossing a coin or rolling a die can be described using the frequentist approach, since the outcomes of these events can be investigated by repeating the same experiment several times and studying the frequency of the outcomes. The Bayesian approach focuses on the concept of uncertainty and this interpretation is common for epistemic uncertainty owing to the lack of knowledge. Statistical events associated with porosity or P-wave velocity in the subsurface are often described using the Bayesian approach, because it is not possible to collect enough data or have a large number of controlled identical experiments in geology.

In geophysical modeling, we often quantify uncertainty using different statistics and probability tools, such as probability distributions, statistical estimators, and geostatistical realizations. For example, the uncertainty associated with the prediction of porosity and fluid saturation from seismic data can be represented by the joint probability distributions of porosity and fluid saturation at each location in the reservoir, by a set of statistical estimations such as the mean, the maximum a posteriori estimate, the confidence interval, and the variance, or by an ensemble of multiple realizations obtained by sampling the probability distribution. In general, it is always possible to build the most-likely model of the properties of interest from these statistics and probability tools and present the solution in a deterministic form. For example, we can compute the most-likely value of the probability distribution of porosity and fluid saturation at each location in the reservoir. However, subsurface models are often highly uncertain owing to the lack of direct measurements, the limited quality and resolution of the available geophysical data, the approximations in the physical models, and the natural variability and heterogeneity of subsurface rocks. Therefore, the uncertainty of the predictions should always be considered in any decision-making process associated with subsurface models.

In this chapter, we review the main concepts of probability and statistics. These results are used in the following chapters to build mathematical methodologies for reservoir modeling, such as geostatistical simulations and inverse methods.

1.2 Probability

In this review, E represents a generic event and $P(E)$ represents its probability. For example, E might represent the occurrence of hydrocarbon sand in a reservoir and $P(E)$ the probability of finding hydrocarbon sand at a given location.

Probability theory is based on three axioms:

1) The probability $P(E)$ of an event E is a real number in the interval [0, 1]:

$$0 \leq P(E) \leq 1. \tag{1.1}$$

2) The probability $P(S)$ of the sample space S is 1:

$$P(S) = 1. \tag{1.2}$$

3) If two events, E_1 and E_2, are mutually exclusive, i.e. $E_1 \cap E_2 = \emptyset$ (in other words, if the event E_1 occurs, then the event E_2 does not occur and if the event E_2 occurs, then the event E_1 does not occur), then the probability of the union $E_1 \cup E_2$ of the two events, i.e. the probability of one of two events occurring, is the sum of the probabilities of the two events $P(E_1)$ and $P(E_2)$:

$$P(E_1 \cup E_2) = P(E_1) + P(E_2). \tag{1.3}$$

The axioms in Eqs. (1.1)–(1.3) are the foundations of probability theory. Based on these axioms, the probability of the complementary event $\overline{E} = S \setminus E$, i.e. the probability that the event E does not occur, is given by:

$$P(\overline{E}) = 1 - P(E), \tag{1.4}$$

since $P(E) + P(\overline{E}) = P(S) = 1$. From the axioms, we can also derive that the probability of the union of two generic events, not necessarily mutually exclusive, is:

$$P(E_1 \cup E_2) = P(E_1) + P(E_2) - P(E_1 \cap E_2), \tag{1.5}$$

where $P(E_1 \cap E_2)$ is the probability that both events occur. For mutually exclusive events, the intersection of the two events is the empty set ($E_1 \cap E_2 = \emptyset$), and the probability of the intersection is $P(E_1 \cap E_2) = 0$; therefore, Eq. (1.5) reduces to the third axiom

(Eq. 1.3). The result in Eq. (1.5) can be proven using set theory and the three axioms. The probability $P(E_1 \cap E_2)$ is generally called joint probability and it is often indicated as $P(E_1, E_2)$.

A fundamental concept in probability and statistics is the definition of conditional probability, which describes the probability of an event based on the outcome of another event. In general, the probability of an event E can be defined more precisely if additional information related to the event is available. For example, seismic velocity depends on a number of factors, such as porosity, mineral volumes, and fluid saturations. If one of these factors is known, then the probability of seismic velocity can be estimated more precisely. For instance, if the average porosity of the reservoir is 0.30, it is more likely that the seismic velocity will be relatively low, whereas if the average porosity of the reservoir is 0.05, then it is more likely that the seismic velocity will be relatively high. This idea can be formalized using the concept of conditional probability. Given two events A and B, the conditional probability $P(A|B)$ is defined as:

$$P(A|B) = \frac{P(A, B)}{P(B)}. \tag{1.6}$$

Two events A and B are said to be independent if the joint probability $P(A, B)$ is the product of the probability of the events, i.e. $P(A, B) = P(A)P(B)$. Therefore, given two independent events A and B, the conditional probability $P(A|B)$ reduces to:

$$P(A|B) = P(A). \tag{1.7}$$

This means that the probability of A does not depend on the outcome of the event B. For a more detailed description of probability theory, we refer the reader to Papoulis and Pillai (2002).

In many practical applications, calculating the joint probability $P(A, B)$ in the definition of the conditional probability (Eq. 1.6) is prohibitive; for example, owing to the lack of data. A common tool to estimate conditional probabilities is Bayes' theorem. This result expresses the conditional probability $P(A|B)$ as a function of the conditional probability $P(B|A)$, which might be easier to calculate in some applications.

Bayes' theorem states that the conditional probability $P(A|B)$ is given by:

$$P(A|B) = \frac{P(B|A)P(A)}{P(B)} \propto P(B|A)P(A), \tag{1.8}$$

where $P(A)$ is the probability of A, $P(B|A)$ is the probability of the event B given the event A, and $P(B)$ is the probability of B. The term $P(A)$ is called the prior probability of A since it measures the probability before considering additional information associated with the event B. The term $P(B|A)$ is the likelihood function of the event B to be observed for each possible outcome of the event A. The term $P(B)$ represents the marginal probability of the event B and it is a normalizing constant to ensure that $P(A|B)$ satisfies the axioms. The resulting conditional probability $P(A|B)$ is also called the posterior probability of the event A, since it is computed based on the outcome of the event B.

In the Bayesian setting, the prior probability represents the prior knowledge about the event of interest and the likelihood function is the probabilistic formulation of the relation between the event of interest and the observable event (i.e. the data). The intuitive idea of Bayes' theorem is to reduce the uncertainty in the prior probability by integrating additional information from the data.

For example, in reservoir modeling applications, the event A might represent the occurrence of porous rocks with porosity higher than 0.30 and the event B might represent the measurements of seismic velocity. Typically, in geoscience studies, prior information about rock and fluid properties can be obtained from geological models, analogues, outcrops, or nearby fields, whereas the likelihood function can be calculated using geophysical models or estimated from available datasets. In the porosity–velocity example, the prior probability of the occurrence of high-porosity rocks in a given reservoir can be estimated from core samples or data from nearby fields, whereas the likelihood function can be computed using a rock physics model that predicts the velocity response in porous rocks (Chapter 2).

The calculation of the normalizing constant $P(B)$ in Bayes' theorem (Eq. 1.8) often requires additional probability tools. A useful result in probability theory is the theorem of total probability. We consider an event A and an ensemble of n events $\{E_i\}_{i=1,\dots,n}$, and we assume that the events $\{E_i\}$ are mutually exclusive ($E_i \cap E_j = \emptyset$, for all $i, j = 1, \dots, n$ with $i \neq j$) and collectively exhaustive ($\bigcup_{i=1}^{n} E_i = S$). These assumptions mean that every possible outcome of the event A is an outcome of one and only one of the events $\{E_i\}_{i=1,\dots,n}$. Then, the probability $P(A)$ of the event A is given by:

$$P(A) = \sum_{i=1}^{n} P(A|E_i)P(E_i). \tag{1.9}$$

The theorem of total probability (Eq. 1.9) expresses the probability of an outcome of an event that can be realized via several distinct events as the contribution of the conditional and marginal probabilities.

Example 1.1 We illustrate the application of Bayes' theorem and the theorem of total probability with an example related to the classification of porous rocks. A team of reservoir geologists classifies a group of porous rocks in a reservoir in three categories: shale, silt, and sand. The geologists are interested in rock samples with porosity greater than 0.20. The team wants to answer the following questions: (i) What is the probability of a rock having porosity greater than 0.20? (ii) If a core sample has porosity greater than 0.20, what is the probability that the rock sample is a sand? (iii) If a core sample has porosity less than or equal to 0.20, what is the probability that the rock sample is a shale?

Based on a nearby field, it is estimated that 20% of the reservoir rocks are shale, 50% are silt, and 30% are sand. If the event A represents the rock type occurrence, the available information can be written in terms of the prior probability $P(A)$ of the event A as:

$$P(A = shale) = 0.2;$$
$$P(A = silt) = 0.5;$$
$$P(A = sand) = 0.3.$$

The literature data for the geological environment under study indicate that the probability of observing porosity greater than 0.20 in that specific region is 0.05 in shale, 0.15 in silt, and 0.4 in sand. This information represents the likelihood function. If the event B represents the occurrence of rock samples having porosity greater than 0.20, then the available information can be written in terms of the conditional probability $P(B|A)$:

$$P(B|A = shale) = 0.05;$$

$$P(B|A = silt) = 0.15;$$
$$P(B|A = sand) = 0.4.$$

We first compute the probability $P(B)$ of observing high porosity in a rock sample, using the theorem of total probability (Eq. 1.9):

$$\begin{aligned} P(B) = {} & P(B|A = shale)P(A = shale) \\ & + P(B|A = silt)P(A = silt) \\ & + P(B|A = sand)P(A = sand) = 0.205. \end{aligned}$$

Then, we compute the posterior probability $P(A = sand|B)$ of a sample with porosity greater than 0.20 being a sand using Bayes' theorem (Eq. 1.8):

$$P(A = sand|B) = \frac{P(B|A = sand)P(A = sand)}{P(B)} = \frac{0.4 \times 0.3}{0.205} = 0.58,$$

where the normalizing constant $P(B)$ is computed using the theorem of total probability, as shown in question (i).

Finally, the probability $P(A = shale|\overline{B})$ of a sample with porosity less than or equal to 0.20 (event $\overline{B}$) being a shale can be computed in two steps. First, we compute the marginal probability of the complementary event $P(\overline{B})$ as:

$$P(\overline{B}) = 1 - P(B) = 0.795;$$

then we compute the conditional probability $P(A = shale|\overline{B})$ using Bayes' theorem (Eq. 1.8):

$$P(A = shale|\overline{B}) = \frac{P(\overline{B}|A = shale)P(A = shale)}{P(\overline{B})} = \frac{0.95 \times 0.2}{0.795} = 0.23.$$

1.3 Statistics

We now consider a variable X, whose value is uncertain. In statistics, such a variable is called a random variable, or a stochastic variable. A random variable is a variable whose value is subject to variations and cannot be deterministically assessed. In other words, the specific value cannot be predicted with certainty before an experiment. Random variables can take discrete or continuous values. In geophysics, a number of subsurface properties, such as facies types, porosity, or P-wave velocity, are considered random variables, because they cannot be exactly measured. Direct measurements are also uncertain, and the measurements should be treated as random variables with a distribution that captures the measurement uncertainty.

1.3.1 Univariate Distributions

We first discuss discrete random variables and then generalize to continuous random variables. A discrete random variable is a variable that can only take a finite number of values (i.e. values in a set N of finite cardinality). For example, tossing a coin or rolling a die are examples of experiments involving discrete random variables. When tossing a coin, the outcome of the random variable is either heads or tails, whereas, when rolling a die, the outcome is a positive integer from 1 to 6. Facies, flow units, or rock types are examples of discrete random variables in subsurface modeling. These variables are often called categorical random variables to indicate that the outcomes of the random variable have no intrinsic order.

Although the outcome of a discrete random variable is uncertain, we can generally assess the probability of each outcome. In other words, we can define the probability of a discrete random variable X by introducing a function $p_X : N \to [0, 1]$, where the probability $P(x)$ of an outcome x is given by the value of the function p_X, i.e. $P(x) = p_X(X = x)$. The uppercase symbol generally represents the random variable, whereas the lowercase symbol represents the specific outcome. The function p_X is called probability mass function and it has the following properties:

$$0 \le p_X(x) \le 1, \tag{1.10}$$

for all outcomes $x \in N$; and

$$\sum_{x \in N} p_X(x) = 1. \tag{1.11}$$

Equations (1.10) and (1.11) imply that the probability of each outcome is a real number in the interval $[0, 1]$ and that the sum of the probabilities of all the possible outcomes is 1. The probability mass function p_X is generally represented by a histogram (or a bar chart for categorical variables).

For example, if we roll a die, there are only six possible outcomes, and the probability of each outcome for a fair die is $P(X = i) = 1/6$, for $i = 1, \ldots, 6$. In a reservoir model where three facies have been identified, for example sand, silt, and shale, the probability mass function is represented by the bar chart of the facies proportions, i.e. the normalized frequency of the facies. In Example 1.1, the prior probabilities of the facies are $P(A = shale) = 0.2$; $P(A = silt) = 0.5$; $P(A = sand) = 0.3$. These probabilities form the probability mass function of the facies and can be represented by a bar chart, as in Figure 1.1.

In the continuous case, the probability of a continuous random variable X is defined by introducing a non-negative integrable function $f_X : \mathbb{R} \to [0, +\infty]$. The function f_X is called probability density function (PDF) and must satisfy the following properties:

$$f_X(x) \ge 0; \tag{1.12}$$

$$f_X(x) \le +\infty; \tag{1.13}$$

$$\int_{-\infty}^{+\infty} f_X(x)dx = 1. \tag{1.14}$$

Equations (1.12)–(1.14) imply that the likelihood of each outcome is a real number in the interval $[0, +\infty]$ and that the integral of the PDF in the domain of X is 1. Because a continuous variable can take infinitely many values, there is an infinite number of possible outcomes; therefore, the probability that a continuous variable X takes the exact value x is 0, i.e. $P(X = x) = 0$. For this reason, instead of computing the probability of an exact value x, we

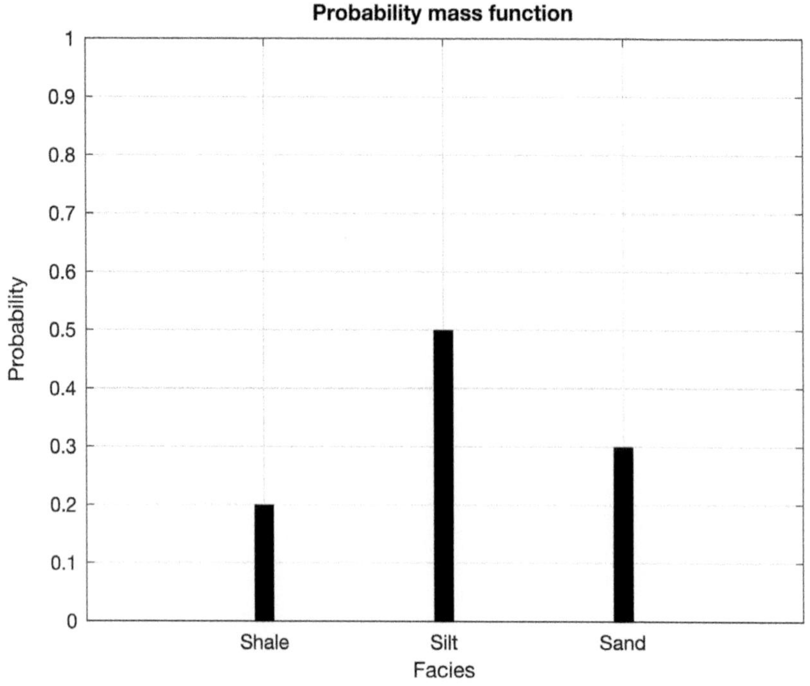

Figure 1.1 Bar chart of the probability mass function of a discrete random variable representing the reservoir facies in Example 1.1.

define the probability of the outcome x to belong to a subset of the domain of the random variable X.

The PDF $f_X(x)$ is then used to define the probability of a subset of values of the random variable X. We define the probability of the outcome x being in the interval $(a, b]$ as the definite integral of the PDF in the interval $(a, b]$:

$$P(a < X \leq b) = \int_a^b f_X(x)dx. \tag{1.15}$$

The interval $(a, b]$ can be infinitesimally small or it can extend to $\pm\infty$. For example, we can study the probability $P(0.20 < \phi \leq 0.21)$ of porosity ϕ being between 0.20 and 0.21 or the probability $P(V_P > 5.5)$ of P-wave velocity V_P being greater than 5.5 km/s. The graphical interpretation of the definition of probability in Eq. (1.15) is shown in Figure 1.2, where the curve represents the PDF of the random variable X, and the area delimited by the PDF and the x-axis of the graph between $a = 2$ and $b = 3$, i.e. the definite integral of the curve in the interval (2, 3], represents the probability $P(2 < X \leq 3)$.

Example 1.2 In this example, we illustrate the calculation of the probability of the random variable X to belong to the interval (2, 3], assuming that X is distributed according to the triangular PDF shown in Figure 1.2. The random variable in Figure 1.2 could represent, for example, the P-wave velocity of a porous rock, expressed in km/s.

The PDF $f_X(x)$ can be written as follows:

$$f_X(x) = \begin{cases} 0 & x \le 1 \\ \frac{1}{4}x - \frac{1}{4} & 1 < x \le 3 \\ -\frac{1}{4}x + \frac{5}{4} & 3 < x \le 5 \\ 0 & x > 5. \end{cases}$$

The triangular function $f_X(x)$ is a non-negative function with integral equal to 1 (i.e. the area of the triangle with base of length 4 and height 0.5); hence, it satisfies the conditions in Eqs. (1.12)–(1.14), and it is a valid PDF. We now compute the probability $P(2 < X \le 3)$ using the definition in Eq. (1.15):

$$P(2 < X \le 3) = \int_2^3 f_X(x)dx = \int_2^3 \left(\frac{1}{4}x - \frac{1}{4}\right)dx = \frac{3}{8} = 0.375.$$

The probability $P(2 < X \le 3)$ represents the area of the region highlighted in Figure 1.2 and can also be computed as the area of a right trapezoid, rotated by 90°.

In many applications, it is useful to describe the distribution of the continuous random variable using the concept of cumulative distribution function (CDF), $F_X : \mathbb{R} \to [0, 1]$. We assume that a random variable X has a PDF $f_X(x)$; then, the CDF $F_X(x)$ is defined as the integral of $f_X(x)$ in the interval $(-\infty, x]$:

$$F_X(x) = \int_{-\infty}^{x} f_X(u)du. \tag{1.16}$$

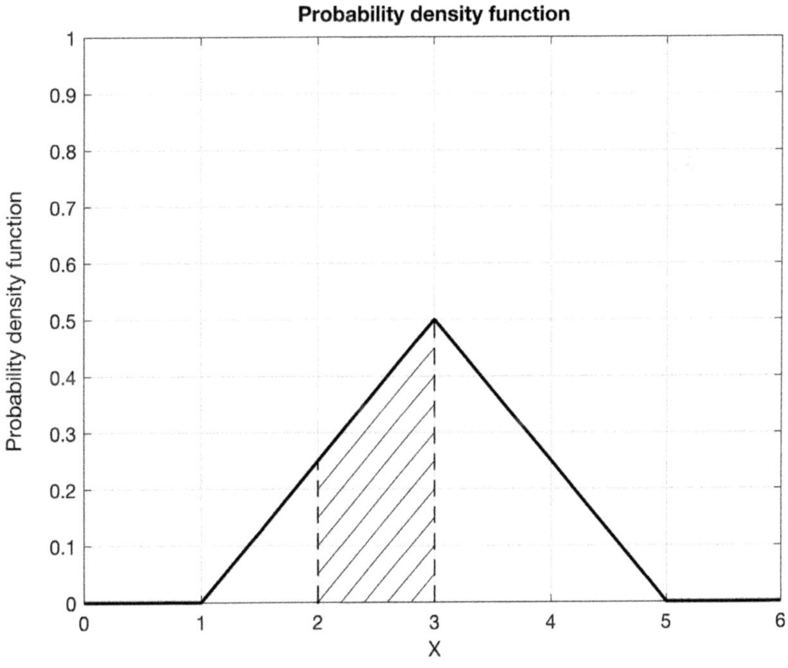

Figure 1.2 Graphical interpretation of the probability $P(2 < X \le 3)$ of a continuous random variable X as the definite integral of the PDF in the interval (2, 3].

Using the definition in Eq. (1.15), we conclude that $F_X(x)$ is the probability of the random variable X being less than or equal to the outcome x, i.e. $F_X(x) = P(X \leq x)$. As a consequence, we can also define the PDF $f_X(x)$ as the derivative of the CDF, i.e. $f_X(x) = \frac{d}{dx}F_X(x)$.

The CDF takes values in the interval [0, 1] and it is non-decreasing, i.e. if $a < b$, then $F_X(a) \leq F_X(b)$. When $F_X(x)$ is strictly increasing, i.e. if $a < b$, then $F_X(a) < F_X(b)$, we can define its inverse function, namely the quantile (or percentile) function. Given a CDF $F_X(x) = p$, the quantile function is $F_X^{-1}(p) = x$ and it associates $p \in [0, 1]$ to the corresponding value x of the random variable X. The definition of percentile is similar to the definition of quantiles, but using percentages rather than fractions in [0, 1]. For example, the 10th percentile (or P10) corresponds to the quantile 0.10. In many practical applications, we often use quantiles and percentiles. The 25th percentile is the value x of the random variable X such that $P(X \leq x) = 0.25$, which means that the probability of taking a value lower than or equal to x is 0.25. The 25th percentile is also called the lower quartile. Similarly, the 75th percentile (or P75) is the upper quartile. The 50th percentile (or P50) is called the median of the distribution, since it represents the value x of the random variable X such that there is an equal probability of observing values lower than or equal to x and observing values greater than x, i.e. $P(X \leq x) = 0.50 = P(X > x)$.

A continuous random variable is completely defined by its PDF (or by its CDF); however, in some special cases, the probability distribution of the continuous random variable can be described by a finite number of parameters. In this case, the PDF is said to be parametric, because it is defined by its parameters. Examples of parameters are the mean and the variance.

The mean is the most common measure used to describe the expected value that a random variable can take. The mean μ_X of a continuous random variable is defined as:

$$\mu_X = \int x f_X(x) dx. \tag{1.17}$$

The mean μ_X is also called the expected value and it is often indicated as $E[X]$. However, the mean itself is not sufficient to represent the full probability distribution, because it does not contain any measure of the uncertainty of the outcomes of the random variable. A common measure for the uncertainty is the variance σ_X^2:

$$\sigma_X^2 = \int (x - \mu_X)^2 f_X(x) dx. \tag{1.18}$$

The variance σ_X^2 describes the spread of the distribution around the mean. The standard deviation σ_X is the square root of the variance, $\sigma_X = \sqrt{\sigma_X^2}$. Other parameters are the skewness coefficient and the kurtosis coefficient (Papoulis and Pillai 2002). The skewness coefficient is a measure of the asymmetry of the probability distribution with respect to the mean, whereas the kurtosis coefficient is a measure of the weight of the tails of the probability distribution.

Other useful statistical estimators are the median and the mode. The median is the 50th percentile (or P50) defined using the CDF and it is the value of the random variable that separates the lower half of the distribution from the upper half. The mode of a random variable is the value of the random variable that is associated with the maximum value of the PDF. For a unimodal symmetric distribution, the mean, the median, and the mode are equal, but, in the general case, they do not necessarily coincide. For example, in a skewed distribution the larger the skewness, the larger the difference between the mode and the median. For a multimodal distribution, the mode can be a more representative statistical estimator, because the mean might fall in a low probability region. A comparison of these three statistical estimators is given in Figure 1.3, for symmetric unimodal, skewed unimodal, and multimodal distributions.

The parameters of a probability distribution can be estimated from a set of n observations $\{x_i\}_{i=1,\ldots,n}$. The sample mean $\overline{\mu}_X$ is an estimate of the mean and it is computed as the average of the data:

$$\overline{\mu}_X = \frac{1}{n}\sum_{i=1}^{n} x_i, \tag{1.19}$$

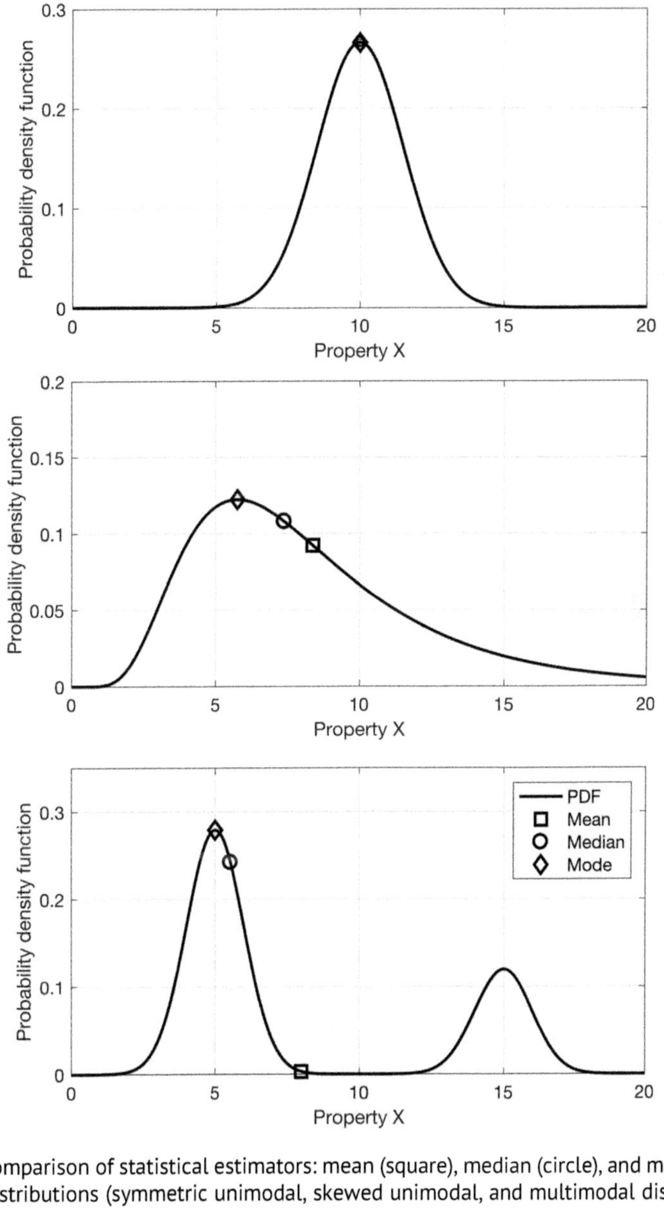

Figure 1.3 Comparison of statistical estimators: mean (square), median (circle), and mode (diamond) for different distributions (symmetric unimodal, skewed unimodal, and multimodal distributions).

whereas the sample variance $\overline{\sigma}_X^2$ is estimated as:

$$\overline{\sigma}_X^2 = \frac{1}{n-1}\sum_{i=1}^{n}(x_i - \overline{\mu}_X)^2, \tag{1.20}$$

where the constant $1/(n-1)$ makes the estimator unbiased (Papoulis and Pillai 2002).

1.3.2 Multivariate Distributions

In many practical applications, we are interested in multiple random variables. For example, in reservoir modeling, we are often interested in porosity and fluid saturation, or P-wave and S-wave velocity. To represent multiple random variables and measure their interdependent behavior, we introduce the concept of joint probability distribution. The joint PDF of two random variables X and Y is a function $f_{X,Y}: \mathbb{R} \times \mathbb{R} \to [0, +\infty]$ such that $0 \le f_{X,Y}(x, y) \le +\infty$ and $\int_{-\infty}^{+\infty}\int_{-\infty}^{+\infty} f_{X,Y}(x,y)dxdy = 1$.

The probability $P(a < X \le b, c < Y \le d)$ of X and Y being in the domain $(a, b] \times (c, d]$ is defined as the double integral of the joint PDF:

$$P(a < X \le b, c < Y \le d) = \int_a^b \int_c^d f_{X,Y}(x,y)dxdy. \tag{1.21}$$

Given the joint distribution of X and Y, we can compute the marginal distributions of X and Y, respectively, as:

$$f_X(x) = \int_{-\infty}^{+\infty} f_{X,Y}(x,y)dy, \tag{1.22}$$

$$f_Y(y) = \int_{-\infty}^{+\infty} f_{X,Y}(x,y)dx. \tag{1.23}$$

In the multivariate setting, we can also introduce the definition of conditional probability distribution. For continuous random variables, the conditional PDF of $X|Y$ is:

$$f_{X|Y}(x) = \frac{f_{X,Y}(x,y)}{f_Y(y)}, \tag{1.24}$$

where the joint distribution $f_{X,Y}(x, y)$ is normalized by the marginal distribution $f_Y(y)$ of the conditioning variable. An analogous definition can be derived for the conditional distribution $f_{Y|X}(y)$ of $Y|X$. All the definitions in this section can be extended to any finite number of random variables.

An example of joint and conditional distributions in a bivariate domain is shown in Figure 1.4. The surface plot in Figure 1.4 shows the bivariate joint distribution $f_{X,Y}(x, y)$ of two random variables X and Y centered at $(1, -1)$. The contour plot in Figure 1.4 shows the probability density contours of the bivariate joint distribution as well as the conditional distribution $f_{Y|X}(y)$ for the conditioning value $x = 1$ and the marginal distributions $f_X(x)$ and $f_Y(y)$.

The conditional probability distribution in Eq. (1.24) can also be computed using Bayes' theorem (Eq. 1.8) as:

$$f_{X|Y}(x) = \frac{f_{Y|X}(y)f_X(x)}{f_Y(y)}. \tag{1.25}$$

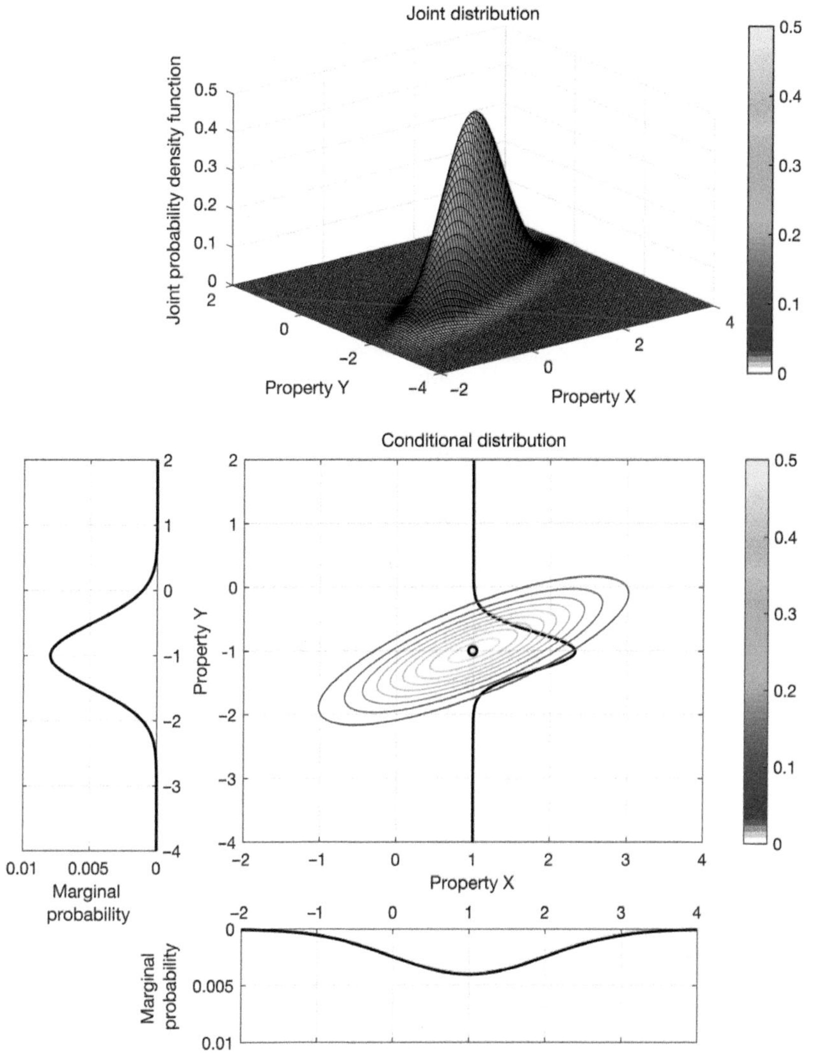

Figure 1.4 Multivariate probability density functions: bivariate joint distribution (surface and contour plots), conditional distribution for $x = 1$, and marginal distributions.

Figure 1.5 shows an example where the uncertainty in the prior probability distribution of a property X is relatively large, and it is reduced in the posterior probability distribution of the property X conditioned on the property Y, by integrating the information from the data contained in the likelihood function. In seismic reservoir characterization, the variable X could represent S-wave velocity and the variable Y could represent P-wave velocity. If a direct measurement of P-wave velocity is available, we can compute the posterior probability distribution of S-wave velocity conditioned on the P-wave velocity measurement. The prior distribution is assumed to be unimodal with relatively large variance. By integrating the likelihood function, we reduce the uncertainty in the posterior distribution.

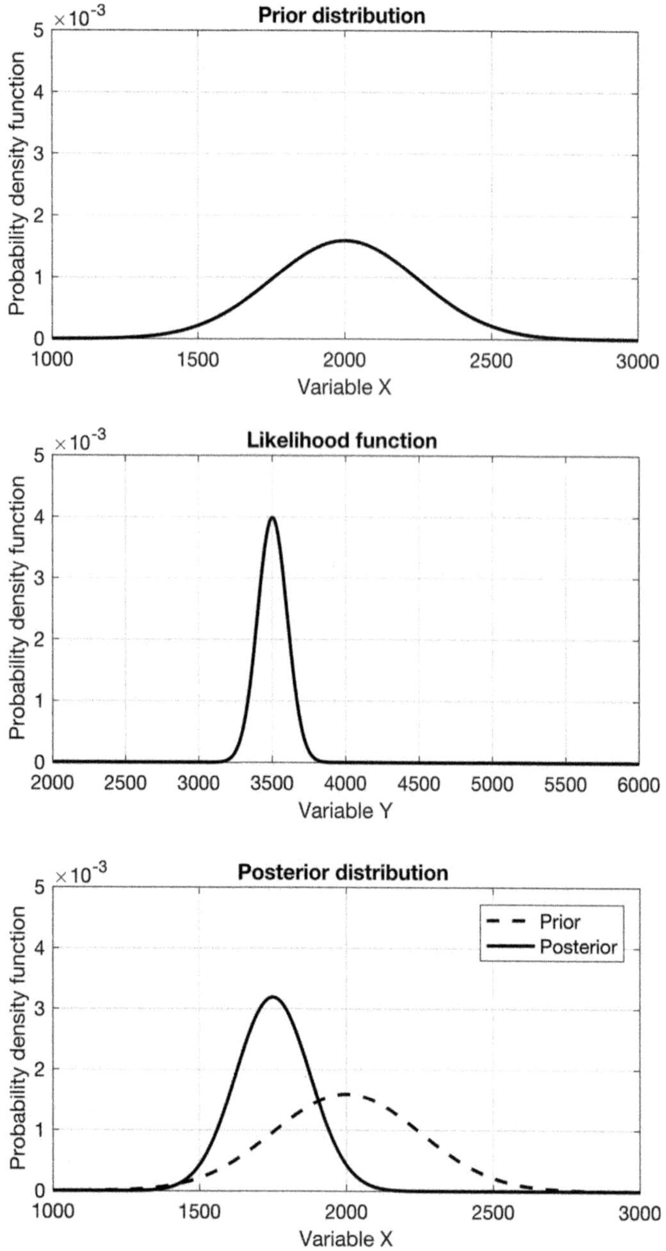

Figure 1.5 Bayes' theorem: the posterior probability is proportional to the product of the prior probability and the likelihood function.

We can also extend the definitions of mean and variance to multivariate random variables. For the joint distribution $f_{X,Y}(x, y)$ of X and Y, the mean $\boldsymbol{\mu}_{X,Y} = [\mu_X, \mu_Y]^T$ is the vector of the means μ_X and μ_Y of the random variables X and Y. In the multivariate case, however, the variances of the random variables do not fully describe the variability of the joint random variable. Indeed, the variability of the joint random variable also depends on how the two variables are related. We define then the covariance $\sigma_{X,Y}$ of X and Y as:

$$\sigma_{X,Y} = \int_{-\infty}^{+\infty} \int_{-\infty}^{+\infty} (x - \mu_X)(y - \mu_Y) f_{X,Y}(x, y) dx dy. \tag{1.26}$$

The covariance is a measure of the linear dependence between two random variables. The covariance of a random variable with itself is equal to the variance of the variable. Therefore, $\sigma_{X,X} = \sigma_X^2$ and $\sigma_{Y,Y} = \sigma_Y^2$. The information associated with the variability of the joint random variable is generally summarized in the covariance matrix $\mathbf{\Sigma}_{X,Y}$:

$$\mathbf{\Sigma}_{X,Y} = \begin{bmatrix} \sigma_X^2 & \sigma_{X,Y} \\ \sigma_{Y,X} & \sigma_Y^2 \end{bmatrix}, \tag{1.27}$$

where the diagonal of the matrix includes the variances of the random variables, and the elements outside the diagonal represent the covariances. The covariance matrix is symmetric by definition, because $\sigma_{X,Y} = \sigma_{Y,X}$ based on the commutative property of the multiplication under the integral in Eq. (1.26). The covariance matrix of a multivariate probability distribution is always positive semi-definite; and it is positive definite unless one variable is a linear transformation of another variable.

We then introduce the linear correlation coefficient $\rho_{X,Y}$ of two random variables X and Y, which is defined as the covariance normalized by the product of the standard deviations of the two random variables:

$$\rho_{X,Y} = \frac{\sigma_{X,Y}}{\sigma_X \sigma_Y}. \tag{1.28}$$

The correlation coefficient is by definition bounded between −1 and 1 (i.e. $-1 \leq \rho_{X,Y} \leq 1$), dimensionless, and easy to interpret. Indeed, a correlation coefficient $\rho_{X,Y} = 0$ means that X and Y are linearly uncorrelated, whereas a correlation coefficient $|\rho_{X,Y}| = 1$ means that Y is a linear function of X. Figure 1.6 shows four examples of two random variables X and Y with different correlation coefficients. When the correlation coefficient is $\rho_{X,Y} = 0.9$, the samples of the two random variables form an elongated cloud of points aligned along a straight line, whereas, when the correlation coefficient is $\rho_{X,Y} \approx 0$, the samples of the two random variables form a homogeneous cloud of points with no preferential alignment. A positive correlation coefficient means that if the random variable X increases, then the random variable Y increases as well, whereas a negative correlation coefficient means that if the random variable X increases, then the random variable Y decreases. For this reason, when the correlation coefficient is $\rho_{X,Y} = -0.6$, the cloud of samples of the two random variables approximately follows a straight line with negative slope.

If two random variables are independent, i.e. $f_{X,Y}(x, y) = f_X(x) f_Y(y)$, then X and Y are uncorrelated. However, the opposite is not necessarily true. Indeed, the correlation coefficient is a measure of linear correlation; therefore, if two random variables are uncorrelated, then there is no linear relation between the two properties, but it does not necessarily mean that the two variables are independent. For example, if $Y = X^2$, and X takes positive and negative values, then the correlation coefficient is close to 0, but yet Y depends deterministically on X through the quadratic relation (Figure 1.6), and the two variables are not independent.

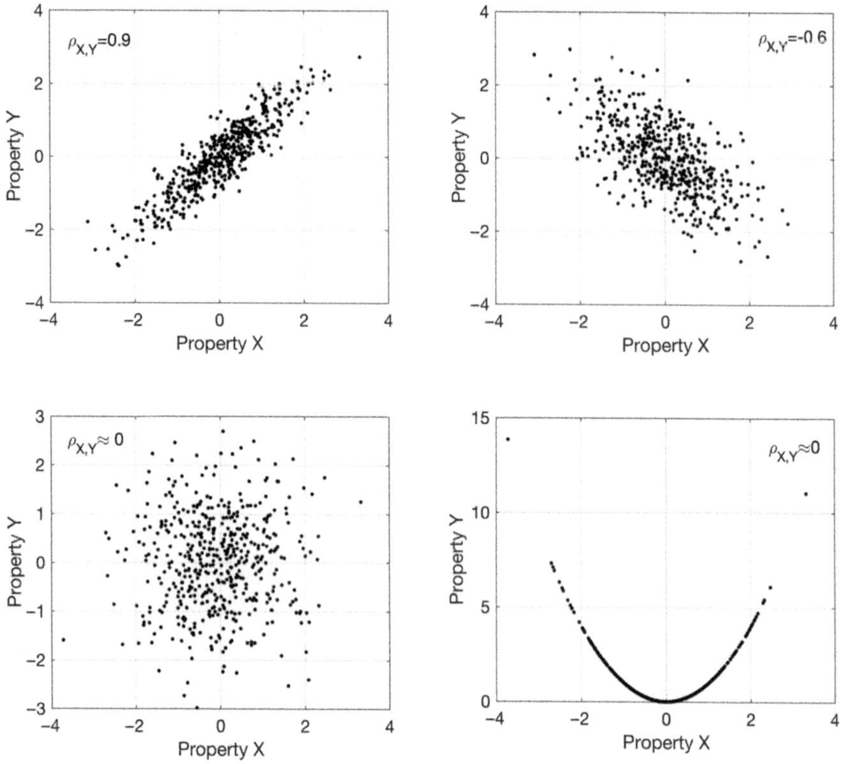

Figure 1.6 Examples of different correlations of the joint distribution of two random variables X and Y. The correlation coefficient $\rho_{X,Y}$ is 0.9 and −0.6 in the top plots and approximately 0 in the bottom plots.

1.4 Probability Distributions

Different probability mass and density functions can be used for discrete and continuous random variables, respectively. For parametric distributions, the function is completely defined by a limited number of parameters (e.g. mean and variance). In this section, we review the most common probability mass and density functions. Probability mass functions are commonly used in geoscience problems for discrete random variables such as facies or rock types, whereas PDFs are used for continuous properties such as porosity, fluid saturations, density, P-wave and S-wave velocity. Some applications in earth sciences include mixed discrete–continuous problems with both discrete and continuous random variables.

1.4.1 Bernoulli Distribution

The simplest probability distribution is the Bernoulli distribution and it is associated with a single experiment with only two possible outcomes. An example of this type of experiment is the toss of a coin. Let X be a random variable representing the experiment, then X is a random variable that takes only two outcomes, 0 and 1, where $X = 1$ means that a favorable event is observed, and $X = 0$ otherwise. We assume that the probability of the favorable event, generally called the probability of success, is a real number p such that $0 \leq p \leq 1$. The probability mass function $p_X(x)$ is then:

$$p_X(x) = \begin{cases} p & x = 1 \\ 1-p & x = 0. \end{cases} \tag{1.29}$$

The mean of the Bernoulli distribution is then $\mu_X = p$ and the variance is $\sigma_X^2 = p(1-p)$.

The Bernoulli distribution has several applications in earth sciences. In reservoir modeling, for example, we can use the Bernoulli distribution for the occurrence of a given facies or rock type. For instance, we define a successful event as finding a high-porosity sand rather than impermeable shale. The probability of success is generally unknown and it depends on the overall proportions of the two facies.

1.4.2 Uniform Distribution

A common distribution for discrete and continuous properties is the uniform distribution on a given interval. According to a uniform distribution, a random variable is equally likely to take any value in the assigned interval. Hence, the PDF is constant within the interval, and 0 elsewhere. If a random variable X is distributed according to a uniform distribution $U([a, b])$ in the interval $[a, b]$, then its PDF $f_X(x)$ can be written as:

$$f_X(x) = \begin{cases} 0 & x < a \\ \dfrac{1}{b-a} & a \leq x \leq b \\ 0 & x > b. \end{cases} \tag{1.30}$$

The mean μ_X of a uniform distribution in the interval $[a, b]$ is:

$$\mu_X = \frac{a+b}{2} \tag{1.31}$$

and it coincides with the median, whereas the variance σ_X^2 is:

$$\sigma_X^2 = \frac{b-a}{12}. \tag{1.32}$$

An example of uniform distribution in the interval $[1, 3]$ is shown in Figure 1.7. The uniform distribution is sometimes called non-informative because it does not provide any additional knowledge other than the interval boundaries.

1.4.3 Gaussian Distribution

Most of the random variables of interest in reservoir modeling are continuous. The most common PDF for continuous variables is the Gaussian distribution, commonly called normal distribution. We say that a random variable X is distributed according to a Gaussian distribution $\mathcal{N}(X; \mu_X, \sigma_X^2)$ with mean μ_X and variance σ_X^2, if its PDF $f_X(x)$ can be written as:

$$f_X(x) = \mathcal{N}(X; \mu_X, \sigma_X^2) = \frac{1}{\sqrt{2\pi\sigma_X^2}} \exp\left(-\frac{1}{2}\frac{(x-\mu_X)^2}{\sigma_X^2}\right). \tag{1.33}$$

A Gaussian distribution $\mathcal{N}(X; \mu_X = 0, \sigma_X^2 = 1)$ with 0 mean and variance equal to 1 is also called standard Gaussian distribution (or normal distribution). The Gaussian distribution is symmetric and unimodal (Figure 1.8) and can be used to describe a number of phenomena

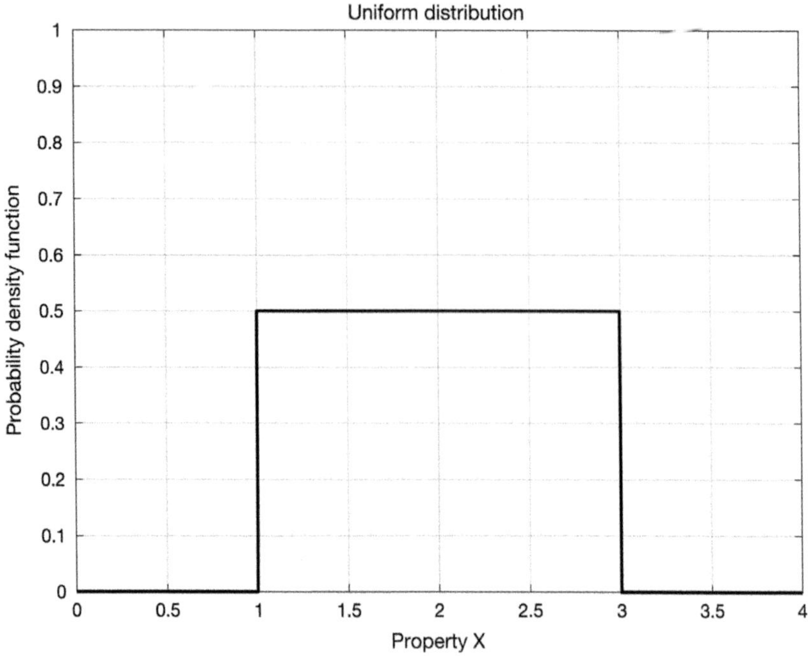

Figure 1.7 Uniform probability density function in the interval [1, 3].

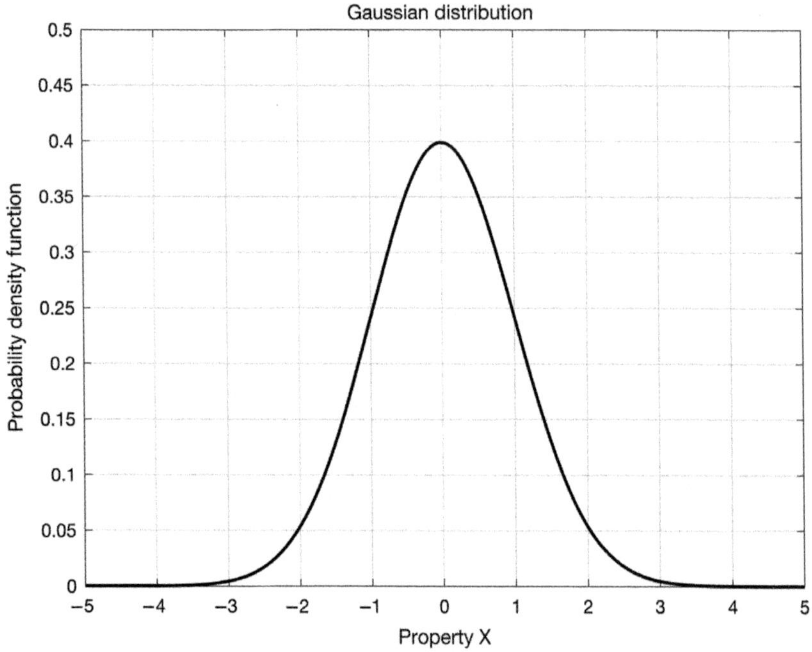

Figure 1.8 Standard Gaussian probability density function with 0 mean and variance equal to 1.

in nature. The Gaussian distribution is defined on the entire set $\mathbb{R}$ of real numbers and it is always positive. For example, the standard Gaussian distribution in Figure 1.8 is always greater than 0, but the probability of the random variable being greater than 3 or less than 3 is close to 0, hence negligible.

As shown in Eq. (1.15), in order to compute any probability associated with the random variable X, we must compute an integral. For a Gaussian distribution, the integral of Eq. (1.33) has no analytical form; therefore, we must use numerical tables of the cumulative density function of the standard Gaussian distribution (Papoulis and Pillai 2002). For example, if X is a random variable distributed according to a standard Gaussian distribution $\mathcal{N}\left(X; \mu_X = 0, \sigma_X^2 = 1\right)$, then we can compute the following probabilities $P(X \leq 1) \cong 0.84$, $P(X \leq 1.5) \cong 0.932$, $P(X \leq 2) \cong 0.977$, using the numerical tables. Similarly, we obtain that $P(-1 \leq X \leq 1) \cong 0.68$, $P(-2 \leq X \leq 2) \cong 0.954$, and $P(-3 \leq X \leq 3) \cong 0.997$.

To compute a probability associated with a non-standard Gaussian distribution $\mathcal{N}\left(Y; \mu_Y, \sigma_Y^2\right)$ with mean μ_Y and variance σ_Y^2, we apply the transformation $X = (Y - \mu_Y)/\sigma_Y$ and we use the numerical tables of the standard Gaussian distribution. Indeed, the random variable X is a standard Gaussian distribution with 0 mean and variance equal to 1, and $P(Y \leq \bar{y}) = P(\mu_Y + \sigma_Y X \leq \bar{y})$. For example, if Y has a Gaussian distribution with mean $\mu_Y = 1$ and variance $\sigma_Y^2 = 4$, then we define a new random variable $X = (Y - 1)/2$. The probability $P(Y \leq 2)$ is then equal to the probability $P(X \leq 0.5) \cong 0.69$. Similarly, $P(-3 \leq Y \leq 3) = P(-2 \leq X \leq 1) \cong 0.82$.

In general, the probability that a Gaussian random variable X takes values in the interval $(\mu_X - \sigma_X, \mu_X + \sigma_X)$ is approximately 0.68; in the interval $(\mu_X - 2\sigma_X, \mu_X + 2\sigma_X)$ is approximately 0.95; and in the interval $(\mu_X - 3\sigma_X, \mu_X + 3\sigma_X)$ is approximately 0.99. Therefore, there is a high probability that the values of the Gaussian random variable $\mathcal{N}\left(X; \mu_X, \sigma_X^2\right)$ are in the interval of length $6\sigma_X$ centered around the mean μ_X, even though the tails of the distributions have non-zero probability for values greater than $\mu_X + 3\sigma_X$ or less than $\mu_X - 3\sigma_X$.

Because a Gaussian PDF takes positive values in the entire real domain, when using Gaussian distributions for bounded random variables, such as volumetric fractions, or positive random variables, such as velocities, one should truncate the distribution to avoid non-physical outcomes or apply a suitable transformation (Papoulis and Pillai 2002), such as the normal score or logit transformations of the original variables.

1.4.4 Log-Gaussian Distribution

We then introduce the log-Gaussian distribution (or log-normal distribution) that is strictly related to the Gaussian distribution. Log-Gaussian distributions are commonly used for positive random variables. For example, suppose that we are interested in the probability distribution of P-wave velocity. To avoid positive values of the PDF for negative (hence non-physical) P-wave velocity values, we can take the logarithm of P-wave velocity and assume a Gaussian distribution in the logarithmic domain. In this case, the distribution of P-wave velocity is said to be log-Gaussian.

We say that a random variable Y is distributed according to a log-Gaussian distribution $Y \sim \text{Log}\mathcal{N}\left(Y; \mu_Y, \sigma_Y^2\right)$ with mean μ_Y and variance σ_Y^2, if $X = \log(Y)$ is distributed according to a Gaussian distribution $X \sim \mathcal{N}\left(\log(Y); \mu_X, \sigma_X^2\right)$. The PDF $f_Y(y)$ can be written as:

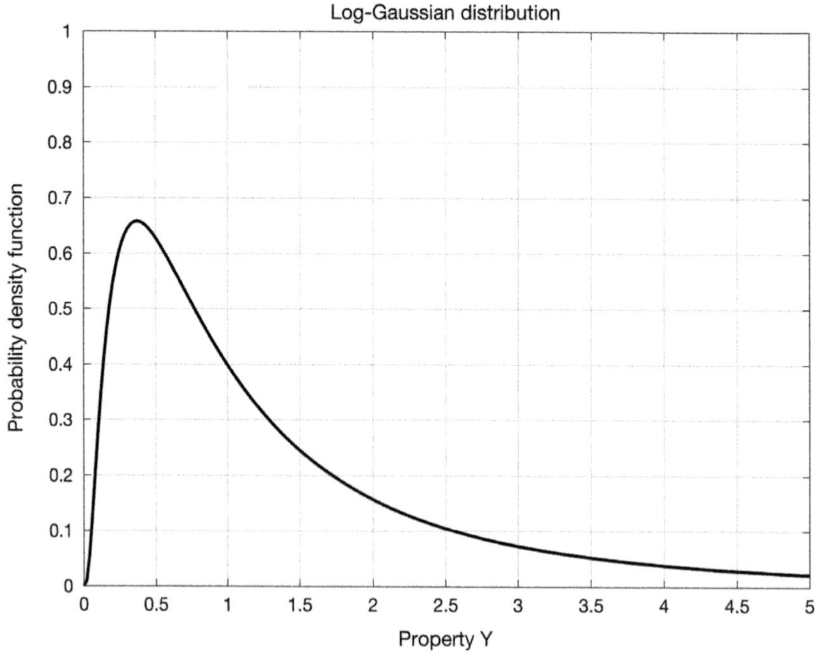

Figure 1.9 Log-Gaussian probability density function associated with the standard Gaussian distribution in Figure 1.8.

$$f_Y(y) = \frac{1}{\sqrt{2\pi\sigma_X^2}\,y} \exp\left(-\frac{1}{2}\frac{(\log(y) - \mu_X)^2}{\sigma_X^2}\right), \tag{1.34}$$

where μ_X and σ_X^2 are the mean and the variance of the random variable in the logarithmic domain.

The mean μ_Y and the variance σ_Y^2 of the log-Gaussian distribution are related to the mean μ_X and variance σ_X^2 of the associated Gaussian distribution, according to the following transformations:

$$\mu_Y = \exp\left(\mu_X + \frac{\sigma_X^2}{2}\right), \tag{1.35}$$

$$\sigma_Y^2 = \mu_Y^2\left[\exp\left(\sigma_X^2\right) - 1\right] = \exp\left(2\mu_X + \sigma_X^2\right)\left[\exp\left(\sigma_X^2\right) - 1\right], \tag{1.36}$$

$$\mu_X = \log\left(\mu_Y\right) - \frac{\sigma_X^2}{2} = \log\left(\frac{\mu_Y^2}{\sqrt{\mu_Y^2 + \sigma_Y^2}}\right), \tag{1.37}$$

$$\sigma_X^2 = \log\left(\frac{\sigma_Y^2}{\mu_Y^2} + 1\right). \tag{1.38}$$

Figure 1.9 shows the log-Gaussian distribution associated with the standard Gaussian distribution $\mathcal{N}\left(X; \mu_X = 0, \sigma_X^2 = 1\right)$ shown in Figure 1.8.

A log-Gaussian distributed random variable takes only positive real values. Its distribution is unimodal but it is not symmetric since the PDF is skewed toward 0. The skewness s of a log-Gaussian distribution is always positive and is given by:

$$s = \left[\exp\left(\sigma_X^2\right) + 2\right]\sqrt{\exp(\sigma_X^2) - 1}. \tag{1.39}$$

For these reasons, the log-Gaussian distribution is a convenient and useful model to describe unimodal positive random variables in earth sciences.

An example of application of log-Gaussian distributions in seismic reservoir characterization can be found in Section 5.3, where we assume a multivariate log-Gaussian distribution of elastic properties in the Bayesian linearized seismic inversion.

1.4.5 Gaussian Mixture Distribution

Gaussian and log-Gaussian distributions are unimodal parametric distributions. However, many rock properties in the subsurface, for example porosity and permeability, are multimodal. Multimodal distributions can be described by non-parametric distributions, but these distributions require a large amount of data to be estimated. In many applications, multimodal distributions can be approximated by Gaussian mixture distributions, i.e. linear combinations of Gaussian distributions.

We say that a random variable X is distributed according to a Gaussian mixture distribution of n components, if the PDF $f_X(x)$ can be written as:

$$f_X(x) = \sum_{k=1}^{n} \pi_k \mathcal{N}\left(X; \mu_{X|k}, \sigma_{X|k}^2\right), \tag{1.40}$$

where the distributions $\mathcal{N}\left(X; \mu_{X|k}, \sigma_{X|k}^2\right)$ represent the Gaussian components with mean $\mu_{X|k}$ and variance $\sigma_{X|k}^2$, and the coefficients π_k are the weights of the linear combination with the condition $\sum_{k=1}^{n} \pi_k = 1$. The condition on the sum of the weights guarantees that $f_X(x)$ is a valid PDF.

Figure 1.10 shows an example of Gaussian mixture PDF with two components with weights equal to 0.4 and 0.6, means equal to −2 and 2, and equal variance of 1. The resulting PDF is multimodal owing to the presence of two components with different mean values and asymmetric due to the larger weight of the second component.

An example of bivariate Gaussian mixture distribution with two components is shown in Figure 1.11. The PDF in Figure 1.11 is estimated from a geophysical dataset of porosity and P-wave velocity. The two components represent two different rock types, namely high-porosity sand and impermeable shale. The mean of porosity in the sand component is 0.22, whereas the mean of porosity in shale is 0.06. Porosity and P-wave velocity are negatively correlated in both components; however, the Gaussian mixture distribution shows a larger variance within the sand component, probably because of the rock heterogeneity and the fluid effect. The distribution is truncated for porosity less than 0, to avoid non-physical values.

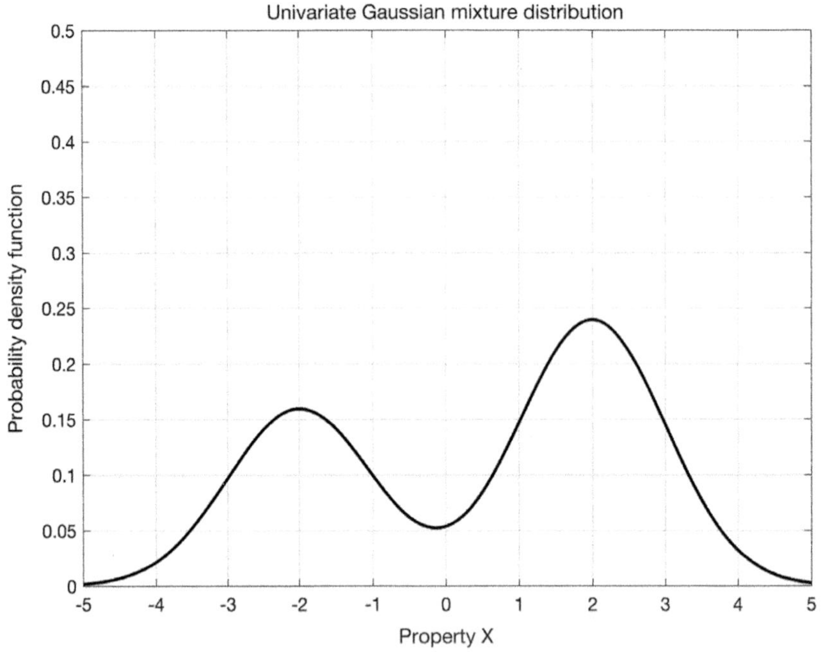

Figure 1.10 Univariate Gaussian mixture probability density function, with two components with weights equal to 0.4 and 0.6, means equal to −2 and 2, and equal variance of 1.

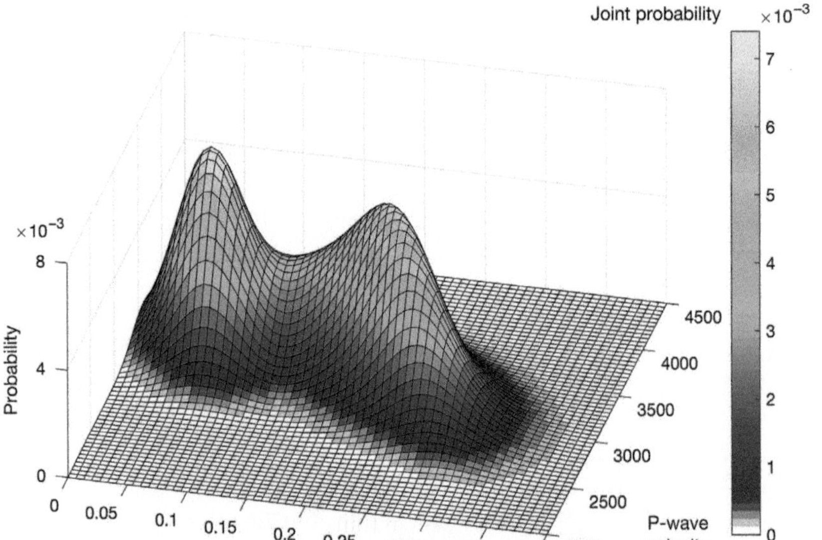

Figure 1.11 Bivariate Gaussian mixture probability density function estimated from a geophysical dataset, in the porosity–velocity domain.

1.4.6 Beta Distribution

The Beta distribution is a PDF for continuous random variables defined in the interval [0, 1]. The PDF is uniquely defined by two positive parameters, namely α and β, that determine the shape of the distribution and its derivatives. The generalization to the multivariate domain is the Dirichlet distribution.

A random variable X is distributed according to a Beta distribution $B(X; \alpha, \beta)$ with parameters $\alpha > 0$ and $\beta > 0$, in the interval [0, 1], if its PDF $f_X(x)$ can be written as:

$$f_X(x) = \frac{x^{\alpha-1}(1-x)^{\beta-1}}{\int_0^1 u^{\alpha-1}(1-u)^{\beta-1}du}, \tag{1.41}$$

where the denominator is a normalizing constant to ensure that Eq. (1.41) is a valid PDF (i.e. the integral of the PDF is equal to 1). The mean μ_X of a Beta distribution is:

$$\mu_X = \frac{\alpha}{\alpha + \beta}, \tag{1.42}$$

whereas the variance σ_X^2 is:

$$\sigma_X^2 = \frac{\alpha\beta}{(\alpha + \beta)^2(\alpha + \beta + 1)}. \tag{1.43}$$

The Beta distribution is a suitable model for random variables describing volumetric fractions, since it is defined on the interval [0, 1]; however, it is not commonly used in practical applications owing to the limited analytical tractability. In geoscience applications, it is often convenient to use the Beta PDF as a prior distribution in Monte Carlo sampling algorithms to sample bounded random variables such as fluid saturations. Figure 1.12 shows two examples of Beta distributions with parameters $\alpha = \beta = 0.1$ (convex curve, solid line) and $\alpha = \beta = 2$ (concave curve, dashed line) that could be used to describe the distribution of fluid saturations and mineral fractions in a hydrocarbon reservoir, respectively. The convex distribution in Figure 1.12 shows high likelihood for values near the boundaries of the interval [0, 1], and low likelihood in the middle of the interval. This behavior is often observed in fluid saturations in the subsurface, where partial saturations are unlikely owing to the gravity effect. The concave distribution in Figure 1.12 shows low likelihood for values near the boundaries of the interval [0, 1], and high likelihood in the middle of the interval. This behavior is often observed in mineral fractions in the subsurface, where rocks are generally made of mixtures of minerals rather than pure minerals.

1.5 Functions of Random Variable

In some applications, we might be interested in transformations of random variables, or, in other words, in the response of a function of a random variable. For example, in reservoir modeling, we might be interested in the permeability distribution, and because of the lack of direct measurements, we estimate the distribution as a function of the distribution of another random variable, for example, porosity.

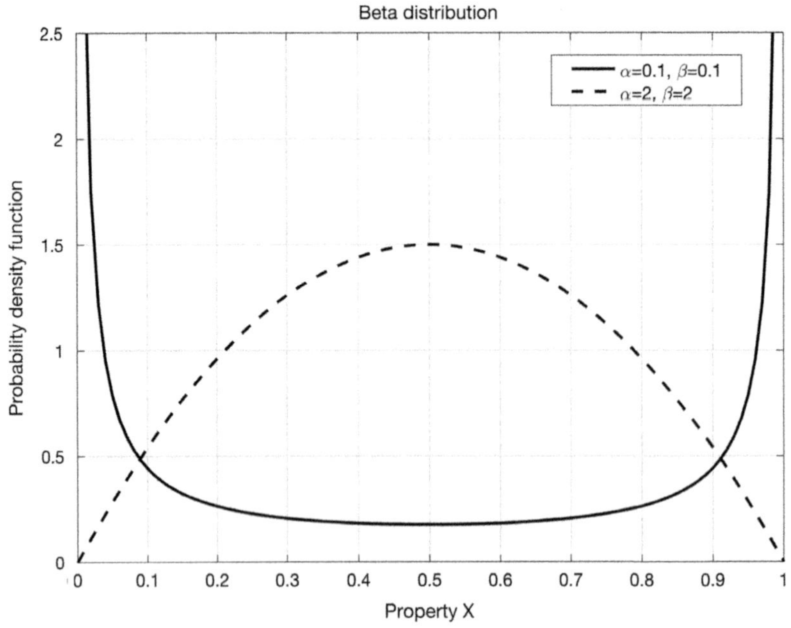

Figure 1.12 Beta probability density function in the interval [0, 1]: the solid line (convex curve) represents a Beta distribution with parameters $\alpha = \beta = 0.1$, whereas the dashed line (concave curve) represents a Beta distribution with parameters $\alpha = \beta = 2$.

If the transformation is linear, then the shape of the distribution of the linear response is the same as the distribution of the initial random variable. For example, if the distribution of the initial random variable is Gaussian, then the distribution of the predicted random variable is also Gaussian. As shown in Section 1.4.3, a Gaussian distribution is fully described by its mean and variance; therefore, to compute the PDF of the transformed variable, we simply compute the mean and the variance of the transformed distribution.

For example, if X is distributed according to a Gaussian distribution $\mathcal{N}(X; \mu_X, \sigma_X^2)$ with mean μ_X and variance σ_X^2, and we apply a linear transformation $Y = g(X) = aX + b$, then Y is still distributed according to a Gaussian distribution $\mathcal{N}(Y; \mu_Y, \sigma_Y^2)$ with mean μ_Y:

$$\mu_Y = a\mu_X + b, \tag{1.44}$$

and variance σ_Y^2:

$$\sigma_Y^2 = a^2\sigma_X^2. \tag{1.45}$$

If X is distributed according to a uniform distribution on the interval $[c, d]$, $X \sim U([c, d])$, and we apply a linear transformation $Y = g(X) = aX + b$, then Y is still distributed according to a uniform distribution $Y \sim U([ac + b, ad + b])$, on the interval $[ac + b, ad + b]$.

These results are intuitive because if we apply a linear transformation to a distribution, we shift the mean and change the variance, but we do not distort the shape of the distribution. For example, if we assume that porosity is distributed according to a Gaussian distribution (neglecting low probability values for porosity values outside the physical bounds), and we apply a linear rock physics model (Chapter 2) to compute the corresponding P-wave velocity distribution, then the distribution of P-wave velocity is still Gaussian and we can compute the mean and the variance using Eqs. (1.44) and (1.45).

If the transformation is not linear, an analytical solution is not always available. A numerical method to obtain an approximation of the probability distribution of the random variable of interest is the Monte Carlo simulation approach. A Monte Carlo simulation consists of three main steps: (i) we generate a set of random samples from the known input distribution; (ii) we apply a physical transformation to each sample; and (iii) we estimate the distribution of the output variable by approximating the histogram of the computed samples. Monte Carlo simulations are often applied in geoscience studies to quantify the propagation of the uncertainty from the input data to the predicted values of a physical model.

1.6 Inverse Theory

In many subsurface modeling problems, we cannot directly measure the properties of interest but can only collect data indirectly related to them. For example, in reservoir modeling, we cannot directly measure porosity far away from the well, but we can acquire seismic data that depend on porosity and other rock and fluid properties. Geophysics generally provides the physical models that link the unknown property, such as porosity, to the measured data, such as seismic velocities. Therefore, the estimation of the unknown model from the measured data is, from the mathematical point of view, an inverse problem.

If $\boldsymbol{m}$ represents the unknown physical variables (i.e. the model), $\boldsymbol{d}$ represents the measurements (i.e. the data), and $\boldsymbol{f}$ is the set of physical equations (i.e. the forward operator) that links the model to the data, then the problem can be formulated as:

$$\boldsymbol{d} = \boldsymbol{f}(\boldsymbol{m}) + \boldsymbol{\varepsilon}, \tag{1.46}$$

where $\boldsymbol{\varepsilon}$ is the measurement error associated with the data. The data $\boldsymbol{d}$ can be a function of time and/or space, or a set of discrete observations. When $\boldsymbol{m}$ and $\boldsymbol{d}$ are vectors of size n_m and n_d, respectively, then $\boldsymbol{f}$ is a function from $\mathbb{R}^{n_m}$ to $\mathbb{R}^{n_d}$. When $\boldsymbol{m}$ and $\boldsymbol{d}$ are functions, then $\boldsymbol{f}$ is an operator. The operator $\boldsymbol{f}$ can be a linear or non-linear system of algebraic equations, ordinary or partial differential equations, or it might involve an algorithm for which there is no explicit analytical formulation. The forward problem is to compute $\boldsymbol{d}$ given $\boldsymbol{m}$. Our focus is on the inverse problem of finding $\boldsymbol{m}$ given $\boldsymbol{d}$ and assessing the uncertainty of the predictions. In other words, we aim to predict the posterior distribution of $\boldsymbol{m}|\boldsymbol{d}$.

In the case of a linear inverse problem with a finite number of measurements $\boldsymbol{d} = \{d_i\}_{i=1,\dots,n_d}$, we can write Eq. (1.46) as a linear system of algebraic equations:

$$\boldsymbol{d} = \mathbf{F}\boldsymbol{m} + \boldsymbol{\varepsilon}, \tag{1.47}$$

where $\mathbf{F}$ is the matrix of size $n_d \times n_m$ associated with the linear operator $\boldsymbol{f}$. A common approach to find the solution of the inverse problem associated with Eq. (1.47) is to estimate the model $\boldsymbol{m}$ that gives the minimum misfit between the data $\boldsymbol{d}$ and the theoretical predictions $\bar{\boldsymbol{d}} = \mathbf{F}\boldsymbol{m}$ of the forward problem, by minimizing the L2-norm (also called the Euclidean norm) $\|\boldsymbol{r}\|_2$ of the residuals $\boldsymbol{r} = \boldsymbol{d} - \mathbf{F}\boldsymbol{m}$:

$$\|\boldsymbol{r}\|_2 = \|\boldsymbol{d} - \mathbf{F}\boldsymbol{m}\|_2 = \sqrt{\sum_{i=1}^{n_d} \left(d_i - (\mathbf{F}\boldsymbol{m})_i\right)^2}. \tag{1.48}$$

The model $\boldsymbol{m}^*$ that minimizes the L2-norm is called the least-squares solution because it minimizes the sum of the squares of the differences of measured and predicted data, and it is given by the following equation, generally called the normal equation (Aster et al. 2018):

$$\boldsymbol{m}^* = \left(\mathbf{F}^T\mathbf{F}\right)^{-1}\mathbf{F}^T\boldsymbol{d}. \tag{1.49}$$

If we consider the data points to be imperfect measurements with random errors, the inverse problem associated with Eq. (1.47) can be seen, from a statistical point of view, as a maximum likelihood estimation problem. Given a model $\boldsymbol{m}$, we assign to each observation d_i a PDF $f_i(d_i|\boldsymbol{m})$ for $i = 1, \dots, n_d$ and we assume that the observations are independent. The joint probability density of the vector of independent observations $\boldsymbol{d}$ is then:

$$f(\boldsymbol{d}|\boldsymbol{m}) = \prod_{i=1}^{n_d} f_i(d_i|\boldsymbol{m}). \tag{1.50}$$

The expression in Eq. (1.50) is generally called likelihood function. In the maximum likelihood estimation, we select the model $\boldsymbol{m}$ that maximizes the likelihood function. If we assume a discrete linear inverse problem with independent and Gaussian distributed data errors ($\varepsilon_i \sim \mathcal{N}(\varepsilon_i; 0, {\sigma_i}^2)$ for $i = 1, \dots, n_d$), then the maximum likelihood solution is equivalent to the least-squares solution. Indeed, under these assumptions, Eq. (1.50) can be written as:

$$f(\boldsymbol{d}|\boldsymbol{m}) = \frac{1}{(2\pi)^{n_d/2} \prod_{i=1}^{n_d} \sigma_i} \prod_{i=1}^{n_d} e^{-\frac{\left(d_i - (\mathbf{F}\boldsymbol{m})_i\right)^2}{2{\sigma_i}^2}} \tag{1.51}$$

and the maximization of Eq. (1.51) is equivalent to the minimization of Eq. (1.48) (Tarantola 2005; Aster et al. 2018).

The L2-norm is not the only misfit measure that can be used in inverse problems. For example, to avoid data points inconsistent with the chosen mathematical model (namely the outliers), the L1-norm is generally preferable to the L2-norm. However, from a mathematical point of view, the L2-norm is preferable because of the analytical tractability of the associated Gaussian distribution.

In science and engineering applications, many inverse problems are not linear; therefore, the analytical solution of the inverse problem might not be available. For non-linear inverse problems, several mathematical algorithms are available, including gradient-based deterministic methods, such as Gauss–Newton, Levenberg–Marquardt, and conjugate gradient; Markov chain Monte Carlo methods, such as Metropolis, Metropolis Hastings, and Gibbs sampling; and stochastic optimization algorithms, such as simulated annealing, particle swarm optimization, and genetic algorithms. For detailed descriptions of these methods we refer the reader to Tarantola (2005), Sen and Stoffa (2013), and Aster et al. (2018).

1.7 Bayesian Inversion

From a probabilistic point of view, the solution of the inverse problem corresponds to estimating the conditional distribution $\boldsymbol{m}|\boldsymbol{d}$. The conditional probability $P(\boldsymbol{m}|\boldsymbol{d})$ can be obtained using Bayes' theorem (Eqs. 1.8 and 1.25):

$$P(\boldsymbol{m}|\boldsymbol{d}) = \frac{P(\boldsymbol{d}|\boldsymbol{m})P(\boldsymbol{m})}{P(\boldsymbol{d})} = \frac{P(\boldsymbol{d}|\boldsymbol{m})P(\boldsymbol{m})}{\int P(\boldsymbol{d}|\boldsymbol{m})P(\boldsymbol{m})d\boldsymbol{m}}, \tag{1.52}$$

where $P(\boldsymbol{d}|\boldsymbol{m})$ is the likelihood function, $P(\boldsymbol{m})$ is the prior distribution, and $P(\boldsymbol{d})$ is the marginal distribution. The probability $P(\boldsymbol{d})$ is a normalizing constant that guarantees that $P(\boldsymbol{m}|\boldsymbol{d})$ is a valid PDF.

In geophysical inverse problems, we often assume that the physical relation $\boldsymbol{f}$ in Eq. (1.46) is linear and that the prior distribution $P(\boldsymbol{m})$ is Gaussian (Tarantola 2005). These two assumptions are not necessarily required to solve the Bayesian inverse problem, but under these assumptions, the inverse solution can be analytically derived. Indeed, in the Gaussian case, the solution to the Bayesian linear inverse problem is well-known (Tarantola 2005). If we assume that: (i) the prior distribution of the model is Gaussian, i.e. $\boldsymbol{m} \sim \mathcal{N}(\boldsymbol{m}; \boldsymbol{\mu}_m, \boldsymbol{\Sigma}_m)$, where $\boldsymbol{\mu}_m$ is the prior mean and $\boldsymbol{\Sigma}_m$ is the prior covariance matrix; (ii) the forward operator $\boldsymbol{f}$ is linear with associated matrix $\mathbf{F}$; and (iii) the measurement errors $\boldsymbol{\varepsilon}$ are Gaussian $\boldsymbol{\varepsilon} \sim \mathcal{N}(\boldsymbol{\varepsilon}; \mathbf{0}, \boldsymbol{\Sigma}_\varepsilon)$, with $\mathbf{0}$ mean and covariance matrix $\boldsymbol{\Sigma}_\varepsilon$, and they are independent of $\boldsymbol{m}$; then, the posterior distribution $\boldsymbol{m}|\boldsymbol{d}$ is also Gaussian $\boldsymbol{m}|\boldsymbol{d} \sim \mathcal{N}(\boldsymbol{m}; \boldsymbol{\mu}_{m|d}, \boldsymbol{\Sigma}_{m|d})$ with conditional mean $\boldsymbol{\mu}_{m|d}$:

$$\boldsymbol{\mu}_{m|d} = \boldsymbol{\mu}_m + \boldsymbol{\Sigma}_m \mathbf{F}^T \left(\mathbf{F}\boldsymbol{\Sigma}_m \mathbf{F}^T + \boldsymbol{\Sigma}_\varepsilon\right)^{-1} (\boldsymbol{d} - \mathbf{F}\boldsymbol{\mu}_m) \tag{1.53}$$

and conditional covariance matrix $\boldsymbol{\Sigma}_{m|d}$:

$$\boldsymbol{\Sigma}_{m|d} = \boldsymbol{\Sigma}_m - \boldsymbol{\Sigma}_m \mathbf{F}^T \left(\mathbf{F}\boldsymbol{\Sigma}_m \mathbf{F}^T + \boldsymbol{\Sigma}_\varepsilon\right)^{-1} \mathbf{F}\boldsymbol{\Sigma}_m. \tag{1.54}$$

For the proof, we refer the reader to Tarantola (2005). This result is extensively used in Chapter 5 for seismic inversion problems.

Example 1.3 We illustrate the Bayesian approach for linear inverse problems in a geophysical application. We assume that the model variable of interest is S-wave velocity V_S and that a measurement of P-wave velocity V_P is available. The goal of this exercise is to predict the conditional probability of S-wave velocity given P-wave velocity.

We assume that S-wave velocity is distributed according to a Gaussian distribution $\mathcal{N}(V_S; \mu_S, \sigma_S^2)$ with prior mean $\mu_S = 2$ km/s and prior standard deviation $\sigma_S = 0.25$ km/s ($\sigma_S^2 = 0.0625$). We assume that the forward operator linking P-wave and S-wave velocity is a linear model of the form:

$$V_P = 2V_s + \varepsilon.$$

We then assume that the measurement error is Gaussian distributed $\mathcal{N}(\varepsilon; \mu_\varepsilon, \sigma_\varepsilon^2)$ with mean $\mu_\varepsilon = 0$ and standard deviation $\sigma_\varepsilon = 0.05$ km/s ($\sigma_\varepsilon^2 = 0.0025$).

If the available measurement of P-wave velocity is $V_P = 3.5$ km/s, then the posterior distribution of S-wave velocity given the P-wave velocity measurement is Gaussian distributed $\mathcal{N}\left(V_S; \mu_{S|P}, \sigma_{S|P}^2\right)$ with mean $\mu_{S|P}$:

$$\mu_{S|P} = 2 + 0.495 \times (3.5 - 2 \times 2) = 1.75 \text{ km/s}$$

and standard deviation $\sigma_{S|P}$:

$$\sigma_{S|P} = \sqrt{0.0625 - 0.495 \times 2 \times 0.0625} = 0.025 \text{ km/s}.$$

If the available measurement of P-wave velocity is $V_P = 4.5$ km/s, then the mean $\mu_{S|P}$ of the posterior distribution is:

$$\mu_{S|P} = 2 + 0.495 \times (4.5 - 2 \times 2) = 2.25 \text{ km/s}$$

and the standard deviation is $\sigma_{S|P} = 0.025$ km/s.

The posterior standard deviation does not depend on the measurement but only on the prior standard deviation of the model variable and the standard deviation of the error.

2

Rock Physics Models

Rock physics is a subdiscipline of geophysics that studies the relations between petrophysical properties, such as porosity, mineral fractions, and fluid saturations, and elastic properties, such as elastic moduli, velocities, density, and impedances (Mavko et al. 2020). There are several factors that affect the elastic behavior of porous rocks, including porosity, mineralogy, fluid volumes, pressure, diagenesis, cementation, and mineral texture. Some of these properties, such as pore volume, can be measured in the laboratory or quantified from well log data, whereas other properties, such as diagenesis, are difficult to quantify. Core samples and well log data, mostly acquired in hydrocarbon reservoirs, show several common trends (Zimmerman 1990; Bourbié et al. 1992; Avseth et al. 2010; Dvorkin et al. 2014; Mavko et al. 2020). For example, P-wave and S-wave velocity and density decrease with increasing porosity. Similarly, if water saturation increases in a mixture of water and hydrocarbon, then P-wave velocity and density generally increase, whereas S-wave velocity slightly decreases.

The MATLAB codes for the examples included in this chapter are provided in the SeReM package and described in the Appendix, Section A.1.

2.1 Rock Physics Relations

We illustrate some of the common rock physics trends using the dataset measured by Han at Stanford University (Han 1986). The dataset includes 64 samples of sandstone from the Gulf of Mexico with measurements of porosity, clay volume, density, and P-wave and S-wave velocity. Figure 2.1 shows a subset of samples with clay volume between 0 and 0.5. As porosity increases, P-wave velocity decreases. For a given porosity value, if clay volume increases, P-wave velocity decreases.

2.1.1 Porosity – Velocity Relations

Several empirical equations have been developed to describe the elastic behavior of porous rocks, in particular for P-wave velocity. Wyllie et al. (1956) propose a linear relation, namely

Seismic Reservoir Modeling: Theory, Examples, and Algorithms, First Edition. Dario Grana, Tapan Mukerji, and Philippe Doyen.

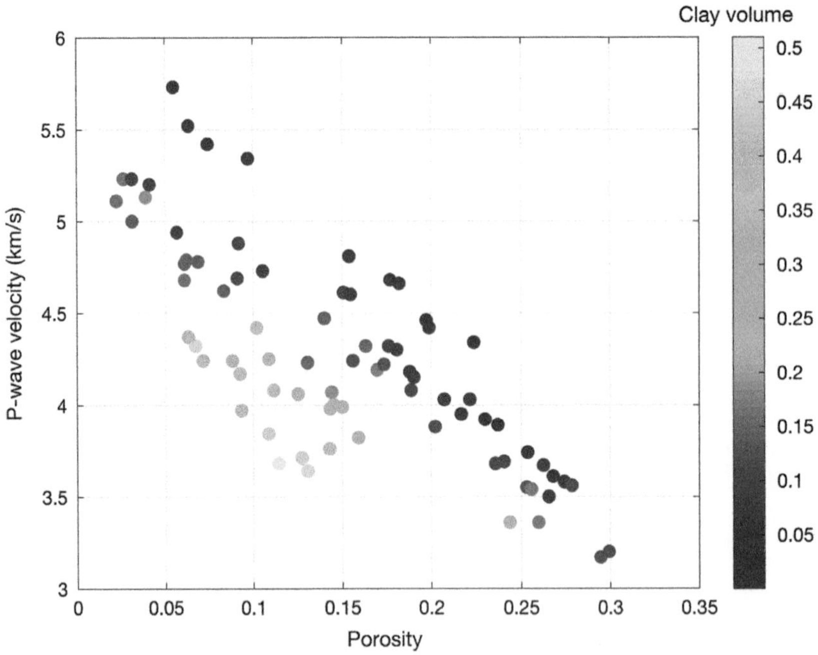

Figure 2.1 Han's dataset: P-wave velocity versus porosity color coded by clay volume. The data are measured at the effective pressure of 40 MPa.

Wyllie's equation, between the travel time of the compressional waves $t_P = 1/V_P$ and porosity ϕ:

$$\frac{1}{V_P} = (1-\phi)\frac{1}{V_{P_{sol}}} + \phi\frac{1}{V_{P_{fl}}}, \tag{2.1}$$

where $V_{P_{sol}}$ and $V_{P_{fl}}$ represent the P-wave velocity of the solid and fluid phases, respectively. Similarly, Raymer et al. (1980) propose a quadratic relation, namely Raymer's equation:

$$V_P = (1-\phi)^2 V_{P_{sol}} + \phi V_{P_{fl}}. \tag{2.2}$$

Raymer's equation generally provides an accurate approximation of P-wave velocity in consolidated sandstones. Equation (2.2) can be extended to S-wave velocity as $V_S = (1-\phi)^2 V_{S_{sol}} \sqrt{(1-\phi)\rho_{sol}/\rho}$, where $V_{S_{sol}}$ and ρ_{sol} are the S-wave velocity and density of the solid phase, respectively, and ρ is the density of the saturated rock (Dvorkin 2008).

Example 2.1 In this example, we study the P-wave velocity of a sandstone with porosity 0.25 saturated with fresh water. We assume that the rock is made of quartz. According to rock physics literature (Mavko et al. 2020), the P-wave velocity of quartz is $V_{P_{sol}} = 6$ km/s and the P-wave velocity of fresh water is $V_{P_{fl}} = 1.5$ km/s. By applying Wyllie's equation (Eq. 2.1), we obtain that the P-wave velocity of the saturated rock is:

$$V_P = \frac{1}{(1-0.25)\times\frac{1}{6} + 0.25\times\frac{1}{1.5}} = 3.43\,\text{km/s}.$$

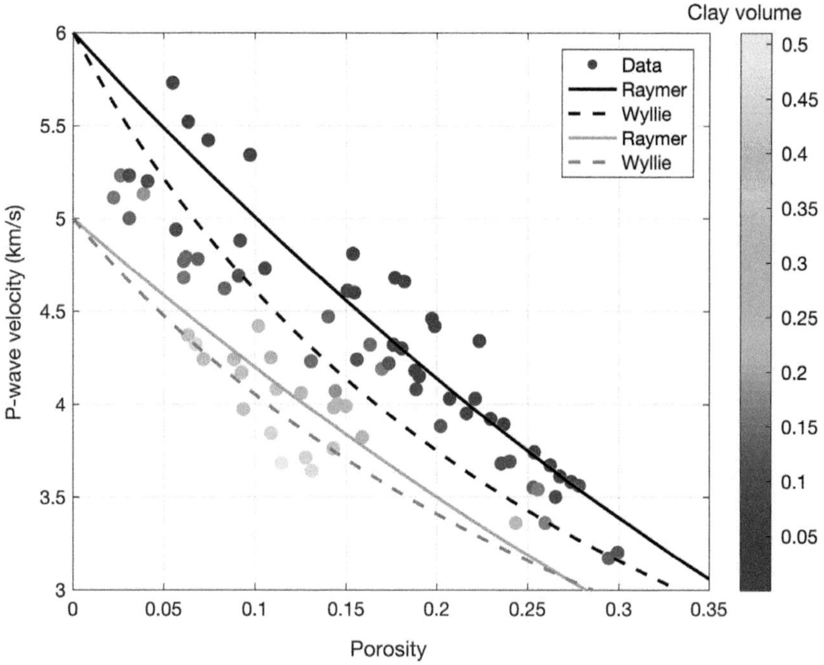

Figure 2.2 Wyllie and Raymer's equations as a function of porosity for clay volume equal to 0 (black lines) and clay volume equal to 0.33 (gray lines). The data are color coded by the clay volume of the measured samples as in Figure 2.1.

Whereas using Raymer's equation (Eq. 2.2), we obtain the following P-wave velocity:

$$V_P = (1 - 0.25)^2 \times 6 + 0.25 \times 1.5 = 3.75\,\text{km/s}.$$

Figure 2.2 shows Wyllie and Raymer's equations (black lines) for a clean sandstone ($V_{P_{sol}} = 6$ km/s) with variable porosity, saturated with fresh water. At porosity equal to 0, the P-wave velocity of the rock is equal to the P-wave velocity of the solid phase for both equations. Indeed, both Eqs. (2.1) and (2.2) reduce to $V_P = V_{P_{sol}}$. At porosity equal to 0.25, the two equations provide the P-wave velocity values in Example 2.1, i.e. $V_P = 3.43$ km/s according to Wyllie's equation and $V_P = 3.75$ km/s according to Raymer's equation. The gray lines in Figure 2.2 show the predictions of Wyllie and Raymer's equations for a shaley sandstone with quartz volume equal to 0.67 and clay volume equal to 0.33. For this mixture of quartz and clay, the P-wave velocity of the solid phase is $V_{P_{sol}} = 5$ km/s. Figure 2.2 compares the model predictions to Han's data. For clean sandstones, Raymer's equation better follows the trend in Han's dataset (dark blue points), whereas for shaley sandstones, both equations provide an accurate approximation.

2.1.2 Porosity – Clay Volume – Velocity Relations

Multilinear models have been proposed to describe the relation between porosity ϕ, clay volume v_c, and P-wave and S-wave velocity V_P and V_S. For a mixture of quartz and clay, these relations can be expressed in the general form as:

$$V_P = a_P\phi + b_P v_c + c_P \tag{2.3}$$

$$V_S = a_S\phi + b_S v_c + c_S \tag{2.4}$$

where (a_P, b_P, c_P) and (a_S, b_S, c_S) are empirical coefficients. The constants c_P and c_S represent the P-wave and S-wave velocity of the solid phase ($\phi = 0$) of a clean sandstone ($v_c = 0$). The coefficients a_P, b_P, a_S, and b_S are generally negative, since both P-wave and S-wave velocities decrease for increasing porosity and clay volume (Figure 2.1). Han (1986) calibrates a set of coefficients based on laboratory measurements at different effective pressure values ranging from 5 to 40 MPa. Similarly, based on the dataset in Figure 2.1, we calibrate a rock physics model using the multilinear regression in Eqs. (2.3) and (2.4). The dataset in Figure 2.1 is measured at 40 MPa. The resulting P-wave velocity equation is given by $V_P = -6.93\phi - 2.18v_c + 5.59$, and the S-wave velocity equation is $V_S = -4.91\phi - 1.89v_c + 3.52$.

Eberhart-Phillips et al. (1989) modify Han's relations (Eqs. 2.3 and 2.4) and present a non-linear multivariate analysis to investigate the combined effect of effective pressure P, porosity, and clay volume based on Han's measurements of velocities in water-saturated shaley sandstones. Eberhart-Phillips's equations are given by:

$$V_P = a_P\phi + b_P\sqrt{v_c} + d_P(P - \exp(-\lambda_P P)) + c_P \tag{2.5}$$

$$V_S = a_S\phi + b_S\sqrt{v_c} + d_S(P - \exp(-\lambda_S P)) + c_S \tag{2.6}$$

where the additional parameters d_P, d_S, λ_P, and λ_S are empirical constant values that are assumed to be positive. Eberhart-Phillips et al. (1989) calibrate the parameters in Eqs. (2.5) and (2.6) on Han's dataset. The resulting P-wave velocity equation is given by $V_P = -6.94\phi - 1.73\sqrt{v_c} + 0.446(P - \exp(-16.7P)) + 5.77$, and the resulting S-wave velocity equation is $V_S = -4.94\phi - 1.57\sqrt{v_c} + 0.361(P - \exp(-16.7P)) + 3.70$. When applied to Han's dataset, the model accounts for 95% of the variance, with a root mean square error of 0.1 km/s.

2.1.3 P-Wave and S-Wave Velocity Relations

In some geophysical applications, only P-wave velocity measurements might be available. This is the case, for example, when P-wave velocity is estimated from refraction seismic data. In these cases, S-wave velocity can be computed from P-wave velocity using linear regressions:

$$V_S = a_1 V_P + a_0 \tag{2.7}$$

or quadratic functions:

$$V_S = a_2 {V_P}^2 + a_1 V_P + a_0, \tag{2.8}$$

where the polynomial coefficients depend on the mineralogy and fluid content. Greenberg and Castagna (1992) propose empirical relations for computing S-wave velocity from P-wave velocity in brine-saturated rocks based on polynomial approximations. These models can be expressed in the general form as:

$$V_S = \frac{1}{2}\left[\sum_{i=1}^{n} v_i \sum_{j=0}^{d} a_{ij}(V_P)^j + \frac{1}{\sum_{i=1}^{n} \frac{v_i}{\sum_{j=0}^{d} a_{ij}(V_P)^j}}\right], \tag{2.9}$$

where n is the number of mineral components, v_i are the volumetric fractions of the mineral components, d is the polynomial degree, and a_{ij} are the polynomial coefficients. This approximation is valid under the assumption that the sum of the mineral fractions is equal to 1 (i.e. $\sum_{i=1}^{n} v_i = 1$). In a single-mineral porous rock saturated with water, the expression simplifies to:

$$V_S = \sum_{j=0}^{d} a_j (V_P)^j. \tag{2.10}$$

Greenberg and Castagna (1992) provide the empirical coefficients a_{ij} for several lithologies, including sandstone, shale, limestone, and dolomite and for linear and quadratic polynomials. All the coefficients are empirically estimated from a calibration dataset. For example, for sandstone saturated with water, the Greenberg and Castagna's linear approximation ($d = 1$) becomes: $V_S = 0.804V_P - 0.856$.

2.1.4 Velocity and Density

The majority of the theoretical models have been developed in the elastic moduli domain (Mavko et al. 2020). The P-wave and S-wave velocities can be computed, by definition, from the bulk and shear moduli, K and μ, and the density ρ of the saturated rock, as:

$$V_P = \sqrt{\frac{K + \frac{4}{3}\mu}{\rho}} \tag{2.11}$$

$$V_S = \sqrt{\frac{\mu}{\rho}}, \tag{2.12}$$

where the density ρ is generally computed as a linear average of the density of the solid and fluid phases, ρ_{sol} and ρ_{fl}, weighted by porosity:

$$\rho = (1 - \phi)\rho_{sol} + \phi\rho_{fl}. \tag{2.13}$$

The bulk modulus K is defined as the ratio of the hydrostatic stress and the volumetric strain, and it represents the reciprocal of the compressibility. The shear modulus μ is defined as the ratio of the shear stress to the shear strain (Mavko et al. 2020). The stress–strain relation of an isotropic linear elastic medium is fully described by the bulk and shear moduli, K and μ, through Hooke's law (Mavko et al. 2020), or by the Lamé's parameter $\lambda = K - 2/3\mu$ and the shear modulus μ. The quantity at the numerator of Eq. (2.11) is generally called the compressional modulus, $M = K + 4/3\mu$, and it is defined as the ratio of the axial stress to the axial strain in a uniaxial strain state.

If the elastic moduli are in Gigapascal (GPa) and density is in grams per centimeters cubed (g/cm^3), then the velocities in Eqs. (2.11) and (2.12) are in kilometers per seconds (km/s). The bulk modulus of a saturated porous rock generally depends on the bulk moduli of the solid and fluid phases and porosity. The shear modulus of a saturated porous rock instead depends on the shear modulus of the solid and porosity, because the shear modulus of the fluid is always 0.

A rock physics model generally includes several steps: if multiple minerals and/or fluids are present, the elastic moduli of the solid and fluid phases are computed using mixing laws; then the elastic moduli of the dry rock are calculated; and the fluid effect is included to compute the elastic moduli of the saturated rock. P-wave and S-wave velocity and density are then computed using Eqs. (2.11)–(2.13).

Example 2.2 We compute the density, P-wave and S-wave velocity of a sandstone with porosity 0.25 saturated with brine, using the definitions in Eqs. (2.11)–(2.13). We first compute the density. The density of quartz is $\rho_{sol} = 2.65 \text{ g/cm}^3$. In this example, we assume that the rock is saturated with brine, hence, the density of water is $\rho_{fl} = 1.01 \text{ g/cm}^3$, owing to the salinity. In these conditions, the density of the brine-saturated sandstone is:

$$\rho = (1-0.25) \times 2.65 + 0.25 \times 1.01 = 2.24 \, \text{g/cm}^3.$$

We then compute the P-wave and S-wave velocity. We assume that the elastic moduli of the brine-saturated sandstone with porosity 0.25 are available from literature (Mavko et al. 2020). The bulk modulus is $K = 17$ GPa and the shear modulus is $\mu = 10$ GPa. The P-wave velocity is then:

$$V_P = \sqrt{\frac{17 + \frac{4}{3} \times 10}{2.24}} = 3.68 \text{ km/s};$$

and the S-wave velocity is:

$$V_S = \sqrt{\frac{10}{2.24}} = 2.11 \text{ km/s}.$$

In the following sections, we present rock physics models to predict the elastic moduli of porous rocks based on porosity, mineral fractions, and fluid saturations.

2.2 Effective Media

We first illustrate the most common mixing laws to compute the density and elastic moduli of mixtures of minerals and mixtures of fluids.

2.2.1 Solid Phase

Most rock physics models assume homogeneous mineralogy; however, porous rocks generally contain several mineral components. The effective medium properties, such as density and elastic moduli, of a mixture of multiple minerals are generally computed using mixing laws that average the properties of the minerals based on the volumetric fractions. The elastic moduli and densities of the mineral components vary from low values for water-rich clays to high values for carbonates. Some of the most commonly used values are reported in Table 2.1. Especially for clay minerals, the variability in the elastic moduli is relatively large, and it depends on the clay composition and the associated volume of

Table 2.1 Density, bulk modulus, and shear modulus of quartz, clay, calcite, and dolomite.

Mineral	Density (g/cm^3)	Bulk modulus (GPa)	Shear modulus (GPa)
Quartz	2.65	36	45
Clay (illite)	2.55	21	7
Clay (kaolinite)	1.58	1.5	1.4
Calcite	2.71	76	32
Dolomite	2.87	95	45

clay-bound water. For example, the elastic moduli and density of kaolinite are lower than the values measured in illite, owing to the large amount of clay-bound water that reduces the density and stiffness of kaolinite. Reference values for a large variety of minerals can be found in Mavko et al. (2020).

If the volumetric fractions of the mineral components are known, for example based on measurements from core samples or well logs, then the density of the solid phase can be calculated as the arithmetic average of the densities of the minerals ρ_i weighted by their volumetric fractions v_i:

$$\rho_{sol} = \sum_{i=1}^{n} v_i \rho_i, \tag{2.14}$$

where n is the number of the mineral components of the solid phase, assuming that the sum of the mineral volumes is 1 (i.e. $\sum_{i=1}^{n} v_i = 1$).

The elastic moduli of the effective solid medium depend on the geometry describing how the mineral components are arranged relative to each other. For example, in a shaley sandstone made of a mixture of quartz and clay, the mineral grains of quartz and clay can be arranged according to different spatial distributions, generating laminated, dispersed, or structural shale distributions (Mavko et al. 2020). In mixtures with laminated, dispersed, or structural shale distributions with the same clay volume, the density of the mixture is the same for all the mixtures, but the elastic moduli might vary owing to the varying stiffness of the mineral texture.

The geometric details of the spatial distribution of the different minerals in a solid mixture cannot be expressed in simple mathematical models owing to the heterogeneity of porous rocks. However, based on the elastic moduli of the mineral components and their volumetric fractions, it is possible to define the upper and lower bounds of the elastic moduli, namely the elastic bounds, of the effective solid phase.

The Voigt mixing law is the arithmetic average of the elastic moduli of the minerals weighted by their volumetric fractions and represents the upper bound for the elastic moduli of the effective solid medium. According to the Voigt average, the bulk and shear moduli of the effective solid medium can be then computed as:

$$K_V = \sum_{i=1}^{n} v_i K_i \tag{2.15}$$

$$\mu_V = \sum_{i=1}^{n} v_i \mu_i, \tag{2.16}$$

where K_i and μ_i are the bulk and shear moduli of the mineral components for $i = 1, \ldots, n$. The Voigt mixing law represents the isostrain average, because it provides the ratio of the average stress to the average strain when all components have the same strain.

The Reuss mixing law is the geometric average of the elastic moduli of the minerals and represents the lower bound for the elastic moduli of the effective solid medium. Using the Reuss average, the bulk and shear moduli of the effective solid medium are then calculated as:

$$K_R = \frac{1}{\sum_{i=1}^{n} \frac{v_i}{K_i}} \tag{2.17}$$

$$\mu_R = \frac{1}{\sum_{i=1}^{n} \frac{v_i}{\mu_i}}. \tag{2.18}$$

The Reuss mixing law represents the isostress average, because it provides the ratio of the average stress to the average strain when all components have the same stress.

Because the Voigt and Reuss bounds provide estimates of the elastic moduli in ideal isostrain and isostress conditions, it is a common practice to use their average to predict the effective properties. The Voigt–Reuss–Hill average is the average of the Voigt and Reuss bounds and it is often used to approximate the bulk and shear moduli of the effective solid medium, K_{sol} and μ_{sol}:

$$K_{sol} = \frac{1}{2}\left(\sum_{i=1}^{n} v_i K_i + \frac{1}{\sum_{i=1}^{n} \frac{v_i}{K_i}}\right) \tag{2.19}$$

$$\mu_{sol} = \frac{1}{2}\left(\sum_{i=1}^{n} v_i \mu_i + \frac{1}{\sum_{i=1}^{n} \frac{v_i}{\mu_i}}\right). \tag{2.20}$$

Example 2.3 We consider a mixture of two minerals, namely quartz and clay, and we illustrate the estimation of the density, bulk and shear moduli of the effective solid medium. The clay type is illite. We assume that clay and quartz are present in the mixture with equal proportions, i.e. the volume of clay is equal to the volume of quartz, $v_c = v_q = 0.5$. The elastic properties of quartz and clay can be found in Table 2.1. We first compute the density of the mixture using Eq. (2.14):

$$\rho_{sol} = 0.5 \times 2.65 + 0.5 \times 2.55 = 2.60 \,\mathrm{g/cm^3}.$$

We then compute the bulk and shear moduli of the mixture using Eqs. (2.15)–(2.20). The Voigt bulk and shear moduli are given by:

$$K_V = 0.5 \times 36 + 0.5 \times 21 = 28.5\,\text{GPa}$$

$$\mu_V = 0.5 \times 45 + 0.5 \times 7 = 26\,\text{GPa}.$$

The Reuss bulk and shear moduli are given by:

$$K_R = \frac{1}{\dfrac{0.5}{36} + \dfrac{0.5}{21}} = 26.5\,\text{GPa}$$

$$\mu_R = \frac{1}{\dfrac{0.5}{45} + \dfrac{0.5}{7}} = 12\,\text{GPa}.$$

We then compute the Voigt–Reuss–Hill average:

$$K_{sol} = \frac{1}{2}(28.5 + 26.5) = 27.5\,\text{GPa}$$

$$\mu_{sol} = \frac{1}{2}(26 + 12) = 19\,\text{GPa}.$$

The elastic bounds for a mixture of quartz and clay with variable clay volume are shown in Figure 2.3. The x-axis represents the clay volume; therefore, clay volume of 0 corresponds to pure quartz, and clay volume of 1 corresponds to pure clay. If the difference between the elastic moduli of the mineral components is small, the upper and lower bounds are relatively narrow; however, if the difference between the elastic moduli is large, the bounds are relatively wide.

Narrower bounds are provided by the Hashin–Shtrikman upper and lower bounds (Mavko et al. 2020). The Hashin–Shtrikman upper bound is given by:

$$K_{HS+} = \left(\sum_{i=1}^{n} \frac{v_i}{K_i + \frac{4}{3}\mu_M}\right)^{-1} - \frac{4}{3}\mu_M \tag{2.21}$$

$$\mu_{HS+} = \left(\sum_{i=1}^{n} \frac{v_i}{\mu_i + \dfrac{\mu_M}{6}\dfrac{9K_M + 8\mu_M}{K_M + 2\mu_M}}\right)^{-1} - \frac{\mu_M}{6}\frac{9K_M + 8\mu_M}{K_M + 2\mu_M}, \tag{2.22}$$

where $K_M = \max_{i=1,\ldots,n}\{K_i\}$ and $\mu_M = \max_{i=1,\ldots,n}\{\mu_i\}$ are the maximum bulk and shear moduli of the mineral components, respectively. The Hashin–Shtrikman lower bound is given by:

$$K_{HS-} = \left(\sum_{i=1}^{n} \frac{v_i}{K_i + \frac{4}{3}\mu_m}\right)^{-1} - \frac{4}{3}\mu_m \tag{2.23}$$

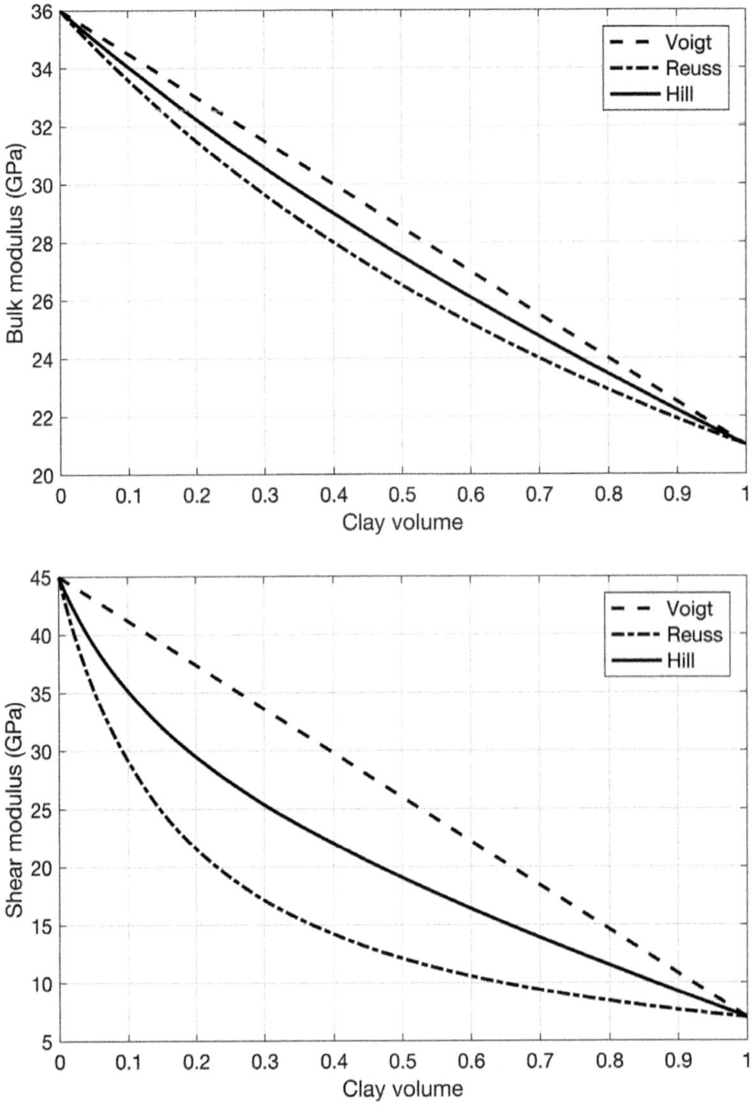

Figure 2.3 Voigt, Reuss, and Voigt–Reuss–Hill averages for the bulk and shear moduli of a mixture of quartz and clay: dashed lines show the Voigt average, dashed-dotted lines show the Reuss average, and solid lines show the Voigt–Reuss–Hill average.

$$\mu_{HS-} = \left(\sum_{i=1}^{n} \frac{v_i}{\mu_i + \frac{\mu_m}{6} \frac{9K_m + 8\mu_m}{K_m + 2\mu_m}} \right)^{-1} - \frac{\mu_m}{6} \frac{9K_m + 8\mu_m}{K_m + 2\mu_m}, \tag{2.24}$$

where $K_m = \min_{i=1,\ldots,n}\{K_i\}$ and $\mu_m = \min_{i=1,\ldots,n}\{\mu_i\}$ are the minimum bulk and shear moduli of the mineral components, respectively. Similar to the Voigt–Reuss–Hill average, the average of the Hashin–Shtrikman upper and lower bounds can be used to approximate the elastic moduli of the effective solid medium.

Example 2.4 For the mixture of quartz and clay in Example 2.3, K_M and μ_M are equal to the bulk and shear moduli of quartz, and K_m and μ_m are equal to the bulk and shear moduli of clay. In this example, we compute the Hashin–Shtrikman elastic moduli for a mixture with equal proportions of quartz and clay. The upper bound bulk modulus is:

$$K_{HS+} = \left(\frac{0.5}{36 + \frac{4}{3} \times 45} + \frac{0.5}{21 + \frac{4}{3} \times 45} \right)^{-1} - \frac{4}{3} \times 45 = 27.8 \text{ GPa},$$

the upper bounds shear modulus is:

$$\mu_{HS+} = \left(\frac{0.5}{45 + \frac{45}{6} \times 5.43} + \frac{0.5}{7 + \frac{45}{6} \times 5.43} \right)^{-1} - \frac{45}{6} \times 5.43 = 20.6 \text{ GPa},$$

the lower bound bulk modulus is:

$$K_{HS-} = \left(\frac{0.5}{36 + \frac{4}{3} \times 7} + \frac{0.5}{21 + \frac{4}{3} \times 7} \right)^{-1} - \frac{4}{3} \times 7 = 27.0 \text{ GPa}.$$

The upper bounds shear modulus is:

$$\mu_{HS-} = \left(\frac{0.5}{45 + \frac{7}{6} \times 7} + \frac{0.5}{7 + \frac{7}{6} \times 7} \right)^{-1} - \frac{7}{6} \times 7 = 15.4 \text{ GPa}.$$

We can then calculate the elastic properties of the effective solid medium by taking the average of the upper and lower bounds. The Hashin–Shtrikman average bulk modulus is $K_{sol} = 27.4$ GPa, and the Hashin–Shtrikman average shear modulus is $\mu_{sol} = 18$ GPa.

The Hashin–Shtrikman bounds for a mixture of quartz and clay are shown in Figure 2.4. The so-obtained Hashin–Shtrikman bounds are narrower than the Voigt and Reuss bounds in Figure 2.3. The Voigt–Reuss–Hill average falls in between the Hashin–Shtrikman bounds.

2.2.2 Fluid Phase

A similar approach based on volumetric averages can be adopted to compute the elastic properties of the effective fluid phase of a mixture of different fluid components. The density of the fluid mixture depends on the fluid saturations and the densities of the fluid components. The bulk modulus of the fluid mixture depends on the fluid saturations and the bulk moduli of the fluid components as well as the spatial distribution of the fluids in the pore space. The shear modulus of the fluid phase is 0 by definition.

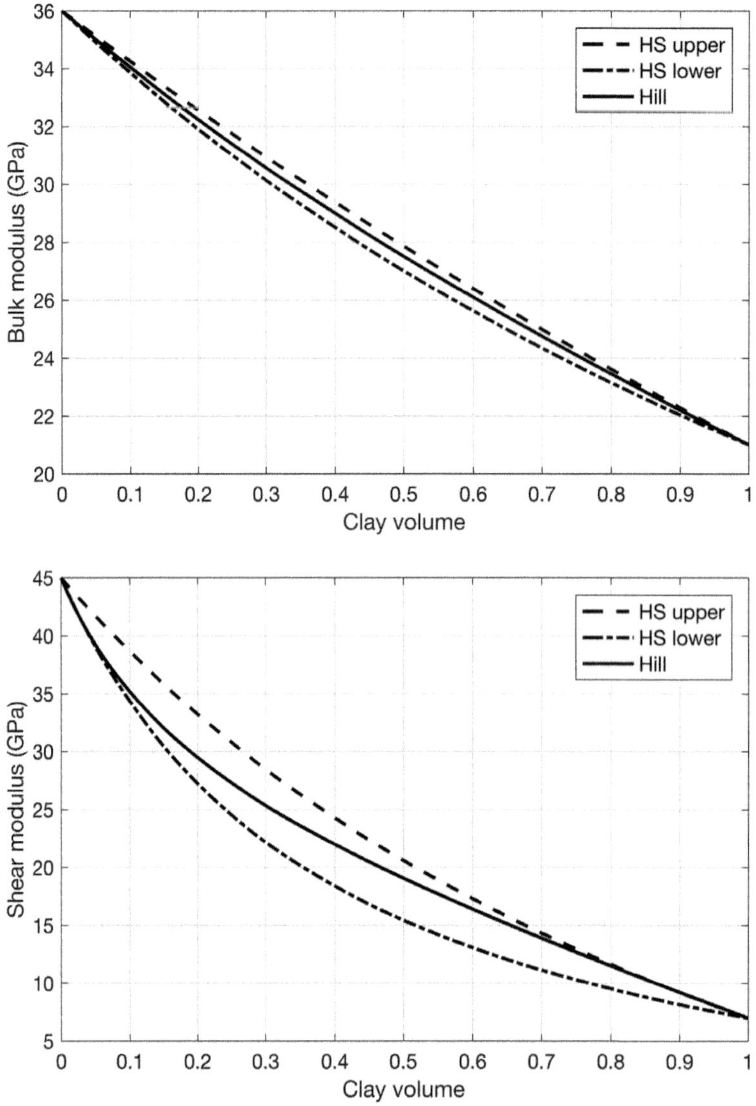

Figure 2.4 Hashin–Shtrikman elastic bounds for the bulk and shear moduli of a mixture of quartz and clay: dashed lines show the Hashin–Shtrikman upper bounds and dashed-dotted lines show the Hashin–Shtrikman lower bounds. The Voigt–Reuss–Hill averages (solid lines) are shown for comparison.

The elastic properties of the fluid components (water, oil, and gas) are not constant and depend on reservoir conditions, such as temperature and pressure, as well as on the fluid composition and characteristics, such as water salinity, oil gravity, gas gravity, and gas–oil ratio (i.e. the amount of gas dissolved in oil). The density and the bulk modulus of the fluid components can be computed using empirical relations, namely the Batzle–Wang

Table 2.2 Density and bulk modulus of water, oil, and gas. The ranges of values depend on reservoir conditions and fluid properties.

Fluid	Density (g/cm^3)	Bulk modulus (GPa)
Water	0.95–1.15	2–3
Oil	0.65–0.95	0.8–1.8
Gas (methane)	0.05–0.25	0.01–0.15

equations (Batzle and Wang 1992), that account for reservoir conditions and fluid characteristics. Table 2.2 shows possible ranges of the elastic properties of water, oil, and gas, for pressure values between 10 and 50 MPa and temperature values between 30 and 80°C. For gas, we report the values of methane, which is the largest component of natural gas. Water is generally denser than oil, and oil is denser than gas. For this reason, gas is more compressible than oil, and oil is more compressible than water, which implies that the bulk modulus of water is higher than the bulk modulus of oil, and the bulk modulus of oil is higher than the bulk modulus of gas.

If the fluid saturations are known, then the density of the effective fluid phase can be calculated as the arithmetic average of the densities of the fluids weighted by their saturations:

$$\rho_{fl} = s_w \rho_w + s_o \rho_o + s_g \rho_g, \tag{2.25}$$

where ρ_w, ρ_o, and ρ_g are the densities of water, oil, and gas, and s_w, s_o, and s_g are the corresponding saturations, with the assumption that $s_w + s_o + s_g = 1$.

In an isostress condition, we can assume that the fluid components are homogeneously mixed together. For a homogeneous mixture of fluids, the Reuss average provides the effective bulk modulus K_{fl} of the fluid mixture:

$$K_{fl} = \frac{1}{\dfrac{s_w}{K_w} + \dfrac{s_o}{K_o} + \dfrac{s_g}{K_g}}. \tag{2.26}$$

In a "patchy" saturation condition, where the fluid mixture is characterized by spatially variable saturations, the Voigt average provides an approximation of the effective bulk modulus K_{fl} of the fluid mixture:

$$K_{fl} = s_w K_w + s_o K_o + s_g K_g. \tag{2.27}$$

Example 2.5 We consider a mixture of water and gas where the water saturation is $s_w = 0.2$ and the gas saturation is $s_g = 0.8$. We assume fresh water with the following density and bulk modulus: $\rho_w = 1$ g/cm^3 and $K_w = 2.25$ GPa; and methane with the following properties: $\rho_g = 0.1$ g/cm^3 and $K_g = 0.1$ GPa. The effective fluid density of the mixture (Eq. 2.25) is then:

$$\rho_{fl} = 0.2 \times 1 + 0.8 \times 0.1 = 0.28\ \text{g/cm}^3.$$

If we assume a homogeneous fluid mixture, the effective fluid bulk modulus (Eq. 2.26) is:

$$K_{fl} = \frac{1}{\frac{0.2}{2.25} + \frac{0.8}{0.1}} = 0.12\,\text{GPa},$$

whereas, if we assume a patchy fluid mixture, the effective fluid bulk modulus (Eq. 2.27) is:

$$K_{fl} = 0.2 \times 2.25 + 0.8 \times 0.1 = 0.53\,\text{GPa}.$$

Similar to the solid phase case, the average of Voigt and Reuss bounds (i.e. patchy and homogeneous mixtures) can also be adopted to describe intermediate cases. Alternatively, Brie et al. (1995) propose a fluid mixing law for patchy mixtures of water and hydrocarbon, namely Brie's equation:

$$K_{fl} = \left(\frac{1}{\frac{s_w}{K_w} + \frac{s_o}{K_o}} - K_g\right)\left(1 - s_g\right)^e + K_g, \tag{2.28}$$

where e is an empirical constant, equal to 3 in the original experiments.

Voigt, Reuss, and Brie's mixing laws are shown in Figure 2.5 for a mixture of water and gas. The figure shows the variations of the bulk modulus as a function of water saturation. If the water saturation increases, the bulk modulus of the fluid mixture increases.

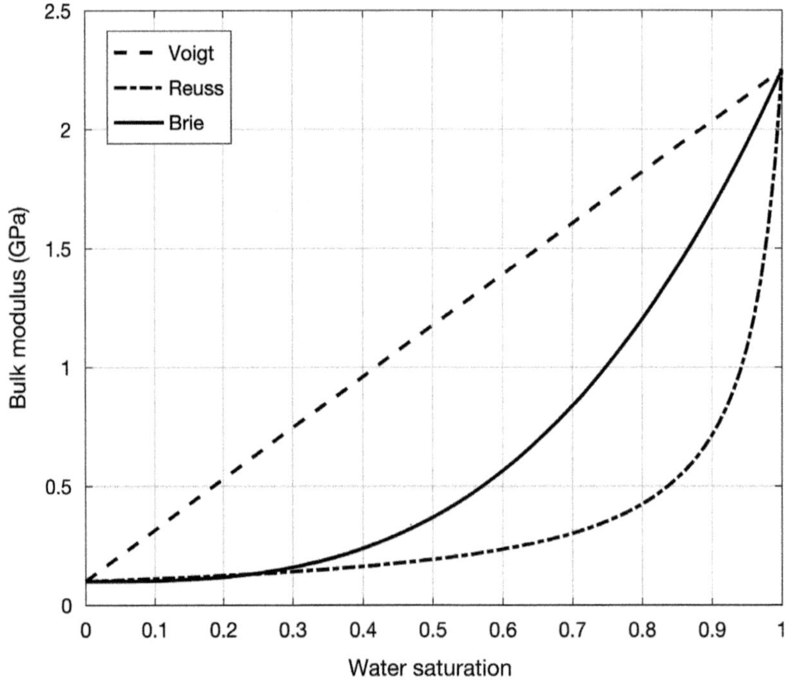

Figure 2.5 Fluid mixing laws for a mixture of water and gas: the dashed line shows Voigt average for patchy saturation mixtures, the dashed-dotted line shows Reuss average for homogeneous mixtures, and the solid line shows Brie's equation.

2.3 Critical Porosity Concept

The effect of porosity on elastic moduli can be modeled according to different rock physics models. Many of these models are based on the concept of critical porosity (Nur et al. 1991). The critical porosity ϕ_c is the porosity value that separates load-bearing sediments, with porosity lower than the critical porosity, and suspensions, with porosity greater than the critical value. The critical porosity value depends on mineral composition, sorting, and angularity at deposition. In a granular material, for example a mixture of quartz and clay, the suspension state at critical porosity describes the rock before compaction and diagenesis take place. Subsequent compaction and diagenesis processes reduce the porosity and increase the elastic stiffness and the elastic properties.

When porosity is lower than the critical porosity, the elastic moduli rapidly decrease from the solid phase values at zero porosity to the suspension values at the critical porosity. As a first approximation, the elastic behavior of a dry rock (i.e. a porous rock with no fluid) can be modeled using a linear trend between zero porosity and the critical porosity (Nur et al. 1991). According to a modified Voigt bound, the dry-rock bulk and shear moduli, K_{dry} and μ_{dry}, can be approximated as:

$$K_{dry} = \left(1 - \frac{\phi}{\phi_c}\right) K_{sol} \tag{2.29}$$

$$\mu_{dry} = \left(1 - \frac{\phi}{\phi_c}\right) \mu_{sol}. \tag{2.30}$$

Non-linear approximations have also been formulated. Krief et al. (1990) propose the following model:

$$K_{dry} = (1 - \beta) K_{sol} \tag{2.31}$$

$$\mu_{dry} = (1 - \beta) \mu_{sol}, \tag{2.32}$$

where β is the Biot's coefficient (Mavko et al. 2020) and it is approximated as $(1 - \beta) = (1 - \phi)^{m(\phi)}$, with $m(\phi) = 3/(1 - \phi)$.

Example 2.6 We consider a porous rock made of quartz ($K_{sol} = 36$ GPa; $\mu_{sol} = 45$ GPa), with porosity $\phi = 0.2$. We assume that the critical porosity is $\phi_c = 0.4$. Then, according to the critical porosity model (Eqs. 2.29 and 2.30), the dry-rock bulk modulus is:

$$K_{dry} = \left(1 - \frac{0.2}{0.4}\right) \times 36 = 18 \text{ GPa}$$

and the dry-rock shear modulus is:

$$\mu_{dry} = \left(1 - \frac{0.2}{0.4}\right) \times 45 = 22.5 \text{ GPa}.$$

Similarly, according to the Krief's model (Eqs. 2.31 and 2.32), the dry-rock bulk modulus is:

$$K_{dry} = (1 - 0.2)^{\frac{3}{(1-0.2)}} \times 36 = 15.6 \text{ GPa}$$

and the dry-rock shear modulus is:

$$\mu_{dry} = (1-0.2)^{\frac{3}{(1-0.2)}} \times 45 = 19.5\ \text{GPa}.$$

2.4 Granular Media Models

Granular media models include several rock physics relations that describe the porous rock as a random pack of spherical grains (Mavko et al. 2020). Spheres are used as idealized representations of grains in sandstone. When porosity increases, the grains lose contact and the porous rock loses its rigidity and its stiffness decreases. The average number of contacts per sphere is generally called the coordination number c. The porosity of a random pack of spheres decreases as the coordination number increases, as a result of the random pack becoming tighter. Random packs of identical spheres have coordination numbers ranging from 6 to 9, whereas in granular media the coordination number varies widely throughout different samples, from 4 to 12.

Several granular media models are based on Hertz–Mindlin contact theory (Mavko et al. 2020). According to Hertz–Mindlin theory, the bulk modulus K_{HM} of a random pack of spherical grains with coordination number c and porosity ϕ is given by:

$$K_{HM} = \sqrt[3]{\frac{c^2(1-\phi)^2 \mu_{sol}^2}{18\pi^2(1-\nu_{sol})^2} P}, \tag{2.33}$$

where μ_{sol} is the shear modulus of the solid phase, ν_{sol} is the Poisson's ratio of the solid phase, and P is the effective pressure. The Poisson's ratio can be computed from the elastic moduli as $\nu = (3K - 2\mu)/(6K + 2\mu)$. The shear modulus μ_{HM} of a random pack of spherical grains is:

$$\mu_{HM} = \frac{5-4\nu_{sol}}{5(2-\nu_{sol})} \sqrt[3]{\frac{3c^2(1-\phi)^2 \mu_{sol}^2}{2\pi^2(1-\nu_{sol})^2} P}. \tag{2.34}$$

The shear modulus expression can also be modified to account for the effect of friction between particles, by introducing the friction coefficient $0 \le f \le 1$, as:

$$\mu_{HM} = \frac{2+3f-\nu_{sol}(1+3f)}{5(2-\nu_{sol})} \sqrt[3]{\frac{3c^2(1-\phi)^2 \mu_{sol}^2}{2\pi^2(1-\nu_{sol})^2} P}, \tag{2.35}$$

where $f = 1$ corresponds to the case where the grains have perfect adhesion and $f = 0$ corresponds to the case where grains have no friction.

Several rock physics models have been derived by combining Hertz–Mindlin theory with the modified Hashin–Shtrikman bounds (Dvorkin et al. 2014). The unconsolidated sand model or soft sand model (Dvorkin and Nur 1996), and the consolidated sand model or stiff sand model (Gal et al. 1998) are commonly used to model dry-rock elastic moduli in sandstone and shaley sandstone. Both models apply Hertz–Mindlin theory (Eqs. 2.33 and 2.34) to compute the dry-rock bulk moduli at the critical porosity and interpolate for intermediate porosities between 0 and the critical porosity ϕ_c using the modified Hashin–Shtrikman bounds.

The soft sand model assumes a dense random pack of identical spherical grains with porosity between 0 and the critical porosity ϕ_c and with coordination number c, typically between 4 and 9, at a given effective pressure P. When porosity is 0 ($\phi = 0$), the dry-rock bulk and shear moduli are equal to the bulk and shear moduli of the solid phase, K_{sol} and μ_{sol}. When porosity equals the critical porosity ($\phi = \phi_c$), the dry-rock bulk and shear moduli are equal to the Hertz–Mindlin bulk and shear moduli, K_{HM} and μ_{HM} (Eqs. 2.33 and 2.34). The model interpolates the values between $\phi = 0$ and $\phi = \phi_c$, using the modified Hashin–Shtrikman lower bounds. The dry-rock bulk and shear moduli, K_{dry} and μ_{dry}, are then given by:

$$K_{dry} = \left(\frac{\dfrac{\phi}{\phi_c}}{K_{HM} + \dfrac{4}{3}\mu_{HM}} + \frac{1 - \dfrac{\phi}{\phi_c}}{K_{sol} + \dfrac{4}{3}\mu_{HM}} \right)^{-1} - \frac{4}{3}\mu_{HM} \tag{2.36}$$

$$\mu_{dry} = \left(\frac{\dfrac{\phi}{\phi_c}}{\mu_{HM} + \dfrac{\mu_{HM}}{6}\xi_{HM}} + \frac{1 - \dfrac{\phi}{\phi_c}}{\mu_{sol} + \dfrac{\mu_{HM}}{6}\xi_{HM}} \right)^{-1} - \frac{\mu_{HM}}{6}\xi_{HM}, \tag{2.37}$$

where $\xi_{HM} = (9K_{HM} + 8\mu_{HM})/(K_{HM} + 2\mu_{HM})$.

Similarly, the stiff sand model interpolates the elastic moduli values for porosity between 0 and the critical porosity ϕ_c using the modified Hashin–Shtrikman upper bounds. The dry-rock elastic moduli are then given by:

$$K_{dry} = \left(\frac{\dfrac{\phi}{\phi_c}}{K_{HM} + \dfrac{4}{3}\mu_{sol}} + \frac{1 - \dfrac{\phi}{\phi_c}}{K_{sol} + \dfrac{4}{3}\mu_{sol}} \right)^{-1} - \frac{4}{3}\mu_{sol} \tag{2.38}$$

$$\mu_{dry} = \left(\frac{\dfrac{\phi}{\phi_c}}{\mu_{HM} + \dfrac{\mu_{sol}}{6}\xi_{sol}} + \frac{1 - \dfrac{\phi}{\phi_c}}{\mu_{sol} + \dfrac{\mu_{sol}}{6}\xi_{sol}} \right)^{-1} - \frac{\mu_{sol}}{6}\xi_{sol}, \tag{2.39}$$

where $\xi_{sol} = (9K_{sol} + 8\mu_{sol})/(K_{sol} + 2\mu_{sol})$.

Example 2.7 We consider two porous rocks, a consolidated sand and an unconsolidated sand, made of quartz with critical porosity $\phi_c = 0.4$ and coordination number $c = 7$, at the effective pressure $P = 20$ MPa. For quartz ($K_{sol} = 36$ GPa; $\mu_{sol} = 45$ GPa), the Poisson's ratio of the mineral is $\nu_{sol} = 0.06$. Then, for both rocks, the dry-rock bulk modulus at the critical porosity (Eq. 2.33) is:

$$K_{HM} = \sqrt[3]{\frac{7^2 \times (1-0.4)^2 \times 45^2}{18 \times \pi^2 \times (1-0.06)^2}} \times 0.02 = 1.65\ \text{GPa}$$

and the dry-rock shear modulus (Eq. 2.34) is:

$$\mu_{HM} = \frac{5 - 4 \times 0.06}{5(2-0.06)} \times \sqrt[3]{\frac{3 \times 7^2 \times (1-0.4)^2 \times 45^2}{2 \times \pi^2 \times (1-0.06)^2}} \times 0.02 = 2.44\ \text{GPa}.$$

For a soft sand with porosity $\phi = 0.2$, the dry-rock bulk modulus (Eq. 2.36) is:

$$K_{dry} = \left(\frac{\frac{0.2}{0.4}}{1.65 + \frac{4}{3} \times 2.44} + \frac{1 - \frac{0.2}{0.4}}{36 + \frac{4}{3} \times 2.44} \right)^{-1} - \frac{4}{3} \times 2.44 = 5.47 \text{ GPa}$$

and the dry-rock shear modulus (Eq. 2.37) is:

$$\mu_{dry} = \left(\frac{\frac{0.2}{0.4}}{2.44 + \frac{2.44}{6} \times 5.26} + \frac{1 - \frac{0.2}{0.4}}{45 + \frac{2.44}{6} \times 5.26} \right)^{-1} - \frac{2.44}{6} \times 5.26 = 6.20 \text{ GPa}.$$

For a stiff sand with porosity $\phi = 0.2$, the dry-rock bulk modulus (Eq. 2.38) is:

$$K_{dry} = \left(\frac{\frac{0.2}{0.4}}{1.65 + \frac{4}{3} \times 45} + \frac{1 - \frac{0.2}{0.4}}{36 + \frac{4}{3} \times 45} \right)^{-1} - \frac{4}{3} \times 45 = 15.09 \text{ GPa}$$

and the dry-rock shear modulus (Eq. 2.39) is:

$$\mu_{dry} = \left(\frac{\frac{0.2}{0.4}}{2.44 + \frac{45}{6} \times 5.43} + \frac{1 - \frac{0.2}{0.4}}{45 + \frac{45}{6} \times 5.43} \right)^{-1} - \frac{45}{6} \times 5.43 = 16.70 \text{ GPa}.$$

The dry-rock elastic moduli of a porous rock made of quartz with critical porosity 0.4 and coordination number 7, at the effective pressure of 20 MPa, are shown in Figure 2.6 as a function of porosity, based on the soft sand (solid lines) and stiff sand (dashed lines) models.

Granular media models assume that a porous rock can be represented by a random pack of solid grains. Porosity variations depend on grain-contact alteration (diagenesis) or deposition of smaller grains in the pore space between larger grains (sorting). Several other models have been developed for granular materials. The contact cement model (Dvorkin and Nur 1996) predicts the dry-rock elastic moduli of cemented sandstones with mid-high porosity values. The constant cement model (Avseth et al. 2010) predicts the dry-rock elastic moduli of a high-porosity grain pack with a given initial cementation, where further porosity reduction is due to non-cementing material into the pore space. In the constant cement model, the elastic behavior for high porosity is described by the contact cement model, whereas the elastic behavior for low porosity is described by the unconsolidated sand model. Other models based on granular media theory include Walton, Digby, Jenkins, Brandt, and Johnson relations (Mavko et al. 2020).

2.5 Inclusion Models

Inclusion models provide an alternative approach to granular media models for the prediction of the elastic moduli of porous rocks with variable porosity. Inclusion models assume that a porous rock can be modeled by placing inclusions within the solid matrix (Mavko

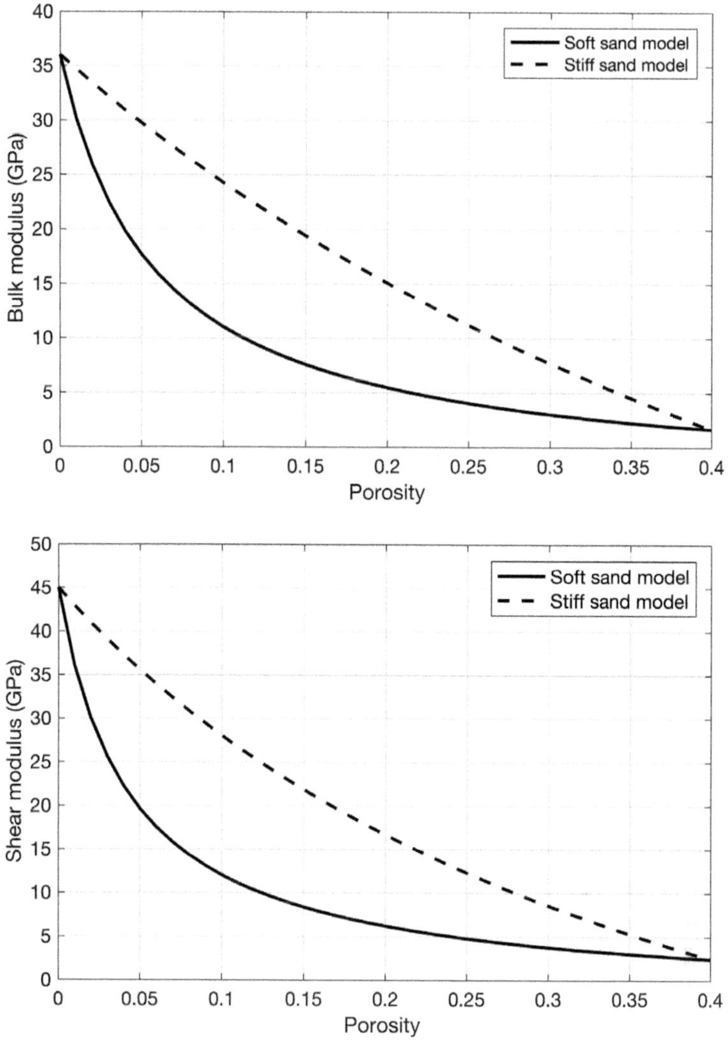

Figure 2.6 Soft sand and stiff sand models for the bulk and shear moduli of a porous rock made of quartz with porosity that varies between 0 and critical porosity 0.4. The solid lines represent the soft sand model; the dashed lines represent the stiff sand model. The models are calculated assuming a coordination number equal to 7 at the effective pressure of 20 MPa.

et al. 2020). These models are suitable for carbonate rocks where the pores appear as inclusions in a calcite or dolomite matrix, but can also be applied to clastic rocks. Kuster and Toksöz (1974) propose expressions for the effective elastic moduli with inclusions of different shapes. The formulations for the effective bulk and shear moduli, K_{KT} and μ_{KT}, are given by:

$$(K_{KT} - K_{sol})\left(\frac{K_{sol} + \frac{4}{3}\mu_{sol}}{K_{KT} + \frac{4}{3}\mu_{sol}}\right) = \sum_{i=1}^{n_i} f_i(K_i - K_{sol})P_i \tag{2.40}$$

$$(\mu_{KT} - \mu_{sol}) \left(\frac{\mu_{sol} + \frac{\mu_{sol}}{6}\xi_{sol}}{\mu_{KT} + \frac{\mu_{sol}}{6}\xi_{sol}} \right) = \sum_{i=1}^{n_i} f_i(\mu_i - \mu_{sol})Q_i \tag{2.41}$$

where n_i is the number of inclusions, the coefficients $\{f_i\}_{i=1,\ldots,n_i}$ represent the volumetric fractions of the inclusions, K_i and μ_i are the bulk and shear moduli of the ith inclusion, P_i and Q_i are the corresponding geometrical factors, and ξ_{sol} is a constant term of the form $\xi_{sol} = (9K_{sol} + 8\mu_{sol})/(K_{sol} + 2\mu_{sol})$. Berryman (1995) proposes several expressions of the geometrical factors for different shapes. For example, for spheres, the expressions of the geometrical factors are:

$$\begin{cases} P_i = \dfrac{K_{sol} + \frac{4}{3}\mu_{sol}}{K_i + \frac{4}{3}\mu_{sol}} \\ \\ Q_i = \dfrac{\mu_{sol} + \frac{\mu_{sol}}{6}\xi_{sol}}{K_i + \frac{\mu_{sol}}{6}\xi_{sol}} \end{cases} \tag{2.42}$$

where $\xi_{sol} = (9K_{sol} + 8\mu_{sol})/(K_{sol} + 2\mu_{sol})$, whereas, for disks, the geometrical factor expressions are:

$$\begin{cases} P_i = \dfrac{K_{sol} + \frac{4}{3}\mu_i}{K_i + \frac{4}{3}\mu_i} \\ \\ Q_i = \dfrac{\mu_{sol} + \frac{\mu_i}{6}\xi_i}{K_i + \frac{\mu_i}{6}\xi_i} \end{cases} \tag{2.43}$$

where $\xi_i = (9K_i + 8\mu_i)/(K_i + 2\mu_i)$.

For a porous rock with a single type of inclusion, the inclusion fraction f_i is equal to the porosity fraction ϕ, and the dry-rock elastic moduli can be computed assuming $K_i = 0$ and $\mu_i = 0$. Assuming spherical inclusions, the dry-rock elastic moduli K_{dry} and μ_{dry} become:

$$K_{dry} = \frac{4K_{sol}\mu_{sol}(1-\phi)}{3K_{sol}\phi + 4\mu_{sol}} \tag{2.44}$$

$$\mu_{dry} = \frac{\mu_{sol}(9K_{sol} + 8\mu_{sol})(1-\phi)}{9K_{sol} + 8\mu_{sol} + 6(K_{sol} + 2\mu_{sol})\phi}. \tag{2.45}$$

Similarly, the saturated-rock bulk modulus K_{sat} can be computed assuming $K_i = K_{fl}$ (i.e. the fluid bulk modulus) and $\mu_i = 0$:

$$K_{sat} = \frac{4K_{sol}\mu_{sol} + 3K_{sol}K_{fl} + 4\mu_{sol}K_{fl}\phi - 4K_{sol}\mu_{sol}\phi}{4\mu_{sol} + 3K_{fl} - 3K_{fl}\phi + 3K_{sol}\phi}, \tag{2.46}$$

whereas the saturated-rock shear modulus is equal to the dry-rock shear modulus $\mu_{sat} = \mu_{dry}$ (Eq. 2.45).

A similar approach is used in self-consistent approximation models (Mavko et al. 2020). The self-consistent approximation model proposed in Te Wu (1966) provides the solution for the elastic moduli for a porous rock with a single type of inclusion with volume fraction f_i:

$$K_{SC} = K_{sol} + f_i(K_i - K_{sol})P_i \tag{2.47}$$

$$\mu_{SC} = \mu_{sol} + f_i(\mu_i - \mu_{sol})Q_i. \tag{2.48}$$

Expressions for the geometrical coefficients P_i and Q_i are available for different pore shapes, including spheres, needles, disks, and penny-cracks. A generalization of the self-consistent approximation model for n_i inclusions is given in Berryman (1995):

$$\sum_{i=1}^{n_i} f_i(K_i - K_{SC})P_i = 0 \tag{2.49}$$

$$\sum_{i=1}^{n_i} f_i(\mu_i - \mu_{SC})Q_i = 0, \tag{2.50}$$

where the summation is over all the n_i phases, including minerals and pores. The coefficients P_i and Q_i for ellipsoidal inclusions of arbitrary aspect ratio α are given in Berryman (1995). The coefficients P_i and Q_i are given by:

$$P_i = \frac{F_1}{F_2}$$

$$Q_i = \frac{1}{5}\left(\frac{2}{F_3} + \frac{1}{F_4} + \frac{F_4F_5 + F_6F_7 - F_8F_9}{F_2F_4}\right)$$

$$F_1 = 1 + A\left[\frac{3}{2}(f + \theta) - R\left(\frac{3}{2}f + \frac{5}{2}\theta - \frac{4}{3}\right)\right]$$

$$F_2 = 1 + A\left[1 + \frac{3}{2}(f + \theta) - \frac{1}{2}R(3f + 5\theta)\right] + B(3 - 4R)$$

$$+ \frac{1}{2}A(A + 3B)(3 - 4R)\left[f + \theta - R\left(f - \theta + 2\theta^2\right)\right]$$

$$F_3 = 1 + A\left[1 - \left(f + \frac{3}{2}\theta\right) + R(f + \theta)\right]$$

$$F_4 = 1 + \frac{1}{4}A[f + 3\theta - R(f - \theta)]$$

$$F_5 = A\left[-f + R\left(f + \theta - \frac{4}{3}\right)\right] + B\theta(3 - 4R)$$

$$F_6 = 1 + A[1 + f - R(f + \theta)] + B(1 - \theta)(3 - 4R)$$

$$F_7 = 2 + \frac{1}{4}A[3f + 9\theta - R(3f + 5\theta)] + B\theta(3 - 4R)$$

$$F_8 = A\left[1 - 2R + \frac{1}{2}f(R - 1) + \frac{1}{2}\theta(5R - 3)\right] + B(1 - \theta)(3 - 4R)$$

$$F_9 = A[(R - 1)f - R\theta] + B\theta(3 - 4R)$$

$$A = \frac{\mu_i}{\mu_{sol}} - 1$$

$$B = \frac{1}{3}\left(\frac{K_i}{K_{sol}} - \frac{\mu_i}{\mu_{sol}}\right)$$

$$R = \frac{1-2\nu_{sol}}{2-2\nu_{sol}}$$

$$\theta = \begin{cases} \dfrac{\alpha}{(\alpha^2-1)^{3/2}}\left[\alpha\left(\alpha^2-1\right)^{1/2} - \cosh^{-1}\alpha\right] & \alpha > 1 \\ \dfrac{\alpha}{(1-\alpha^2)^{3/2}}\left[\cos^{-1}\alpha - \alpha\left(1-\alpha^2\right)^{1/2}\right] & \alpha < 1 \end{cases}$$

$$f = \frac{\alpha^2}{1-\alpha^2}(3\theta - 2). \tag{2.51}$$

where $\alpha < 1$ represents oblate spheroids and $\alpha > 1$ represents prolate spheroids.

Example 2.8 We consider two porous rocks made of quartz, one with spherical pores (aspect ratio $\alpha = 1$) and one with elliptical pores (oblate spheroids with aspect ratio $\alpha = 0.2$). We first compute the dry-rock elastic properties of the rock with spherical pores using Eqs. (2.44) and (2.45). The dry-rock bulk modulus (Eq. 2.44) is:

$$K_{dry} = \frac{4 \times 36 \times 45 \times (1-0.2)}{3 \times 36 \times 0.2 + 4 \times 45} = 25.71 \text{ GPa}$$

and the dry-rock shear modulus (Eq. 2.45) is:

$$\mu_{dry} = \frac{45 \times (9 \times 36 + 8 \times 45)(1-0.2)}{9 \times 36 + 8 \times 45 + 6 \times (36 + 2 \times 45) \times 0.2} = 29.48 \text{ GPa}.$$

We then compute the dry-rock elastic properties of the rock with elliptical pores using Eqs. (2.49)–(2.51). For a rock with elliptical pores with aspect ratio $\alpha = 0.2$, the dry-rock bulk and shear moduli are $K_{dry} = 19.09$ GPa and $\mu_{dry} = 22.88$ GPa, respectively. The so-obtained moduli are generally higher than the values obtained using the granular media models, namely soft and stiff sand models, in Example 2.7.

The dry-rock elastic moduli computed using inclusion models as a function of porosity are shown in Figure 2.7, for two porous rocks made of quartz, one with spherical pores (solid lines) and one with elliptical pores with aspect ratio $\alpha = 0.2$ (dashed lines). The elastic moduli obtained from the inclusion models are generally higher than the predictions obtained using the granular media models shown in Figure 2.6.

Inclusion models can be applied to compute dry-rock elastic moduli by modeling the pore space as an empty inclusion or saturated-rock elastic moduli by modeling the pore space as a fluid-saturated inclusion. However, these models assume that the pores are isolated with respect to the fluid flow; therefore, pore pressure is unequilibrated. For these reasons, the calculation of the saturated-rock elastic moduli is suitable for high-frequency laboratory conditions. For low frequencies, such as surface reflection seismic measurements, inclusion models can be used to compute the dry-rock elastic moduli and combined with Gassmann's equations (Section 2.6) to calculate the saturated-rock elastic moduli.

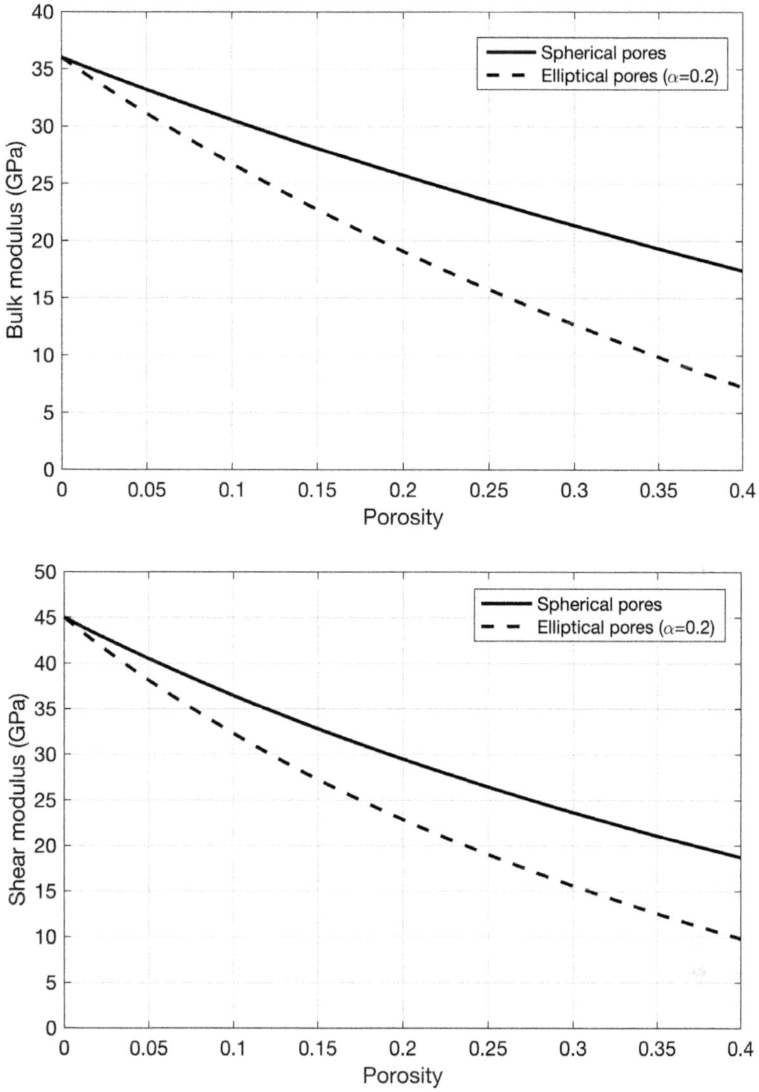

Figure 2.7 Inclusion models for the bulk and shear moduli of a porous rock made of quartz with porosity that varies between 0 and 0.4. The solid lines represent the model for spherical pores; the dashed lines represent the model for elliptical pores with aspect ratio 0.2.

2.6 Gassmann's Equations and Fluid Substitution

Critical porosity, granular media, and inclusion models are generally applied to compute the dry-rock elastic properties as a function of porosity and mineral volumes. The most common approach to include the fluid effect and compute the saturated-rock elastic properties is based on Gassmann's equations (Gassmann 1951).

According to Gassmann's equations, the saturated-rock bulk modulus K_{sat} can be computed as:

$$K_{sat} = K_{dry} + \frac{\left(1 - \frac{K_{dry}}{K_{sol}}\right)^2}{\frac{\phi}{K_{fl}} + \frac{(1-\phi)}{K_{sol}} - \frac{K_{dry}}{K_{sol}^2}}, \tag{2.52}$$

whereas the saturated-rock shear modulus μ_{sat} is equal to the dry-rock shear modulus in Gassmann's static (low-frequency) theory:

$$\mu_{sat} = \mu_{dry}. \tag{2.53}$$

Gassmann's equations, in the proposed formulation, assume that the rock is isotropic and the mineral moduli are homogeneous. These assumptions are generally not verified in porous rocks; however, for isotropic and weakly anisotropic rocks, Gassmann's equations provide an accurate approximation. Extensions of Gassmann's equations for anisotropic rocks are also available in the rock physics literature (Gassmann 1951; Mavko et al. 2020). Brown and Korringa (1975) generalize Eqs. (2.52) and (2.53) for heterogeneous rocks with mixed mineralogy. Gassmann's equations also assume that the pore pressure variations induced in the fluid by seismic waves can rapidly equilibrate through the pore space. This assumption is not valid for high-frequency laboratory conditions, but it is generally verified for field measurements, such as surface seismic data and possibly well logs (Mavko et al. 2020).

Example 2.9 We consider two porous rocks, a consolidated sand and an unconsolidated sand, with the following properties: porosity is $\phi = 0.2$ and the bulk and shear moduli of quartz are $K_{sol} = 36$ GPa and $\mu_{sol} = 45$ GPa, respectively. We first adopt the soft sand model to compute the dry-rock elastic properties, $K_{dry} = 5.47$ GPa and $\mu_{dry} = 6.20$ GPa, as shown in Example 2.7 (Section 2.4). We assume that the porous rock is saturated with water with bulk modulus $K_{fl} = 2.25$ GPa and we apply Gassmann's equations (Eqs. 2.52 and 2.53) to compute the saturated-rock elastic moduli. The saturated-rock bulk modulus (Eq. 2.52) is:

$$K_{sat} = 5.47 + \frac{\left(1 - \frac{5.47}{36}\right)^2}{\frac{0.2}{2.25} + \frac{(1-0.2)}{36} - \frac{5.47}{36^2}} = 12.20 \text{ GPa},$$

and the saturated-rock shear modulus (Eq. 2.53) is equal to the dry-rock shear modulus $\mu_{sat} = \mu_{dry} = 6.20$ GPa. We then adopt the stiff sand model to compute the dry-rock elastic properties, $K_{dry} = 15.09$ GPa and $\mu_{dry} = 16.70$ GPa, as shown in Example 2.7 (Section 2.4), and apply Gassmann's equations to compute the saturated-rock elastic moduli. The saturated-rock bulk modulus is:

$$K_{sat} = 15.09 + \frac{\left(1 - \frac{15.09}{36}\right)^2}{\frac{0.2}{2.25} + \frac{(1-0.2)}{36} - \frac{15.09}{36^2}} = 18.48 \text{ GPa},$$

and the saturated-rock shear modulus is equal to the dry-rock shear modulus $\mu_{sat} = \mu_{dry} = 16.70$ GPa.

Similarly, we can repeat the calculations using the dry-rock elastic moduli computed in Example 2.8 (Section 2.5). For the porous rock with spherical pores, with dry-rock elastic moduli $K_{dry} = 25.71$ GPa and $\mu_{dry} = 29.48$ GPa, the saturated-rock bulk and shear moduli are $K_{sat} = 26.61$ GPa and $\mu_{sat} = 29.48$ GPa, respectively. For the porous rock with elliptical pores with aspect ratio 0.2, with dry-rock elastic moduli $K_{dry} = 19.09$ GPa and $\mu_{dry} = 22.88$ GPa, the saturated-rock bulk and shear moduli are $K_{sat} = 21.28$ GPa and $\mu_{sat} = 22.88$ GPa, respectively.

Example 2.9 assumes that the dry-rock elastic moduli are computed using one of the theoretical rock physics models in Sections 2.4 and 2.5. However, in many practical applications, the dry-rock elastic moduli can be predicted from velocity measurements. If the P-wave and S-wave velocity and density are available, for example, from core samples or well logs, then the saturated-rock elastic moduli can be calculated from the definitions of P-wave and S-wave velocity (Eqs. 2.11 and 2.12) as:

$$K_{sat} = \rho {V_P}^2 - \frac{4}{3}\rho {V_S}^2 \tag{2.54}$$

$$\mu_{sat} = \rho {V_S}^2. \tag{2.55}$$

The dry-rock bulk modulus K_{dry} can then be computed using Gassmann's inverse equation:

$$K_{dry} = \frac{K_{sat}\left(\frac{K_{sol}}{K_{fl}}\phi + 1 - \phi\right) - K_{sol}}{\frac{K_{sol}}{K_{fl}}\phi + \frac{K_{sat}}{K_{sol}} - 1 - \phi}, \tag{2.56}$$

whereas the dry-rock shear modulus μ_{dry} is equal to the saturated-rock shear modulus $\mu_{dry} = \mu_{sat}$ (Eq. 2.53).

A common application of Gassmann's equations is the fluid substitution method, commonly used in rock physics studies to predict the saturated-rock elastic properties under different fluid conditions (Mavko et al. 2020). The fluid substitution method is generally applied to predict the elastic moduli and velocities of a porous rock saturated with a given fluid (fluid 2) from the elastic properties of the same rock saturated with a different fluid (fluid 1). We assume that P-wave and S-wave velocity and density of the porous rock saturated with fluid 1 are available and indicate them with V_{P_1}, V_{S_1}, and ρ_1. The fluid substitution method includes the following steps:

1) We compute the elastic moduli K_{sat_1} and μ_{sat_1} using Eqs. (2.54) and (2.55).
2) We apply Gassmann's inverse equation to compute the dry-rock bulk modulus K_{dry} (Eq. 2.56).
3) We calculate the bulk modulus K_{sat_2} of the rock saturated with fluid 2 using Gassmann's equation (Eq. 2.52).
4) We assume that shear modulus μ_{sat_2} is equal to the dry-rock shear modulus, hence $\mu_{sat_2} = \mu_{dry} = \mu_{sat_1}$, according to Gassmann's assumption (Eq. 2.53).
5) We compute the density of the rock saturated with fluid 2 as:

$$\rho_2 = \rho_1 + \phi\left(\rho_{fl_2} - \rho_{fl_1}\right) \tag{2.57}$$

based on the linearity of the density equation (Eq. 2.13).
6) We calculate the velocities V_{P_2} and V_{S_2} by definition (Eqs. 2.11 and 2.12).

The fluid substitution can also be performed based on P-wave velocity only, by using the compressional modulus $M = (K + 4/3\mu)$ as in Mavko et al. (1995).

Figure 2.8 shows an example of application of Gassmann's equations and fluid substitution to Han's dataset. The original dataset includes rock samples saturated with brine (Figure 2.1). Because Han's measurements were made at ultrasonic frequencies, dispersion effects can have affected velocity measurements in saturated rocks. In this example, we ignore dispersion effects in the calculations. We first compute the bulk modulus of the brine-saturated samples (black squares in Figure 2.8) using Eq. (2.54). We then compute the bulk modulus of the dry samples (gray crosses in Figure 2.8) using Gassmann's inverse relation (Eq. 2.56). We then apply Gassmann's relation (Eq. 2.52) to compute the bulk

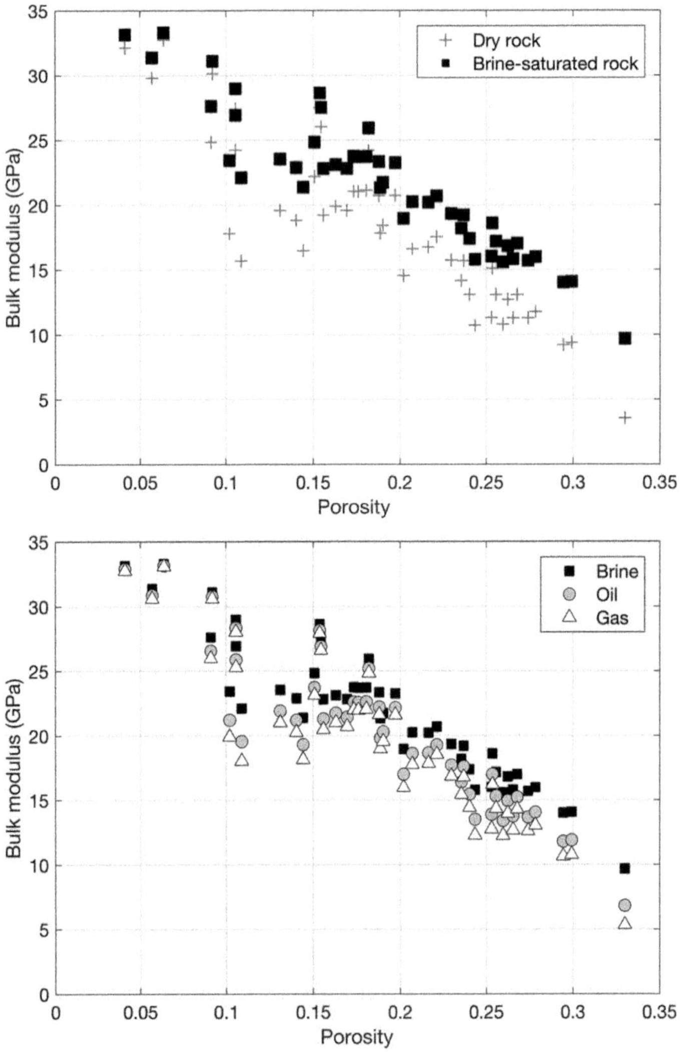

Figure 2.8 Application of Gassmann's equations and fluid substitution to a subset of samples from Han's dataset for the estimation of the bulk modulus of rock samples saturated with different fluids: black squares represent brine-saturated rocks, gray crosses represent dry rocks, gray circles represent oil-saturated rocks, and white triangles represent gas-saturated rocks.

modulus of the samples saturated with oil (gray circles) and gas (white triangles). Owing to the different compressibility of brine, oil, and gas, brine-saturated rocks are stiffer than oil-saturated rocks, and oil-saturated rocks are stiffer than gas-saturated rocks. Therefore, the bulk modulus of brine-saturated samples is greater than the bulk modulus of oil-saturated samples, and the bulk modulus of oil-saturated samples is greater than the bulk modulus of gas-saturated samples. The difference in the bulk moduli is larger for high porosity rocks, where the fluid effect is more significant, and converges to 0 for low porosity rocks (Figure 2.8).

Gassmann's equations can be combined with granular media and inclusion models to predict the saturated-rock bulk moduli. Velocities are then computed by definition using Eqs. (2.11) and (2.12). Figure 2.9 shows the velocity predictions obtained using the soft and stiff sand

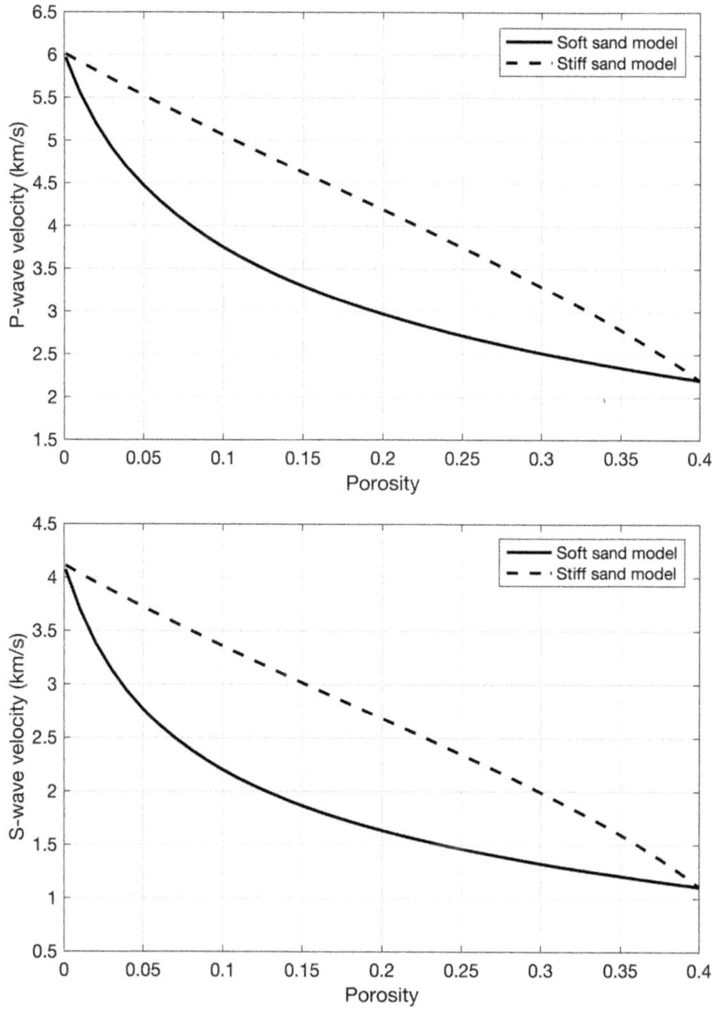

Figure 2.9 P-wave and S-wave velocity predictions as a function of porosity obtained combining Gassmann's equations with the soft sand model (solid lines) and stiff sand model (dashed lines) for a water-saturated sandstone.

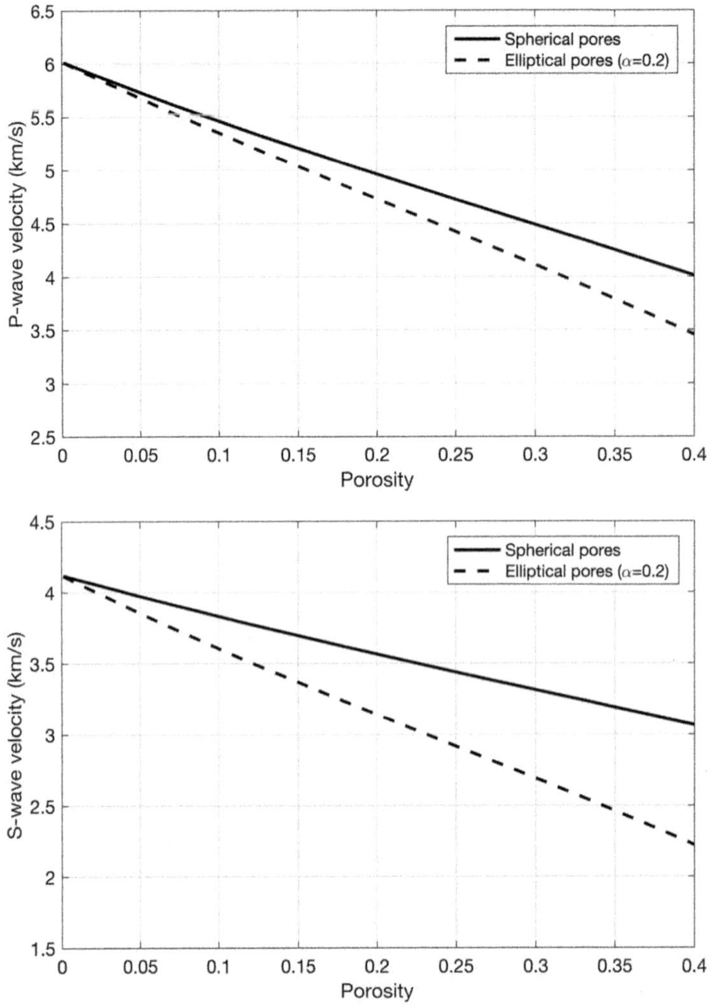

Figure 2.10 P-wave and S-wave velocity predictions as a function of porosity obtained combining Gassmann's equations with inclusion models for spherical pores (solid lines) and elliptical pores with aspect ratio 0.2 (dashed lines) for a water-saturated sandstone.

models (Figure 2.6) combined with Gassmann's equations for a water-saturated sandstone. Figure 2.10 shows the velocity predictions obtained using the inclusion models for spherical and ellipsoidal pore shapes (Figure 2.7) combined with Gassmann's equations.

2.7 Other Rock Physics Relations

The rock physics models in Sections 2.1–2.6 focus on the elastic response, namely P-wave and S-wave velocity and density, of a porous rock with known porosity, mineral fractions, and fluid saturations. However, rock physics includes a broad variety of relations between different rock and fluid properties (Mavko et al. 2020).

Petro-elastic relations linking petrophysical properties to elastic attributes are the most commonly used in reservoir characterization studies, where elastic attributes are estimated from seismic data. In some applications, resistivity logs and controlled-source electromagnetic (CSEM) data might be available (Section 7.3) and provide information about the resistivity of the saturated rocks. Resistivity is defined as the reciprocal of conductivity. Several rock physics models have been developed to link porosity ϕ, water saturation s_w, and resistivity R (Mavko et al. 2020). These models are commonly used to predict saturation at the borehole locations using resistivity well logs and in reservoir models using CSEM data.

The most common rock physics relation to predict the resistivity R of a porous rock partially saturated with water is Archie's law (Archie 1942; Mavko et al. 2020):

$$R = a\frac{R_w}{\phi^m {s_w}^n}, \tag{2.58}$$

where a is an empirical constant generally close to 1, R_w is the resistivity of the pore water that depends on salinity and temperature, and the exponents m and n are two empirical constants generally calibrated using laboratory measurements. The parameter m is called the cementation exponent. It is generally close to 2 for sandstones and varies between approximately 1.3 and 2.5 for most sedimentary rocks. The parameter n is called the saturation exponent. It is generally close to 2 and it depends on the fluid characteristics. Archie's law is commonly applied to clean sandstones; the extension to shaley sandstones requires an additional term to account for the clay resistivity (Worthington 1985). Several formulations have been proposed to predict the resistivity of shaley sandstones partially saturated with water, such as Simandoux and Poupon–Leveaux relations (Simandoux 1963; Poupon and Leveaux 1971; Mavko et al. 2020), which include an additional term at the denominator of Archie's law (Eq. 2.58) that depends on the clay volume v_c. The Simandoux equation can be written as:

$$R = a\frac{1}{\left(\frac{\phi^m}{R_w} + \frac{v_c}{R_c}\right){s_w}^n}, \tag{2.59}$$

where R_c is the intrinsic resistivity of clay and it depends on the clay type. Similarly, the Poupon–Leveaux equation can be formulated as:

$$R = a\frac{1}{\left[\left(\frac{\phi^m}{R_w}\right)^{\frac{1}{2}} + \left(\frac{{v_c}^{\alpha - v_c}}{R_c}\right)^{\frac{1}{2}}\right]^2 {s_w}^n}, \tag{2.60}$$

where α is an empirical parameter generally calibrated using laboratory measurements.

Another rock property of interest in reservoir modeling is permeability. Fluid flow simulation models require the knowledge of the initial fluid pressure and saturation conditions and the spatial distribution of porosity and permeability (Section 7.4). Seismic reservoir characterization methods generally allow the prediction of porosity from seismic data (Chapter 5), whereas permeability is often computed from porosity based on rock physics relations calibrated on rock sample measurements. The Kozeny–Carman equation (Kozeny 1927; Carman 1937; Mavko et al. 2020) provides a relation between permeability and rock and fluid properties, including porosity and particle size. The Kozeny–Carman equation assumes that a porous rock with porosity ϕ can be geometrically described as a random pack of identical spherical grains with diameter d, and its permeability k is given by:

$$k = \frac{1}{B}\frac{\phi^3}{(1-\phi)^2}d^2, \tag{2.61}$$

where B is a constant that depends on a geometrical factor associated with the tortuosity of the pore space and it is often calibrated using laboratory measurements. Equation (2.61) can also be modified to account for a percolation porosity ϕ_p as:

$$k = \frac{1}{B}\frac{\left(\phi-\phi_p\right)^3}{\left[1-\left(\phi-\phi_p\right)\right]^2}d^2, \tag{2.62}$$

to neglect the disconnected pore volume that does not contribute to the fluid flow (Mavko and Nur 1997).

During hydrocarbon reservoir production, changes in fluid saturation and effective pressure affect the elastic and seismic response of the saturated rocks (Christensen and Wang 1985). Isotropic effective pressure P is defined as the difference between the overburden pressure P_o and the pore pressure P_p:

$$P = P_o - \eta P_p, \tag{2.63}$$

where the overburden pressure P_o is generally computed as the integral of density ρ from the surface to the depth of interest z times the acceleration due to gravity g:

$$P_o = g\int_0^z \rho(z)dz. \tag{2.64}$$

The parameter $\eta \leq 1$ in Eq. (2.63) is the effective stress coefficient and it depends on the lithology and the stress conditions (Zimmerman 1990). When $\eta = 1$, the effective stress $P = P_o - P_p$ is known as Terzaghi's effective stress or Terzaghi's law (Terzaghi 1943).

The relation between effective pressure and velocity is non-linear. Generally, both P-wave and S-wave velocities tend to increase as effective pressure increases, until they reach an asymptotic value. The pressure dependence results from the closing of the crack-like pore space and tightening of grain boundaries, which makes the rock stiffer. The magnitude of the velocity variation generally depends on the pore volume and shape. Figure 2.11 shows the P-wave and S-wave velocity measured at different effective pressure values from 5 to 40 MPa, for two samples of shaley sandstones from Han's dataset (Han 1986).

Several models have been proposed to model the effective pressure dependence of elastic properties, including velocities and elastic moduli (Gutierrez et al. 2006; Mavko et al. 2020).

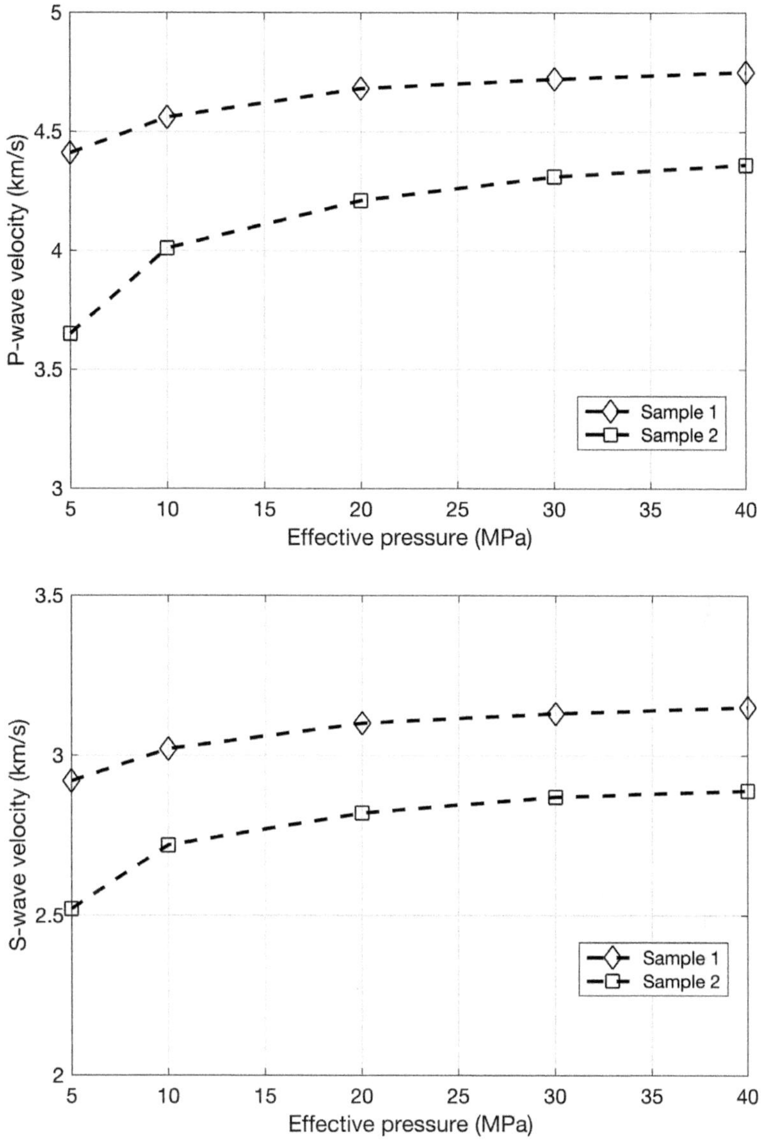

Figure 2.11 P-wave and S-wave velocity variations as a function of effective pressure for two shaley sandstones from Han's dataset.

Eberhart-Phillips et al. (1989) propose a general form for P-wave and S-wave velocities given by:

$$V_P = A_P + B_P(P - \exp(-\lambda_P P)) \tag{2.65}$$

$$V_S = A_S + B_S(P - \exp(-\lambda_S P)) \tag{2.66}$$

where the model parameters A_P, A_S, B_P, B_S, λ_P, and λ_S are empirically calibrated using laboratory measurements. Eberhart-Phillips et al. (1989) show that in sandstone and shaley sandstone, the coefficients A_P and A_S depend on porosity and clay content, as shown in Eqs. (2.5) and (2.6). MacBeth (2004) proposes an analogous equation to link the dry-rock elastic moduli to effective pressure using exponential relations as:

$$K_{dry}(P) = \frac{K^{\infty}}{1 + A_K \exp\left(-\frac{P}{P_K}\right)} \tag{2.67}$$

$$\mu_{dry}(P) = \frac{\mu^{\infty}}{1 + A_\mu \exp\left(-\frac{P}{P_\mu}\right)}, \tag{2.68}$$

where the model parameters K^{∞}, μ^{∞}, A_K, A_μ, P_K, and P_μ are empirically obtained by fitting a set of lab measurements conducted on dry samples of sandstone and shaley sandstone. Equations (2.67) and (2.68) converge to the asymptotic values K^{∞} and μ^{∞} when effective pressure P increases and reaches the characteristic pressure values P_K and P_μ, respectively. As for Eqs. (2.65) and (2.66), in sandstone and shaley sandstone, the asymptotic values K^{∞} and μ^{∞} depend on porosity and clay content, and Eqs. (2.67) and (2.68) can be rewritten as:

$$K_{dry}(P) = \frac{a_K \phi + b_K v_c + c_K}{1 + A_K \exp\left(-\frac{P}{P_K}\right)} \tag{2.69}$$

$$\mu_{dry}(P) = \frac{a_\mu \phi + b_\mu v_c + c_\mu}{1 + A_\mu \exp\left(-\frac{P}{P_\mu}\right)}, \tag{2.70}$$

where c_K and c_μ have the dimensions of elastic moduli and the parameters a_K, b_K, a_μ, and b_μ are generally negative. MacBeth's equations can be combined with Gassmann's equations to predict the elastic response as a function of saturation and pressure in time-lapse seismic studies (Section 7.2).

2.8 Application

We show two different applications of rock physics models to well log data acquired in the Rock Springs Uplift, a potential carbon dioxide (CO_2) sequestration site, in southwest Wyoming (Grana et al. 2017b). Two potential storage units have been identified: the Mississippian Madison Limestone and Middle-Late Pennsylvanian Weber Sandstone. These formations are deep saline aquifers with effective sealing layers that could guarantee long-term CO_2 storage. The well logs were acquired for a pre-injection reservoir characterization study and include P-wave and S-wave velocity, density, and petrophysical curves of porosity, mineral volumes, and water saturation obtained from the formation evaluation analysis.

The Weber Sandstone formation is mostly composed of sand with low porosity values. The average porosity is 0.08 and the average shale content is 0.1 with limited variations within the interval. The solid phase of the porous rocks also includes fractions of dolomite and limestone; in the upper part of the interval, the mineral volumes of dolomite and limestone are approximately constant, whereas in the lower part, several interbedded carbonate layers are present. The fluid is brine with a residual oil saturation between 0.1 and 0.3. The well logs are shown in Figure 2.12.

We adopt the stiff sand model combined with Gassmann's equations to predict the elastic properties, i.e. P-wave and S-wave velocity, and density, and we compare the predictions with the actual measurements in Figure 2.12. We first use Hertz–Mindlin relations (Eqs. 2.33 and 2.34) to compute the dry-rock bulk and shear moduli at the critical porosity of 0.4, assuming a coordination number of 9. Then we apply the modified Hashin–Shtrikman upper bounds (Eqs. 2.38 and 2.39) to compute the dry-rock bulk and shear moduli for porosities between 0 and critical porosity 0.4. To account for the complex mineral composition of the solid phase and local heterogeneities, we apply the stiff-sand model assuming a range of values for the bulk modulus between 39 and 51 GPa and a range of values for the shear modulus between 37 and 49 GPa, based on the variability of the measured mineral fractions. The fluid effect is computed using Gassmann's equations (Eqs. 2.52 and 2.53). The rock physics model in brine-saturated conditions is shown in Figure 2.13, for different values of the solid-phase bulk and shear moduli. The model predictions in the borehole are displayed in Figure 2.12 (dashed red lines) and they are compared with the actual measurements, showing accurate results.

The Madison Limestone formation includes limestone with porosity close to 0 in the upper part of the interval, and dolomite with porosity gradually decreasing from 0.27 to 0.05 in the lower part of the interval. Shale and sand percentages are negligible. The fluid is brine with residual oil saturation. The well logs are shown in Figure 2.14.

We adopt Berryman's inclusion model assuming an aspect ratio of 0.2 (Eqs. 2.49–2.51). The bulk modulus of the solid phase varies between 67 and 77 GPa and the shear modulus varies between 32 and 46 GPa, based on the range of the measured mineral fractions. The fluid effect is computed using Gassmann's equations. The rock physics model in brine-saturated conditions is shown in Figure 2.15. The model predictions in the borehole are displayed in Figure 2.14 and show accurate predictions compared with the actual measurements.

Rock physics models are commonly used in petrophysical inversion studies (Chapter 5) to predict the petrophysical properties of reservoir rocks based on elastic attributes or seismic data. Examples of applications of rock physics models in seismic reservoir characterization studies are shown in Chapter 8.

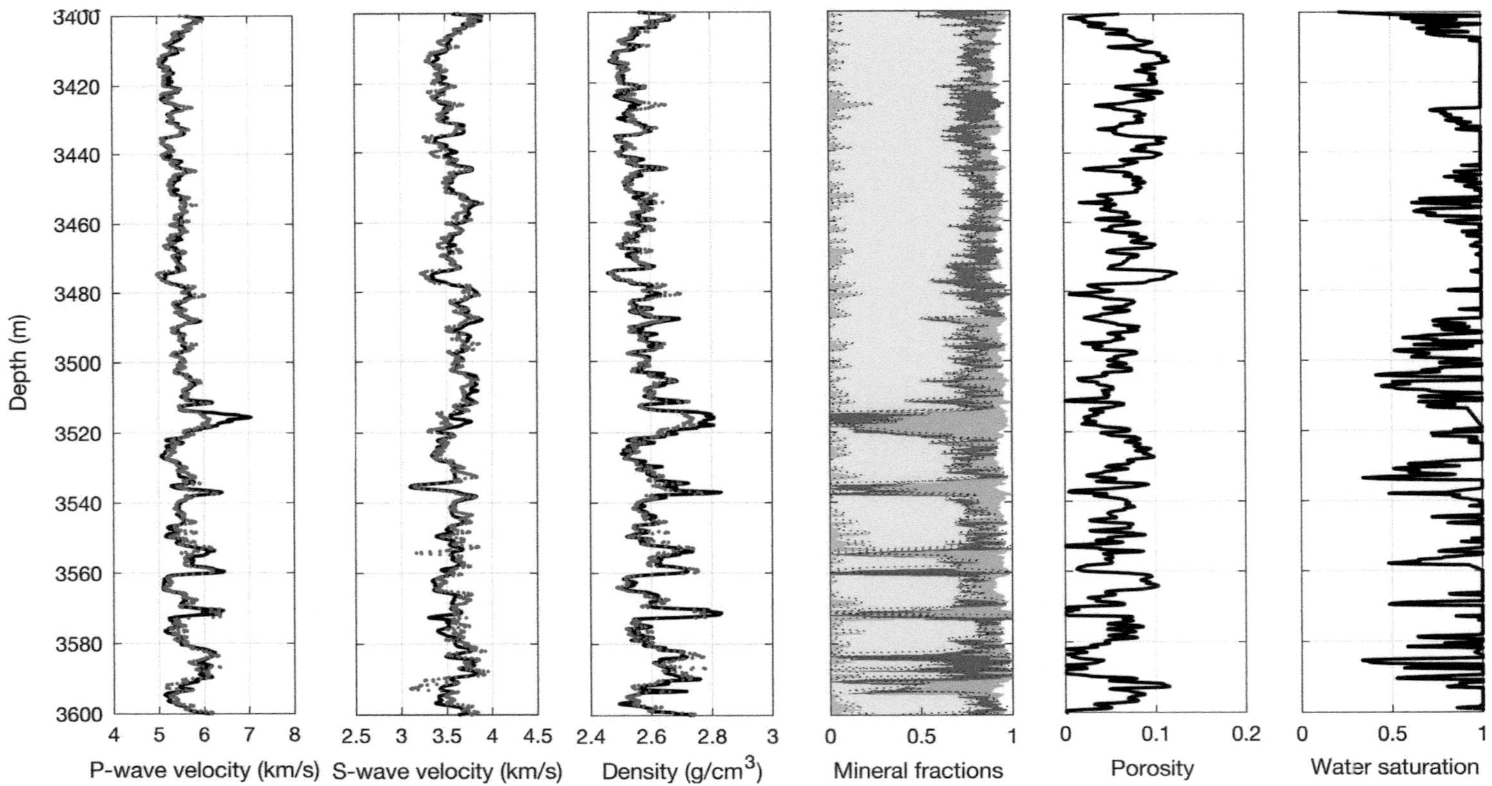

Figure 2.12 Well logs for the Weber Sandstone formation, from left to right: P-wave velocity, S-wave velocity, density, mineral fractions (volume of shale in tan, volume of sand in yellow, volume of limestone in blue, volume of dolomite in green, and pore volume in white), porosity and water saturation. The black curves show the well log data; the dashed red curves show the rock physics model predictions.

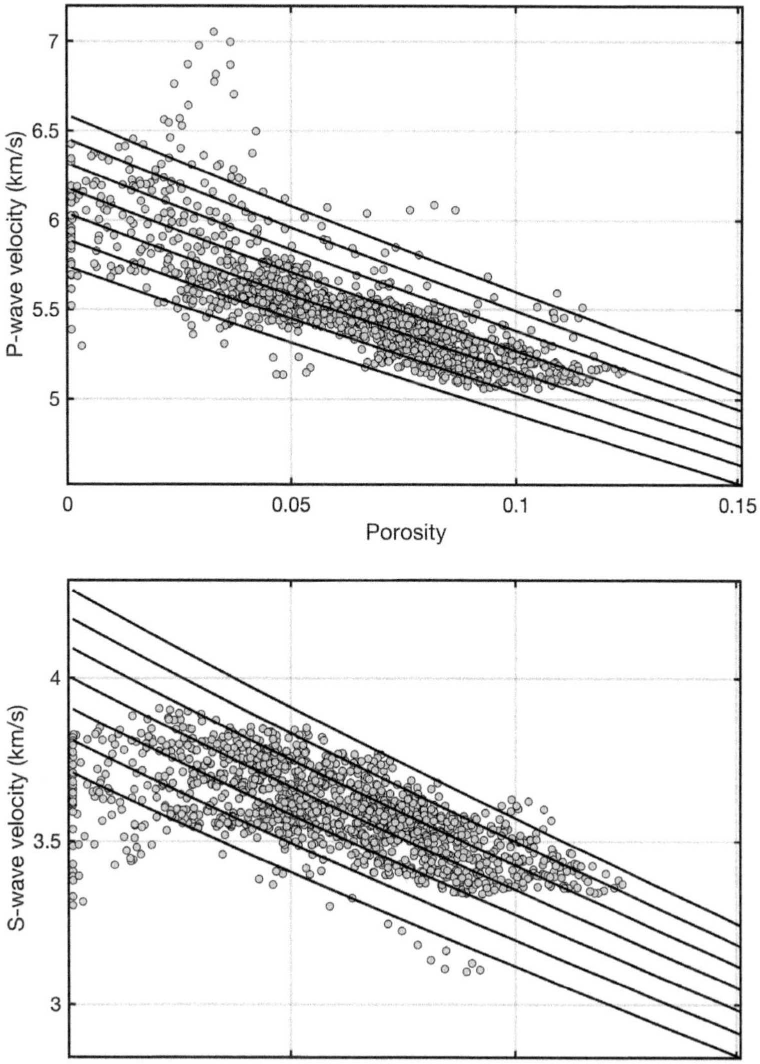

Figure 2.13 Stiff sand rock physics model for the Weber Sandstone formation: P-wave and S-wave velocity versus porosity: the gray dots represent the data from well logs; the black curves represent the rock physics models for different values of the bulk modulus (39–51 GPa) and shear modulus (37–49 GPa) of the solid phase.

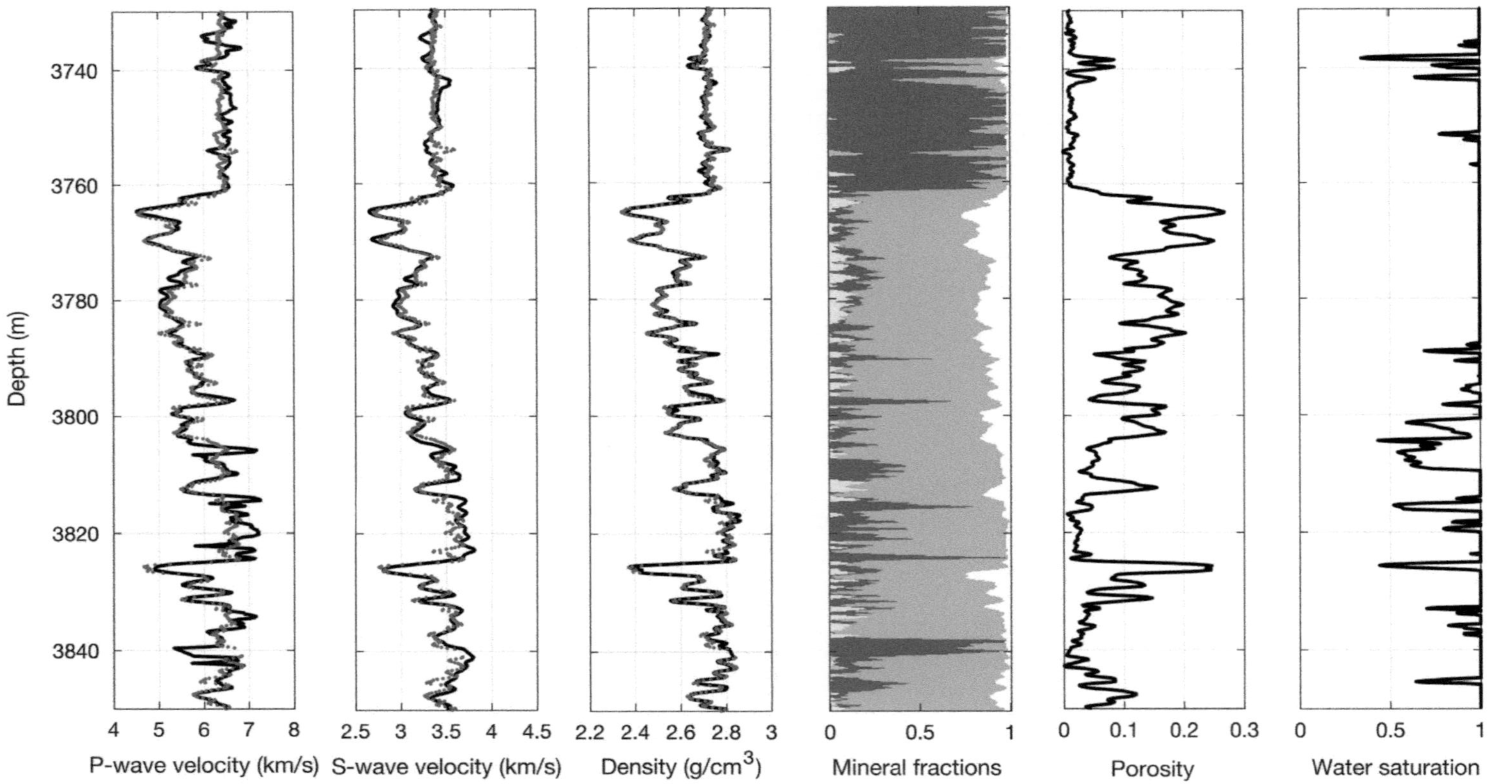

Figure 2.14 Well logs for the Madison Limestone formation, from left to right: P-wave velocity, S-wave velocity, density, mineral fractions (volume of shale in tan, volume of sand in yellow, volume of limestone in blue, volume of dolomite in green, and pore volume in white), porosity and water saturation. The black curves show the well log data; the dashed red curves show the rock physics model predictions.

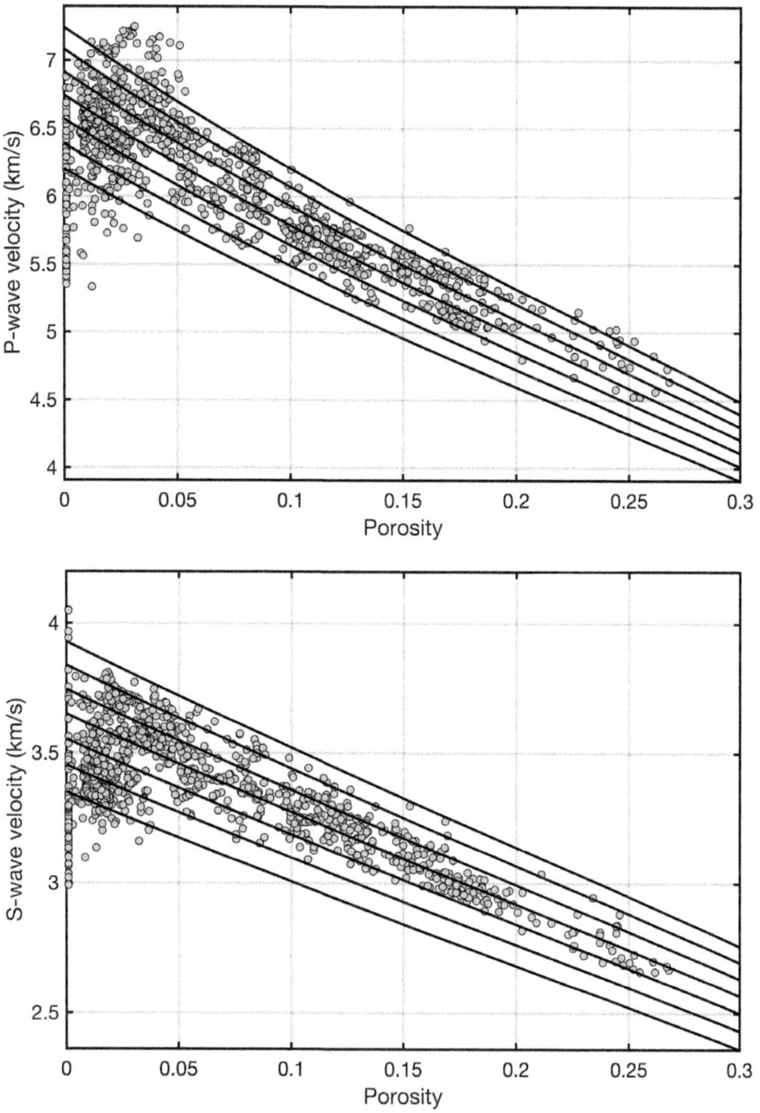

Figure 2.15 Berryman's inclusion rock physics model for the Madison Limestone formation: P-wave and S-wave velocity versus porosity: the gray dots represent the data from well logs; the black curves represent the rock physics models for different values of the bulk modulus (67–77 GPa) and shear modulus (32–46 GPa) for the solid phase.

3

Geostatistics for Continuous Properties

Geostatistics is a discipline that aims to analyze and predict properties associated with spatiotemporal phenomena in geosciences. In subsurface modeling, geostatistical concepts and tools are commonly used to build geological models that account for the spatial and temporal continuity of geological features. Geostatistical methods were initially developed in the 1950s to predict mineral grade distributions in mining engineering problems and then applied to several other domains of geosciences, including geophysics, petroleum engineering, hydrology, environmental sciences, and meteorology. The mathematical foundation of spatial random variables was formalized by Matheron (1970). Several textbooks have been published since then, including Journel and Huijbregts (1978), Isaaks and Srivastava (1989), Kitanidis (1997), Goovaerts (1997), Deutsch and Journel (1998), Deutsch (2002), Wackernagel (2003), Dubrule (2003), Chilès and Delfiner (2009), Caers (2011), Lantuéjoul (2013), Mariethoz and Caers (2014), and Pyrcz and Deutsch (2014). Generally, the goal of geostatistical methods is to predict the value of a property at a given location at a given time, using data measured at other nearby locations. Traditionally geostatistical modeling includes two main families of methods for interpolation and simulation. Interpolation methods aim to predict the optimal values, for example in the mean square sense, whereas simulation methods aim to draw samples from the probability distribution of the variable and provide multiple outcomes for different realizations that represent the model uncertainty. The majority of geostatistical methods, in their most basic form, assume stationarity of the random variable, which means that its statistical properties, such as the mean and the variance, do not change in the entire spatiotemporal domain. However, there are many geostatistical methods that can integrate spatially varying mean models. In this chapter, we focus on spatial problems but the same concepts and methods could be extended to spatiotemporal applications.

The MATLAB codes for the examples included in this chapter are provided in the SeReM package and described in the Appendix, Section A.2.

3.1 Introduction to Spatial Correlation

In geoscience applications, the focus is often on the spatial distribution of rock and fluid properties, to investigate the local variations of subsurface variables and to understand how these properties change in space. In this section, we introduce the basic concepts

Seismic Reservoir Modeling: Theory, Examples, and Algorithms, First Edition. Dario Grana, Tapan Mukerji, and Philippe Doyen.

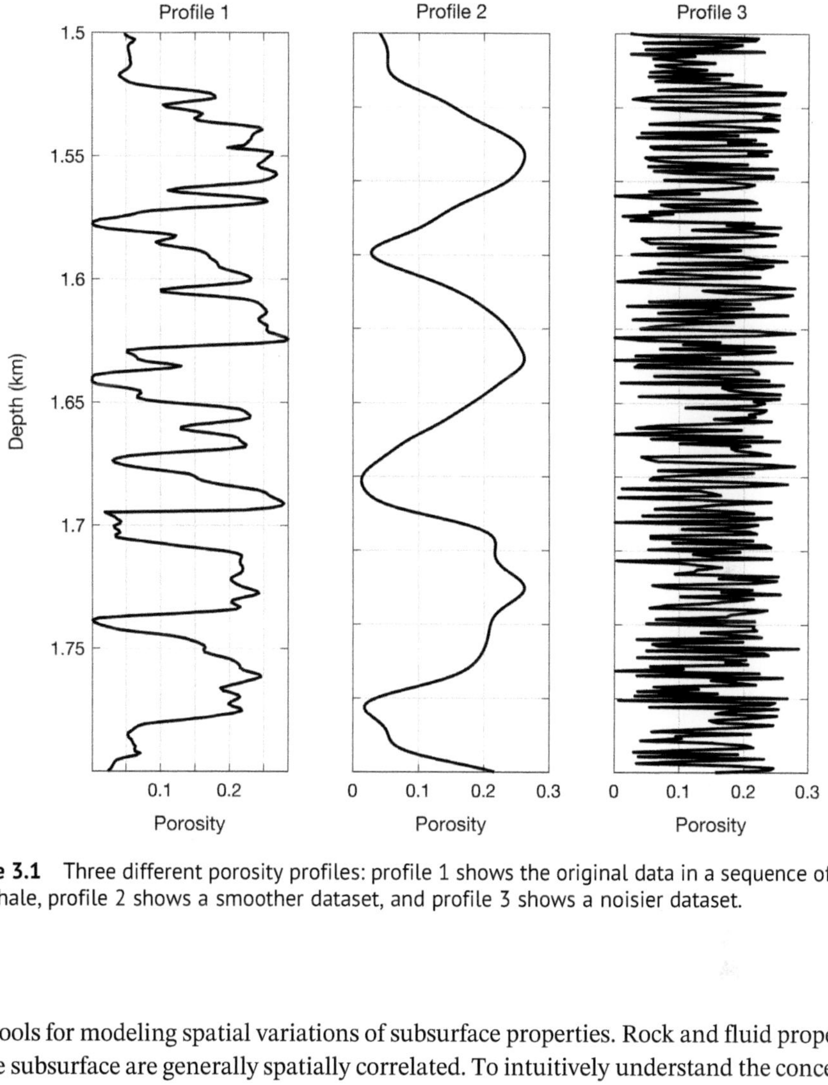

Figure 3.1 Three different porosity profiles: profile 1 shows the original data in a sequence of sand and shale, profile 2 shows a smoother dataset, and profile 3 shows a noisier dataset.

and tools for modeling spatial variations of subsurface properties. Rock and fluid properties in the subsurface are generally spatially correlated. To intuitively understand the concept of spatial correlation, we first discuss an illustrative example. Figure 3.1 shows three different porosity profiles: profile 1 shows the bimodal behavior of a sequence of sand and shale; profile 2 shows a smoother behavior; and profile 3 shows a noisier behavior. These three profiles look different in terms of spatial continuity; however, from a statistical point of view, the marginal distribution of porosity in the three profiles is the same. The mean is 0.14 and the standard deviation is 0.08 in all three profiles and the histograms are almost identical (Figure 3.2).

The difference between the three profiles in Figure 3.1 is in the spatial distribution of the porosity values: profile 1 shows a shorter spatial correlation than profile 2 and a longer spatial correlation than profile 3. We introduce the concept of spatial correlation using profile 1 in Figure 3.1 as an illustrative example. If we select a sample along the profile, it is likely that the samples measured at the adjacent locations show similar

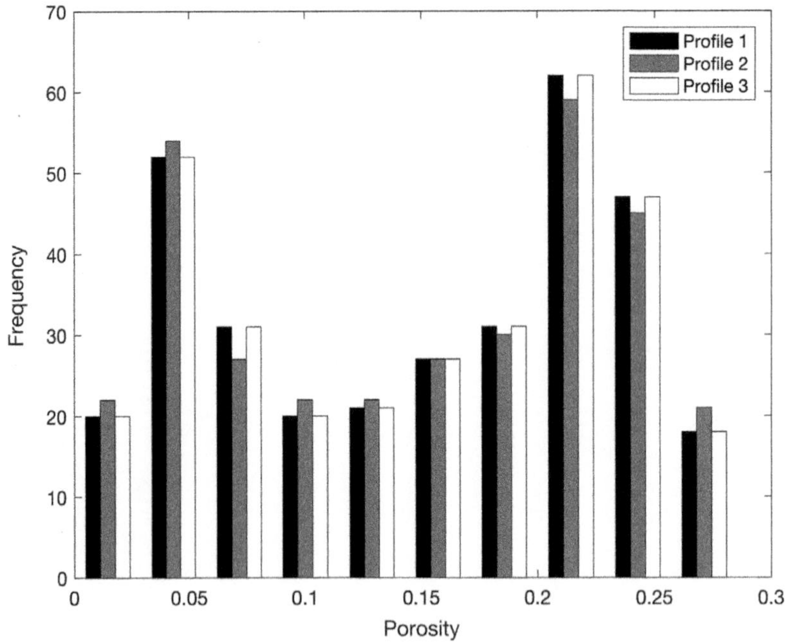

Figure 3.2 Histograms of porosity profiles in Figure 3.1, showing a similar marginal distribution for all three profiles.

values owing to the geological continuity in the subsurface. Conversely, points far from the selected sample might show larger variations in the porosity values. The vertical correlation function allows representing these similarities and dissimilarities due to the distance, in a mathematical form.

In Figure 3.3, we study the spatial correlation by analyzing couples of samples measured at different distances. We first consider all the pairs of samples with distance 0.5 m. We plot the porosity at the deeper location on the y-axis versus the porosity at the shallower location on the x-axis, for each couple of points in the profile (Figure 3.3). When the distance between the samples in a pair is 0.5 m, the two porosity values are likely to be similar (except in the proximity of an abrupt lithological change), and the data points in the plot follow a linear trend. The resulting correlation coefficient (Section 1.3.2) is 0.99. If we increase the distance between the samples in each pair to 1, 2, 3, and 6 m, respectively, then, the spread of the data points around the linear trend becomes larger (Figure 3.3) and the correlation coefficient decreases to 0.96, 0.88, 0.77, and 0.46, respectively. If we increase the distance between the samples in each pair to 12 m, the data points form a homogeneous cloud of points and they are no more correlated. The resulting correlation coefficient becomes 0.09. Conversely, if we reduce the distance between the samples in each pair to 0 m, then both samples in each pair have the same values; therefore, the data points are perfectly aligned along a straight line and the correlation coefficient is exactly 1. In general, when the distance of the samples tends to 0, then the correlation coefficient tends to 1; when the distance becomes large, then the correlation coefficient tends to 0.

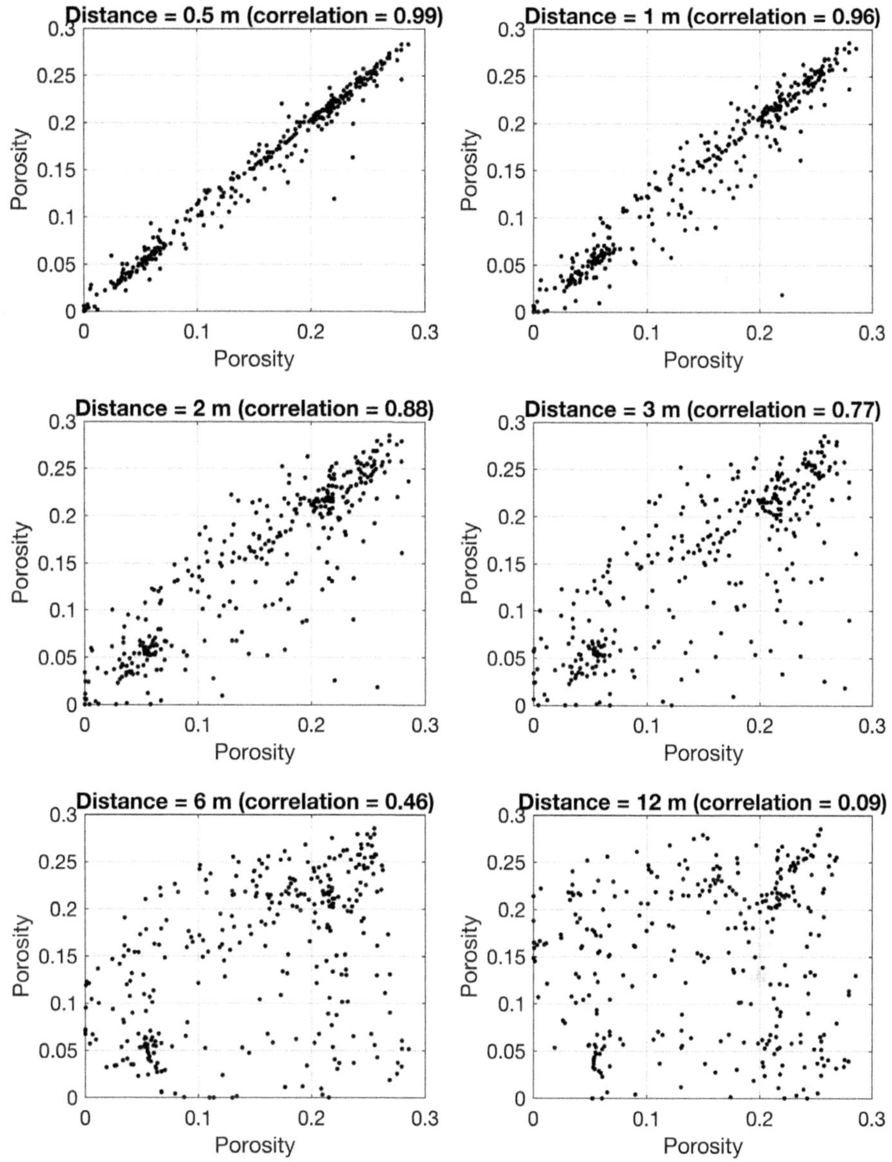

Figure 3.3 Cross-plots of pairs of porosity samples measured at gradually increasing distances of 0.5, 1, 2, 3, 6, and 12 m, along the porosity profile 1 in Figure 3.1.

We then study the correlation coefficient as a function of the distance h. This function is called the experimental spatial correlation function $\rho(h)$. For the three vertical profiles in Figure 3.1, the calculated experimental vertical correlation functions are shown in Figure 3.4. For profile 1, the correlation decreases to 0 in 14 m; for profile 2, the correlation decreases to 0 in 24 m, owing to the smooth behavior of the data; and for profile 3, the correlation decreases to 0 in 1 m, owing to the noisy nature of the data.

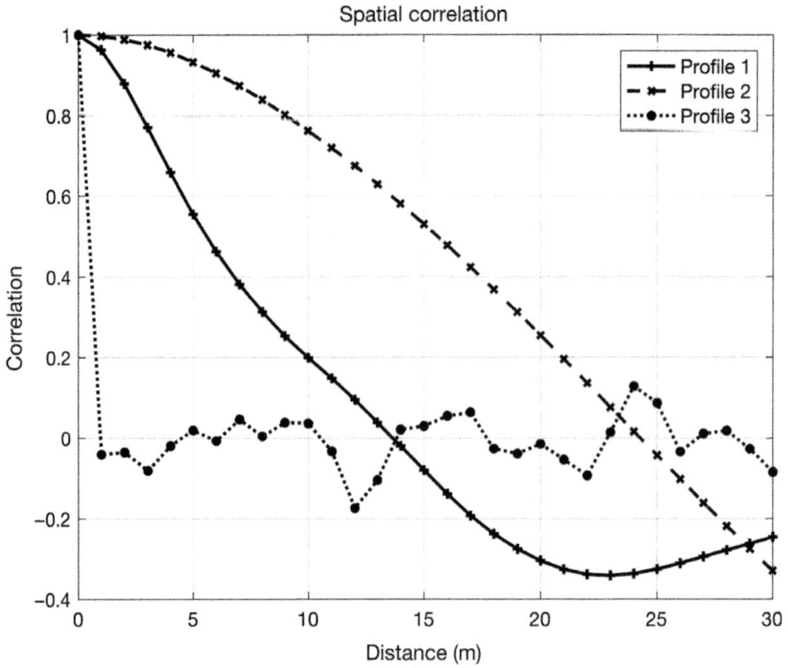

Figure 3.4 Experimental vertical correlation functions for porosity profiles in Figure 3.1.

3.2 Spatial Correlation Functions

The experimental spatial correlation function describes the variations of the correlation coefficient as a function of the distance between the samples. The spatial correlation coefficient generally decreases as the distance increases. The distance at which the correlation coefficient becomes 0 is called the correlation length (or correlation range). Beyond this distance, the variable can be generally considered spatially uncorrelated. In this section, we first review the mathematical definitions of spatial covariance and correlation functions and variograms, then we extend the formulation to multidimensional problems.

We first focus on univariate problems, where the variable of interest varies along one dimension, for example a vertical profile in the subsurface. From a mathematical point of view, given a random variable X with stationary mean μ_X and variance σ_X^2, the spatial covariance $C(h)$ is defined as the expected value (Section 1.3.1) of the product of the residuals of the property values at two locations separated by a distance h:

$$C(h) = E[X(u) - \mu_X][X(u + h) - \mu_X], \tag{3.1}$$

where u represents the spatial location. When the distance $h = 0$, then Eq. (3.1) coincides with the definition of variance (Section 1.3.1), hence $C(0) = \sigma_X^2$. The spatial correlation $\rho(h)$ is then defined as the spatial covariance $C(h)$ normalized by the variance of the property σ_X^2:

$$\rho(h) = \frac{C(h)}{\sigma_X^2} = \frac{C(h)}{C(0)}. \tag{3.2}$$

Given a set of n measured samples $\{x_i\}_{i=1,\dots,n}$ at different locations in space $\{u_i\}_{i=1,\dots,n}$, the experimental spatial covariance $\hat{C}(h)$ can then be computed as:

$$\hat{C}(h) = \frac{1}{n_h}\sum_{j=1}^{n_h}\left(x_j - \hat{\mu}_X\right)\left(x_{j+h} - \hat{\mu}_X\right), \tag{3.3}$$

where n_h is the number of pairs of samples, $\left\{\left(x_j, x_{j+h}\right)\right\}_{j=1,\dots,n_h}$ represent the pairs of samples measured at locations $\left\{\left(u_j, u_{j+h}\right)\right\}_{j=1,\dots,n_h}$ separated by a distance h, and $\hat{\mu}_X = \frac{1}{n}\sum_{i=1}^{n} x_i$ is the empirical mean of the measured samples. When the distance $h = 0$, the experimental spatial covariance $\hat{C}(h = 0)$ becomes:

$$\hat{C}(0) = \frac{1}{n_h}\sum_{j=1}^{n_h}\left(x_j - \hat{\mu}_X\right)^2 \tag{3.4}$$

and it is equal to the empirical variance $\hat{\sigma}_X^2$. The experimental correlation function $\hat{\rho}(h)$ is given by:

$$\hat{\rho}(h) = \frac{\hat{C}(h)}{\hat{\sigma}_X^2} = \frac{\hat{C}(h)}{\hat{C}(0)}. \tag{3.5}$$

When building the experimental covariance and correlation functions, we implicitly assume that the mean is stationary and that the spatial covariance is only a function of the distance.

The variogram $\gamma(h)$ is related to the spatial covariance function by the following relation:

$$\gamma(h) = \frac{1}{2}E[X(u+h) - X(u)]^2 = C(0) - C(h). \tag{3.6}$$

The variogram measures the dissimilarity of the property values as a function of the distance h; hence, the variogram $\gamma(h)$ generally increases and converges to the empirical variance $\hat{\sigma}_X^2$ as the distance h increases. The asymptotic value of the variogram is often called the sill. When the distance $h = 0$, then the variogram $\gamma(h = 0) = 0$. The experimental variogram $\hat{\gamma}(h)$ can be then computed as:

$$\hat{\gamma}(h) = \frac{1}{2n_h}\sum_{j=1}^{n_h}\left(x_{j+h} - x_j\right)^2. \tag{3.7}$$

Experimental variograms might show a discontinuity at $h = 0$, which is pure noise or variability at a scale smaller than the sampling interval. This discontinuity is known as the nugget effect.

Example 3.1 The experimental vertical correlation function and the corresponding variogram of profile 1 in Figure 3.1 are shown in Figure 3.5. The correlation length is approximately 14 m. For distances larger than the correlation length, we do not expect to observe any correlation in the data, therefore, the correlation function converges to 0

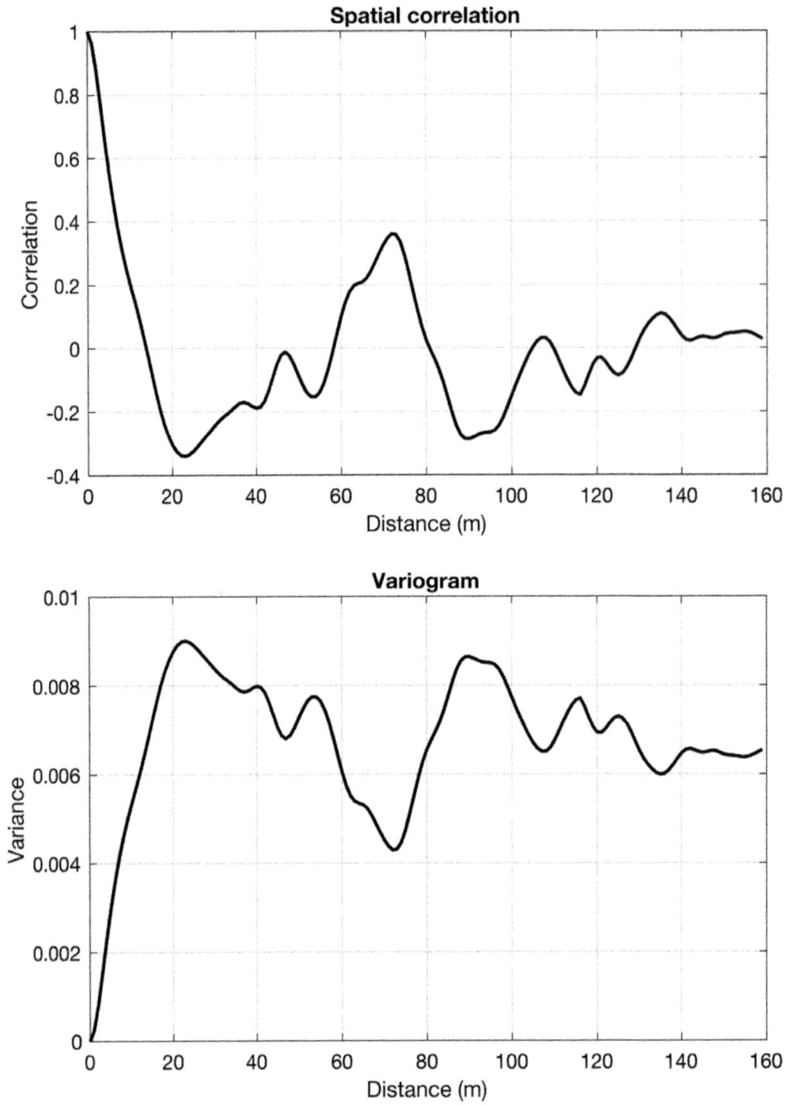

Figure 3.5 Example of experimental vertical correlation function and corresponding experimental vertical variogram function, computed based on profile 1 in Figure 3.1.

and the variogram converges to the variance of the property, i.e. the sill of the variogram. Beyond the correlation length, the variability around 0 in the correlation function and around the sill in the variogram is because of the periodic behavior of the data.

The experimental functions, estimated from real data, can be noisy owing to measurement errors, limited number of samples, and property heterogeneities. It is a common practice to fit the experimental functions using theoretical models. Several parametric models have been proposed to approximate the experimental spatial correlation $\hat{\rho}(h)$, the experimental spatial covariance $\hat{C}(h)$, and experimental variogram $\hat{\gamma}(h)$. Parametric models are generally preferred to experimental functions because they guarantee that their corresponding spatial covariance matrices are positive definite and their inverse can be computed. The three most common models of spatial correlation include the exponential model:

$$\rho_{exp}(h) = \exp\left(-\frac{3h}{l}\right), \tag{3.8}$$

the Gaussian model:

$$\rho_{Gau}(h) = \exp\left(-\frac{3h^2}{l^2}\right), \tag{3.9}$$

and the spherical model:

$$\rho_{sph}(h) = \begin{cases} 1 - \frac{3h}{2l} + \frac{h^3}{2l^3} & h \leq l \\ 0 & h > l \end{cases} \tag{3.10}$$

where l is the correlation length. The corresponding exponential, Gaussian, and spherical variograms can be computed as:

$$\gamma(h) = \sigma_X^2[1 - \rho(h)], \tag{3.11}$$

where $\rho(h)$ represents one of the theoretical spatial correlation functions in Eqs. (3.8)–(3.10). Examples of parametric spatial covariance functions and variograms with correlation length of 5 m, sill (i.e. the variance of the variable) equal to 1, and nugget of 0 are shown in Figure 3.6. Because the Gaussian and exponential models are always positive, we define the correlation length as the distance where the value of the function is equal to 5% of the variance σ_X^2. The Gaussian model is generally used to describe smooth spatial variations. In several applications, a sum of multiple models is adopted to describe different scales of variability in a nested structure.

The analysis of the spatial variability can also be applied to multidimensional problems. For isotropic conditions, the definition of spatial covariance, correlation, and variogram (Eqs. 3.1, 3.2 and 3.6) can be extended to the multidimensional case. However, in two- and three-dimensional applications, the spatial covariance might depend on the direction along which it is calculated. Therefore, the spatial model parameters are generally directionally dependent owing to the anisotropy associated with geological processes such as layering and depositional sequences. In two-dimensional problems, we typically observe a direction of maximum correlation and a direction of minimum correlation that are often orthogonal. In subsurface modeling, the directions of maximum and minimum correlation often coincide with the dominant direction of the geological structures and its orthogonal component. For example, in a seismic section, the horizontal direction often shows a larger spatial correlation than the vertical direction owing to the geological continuity of the sedimentary layers. In general, the correlation length is a function of the spatial direction and its variations can be described by an ellipse, where the directions of the major and minor axes correspond to the directions of the maximum and minimum correlation length, respectively. The geological anisotropy can be then described by the orientation of the ellipse, represented by the azimuthal angle α indicating the counterclockwise angle of the ellipse major axis from the Cartesian horizontal axis, and by the maximum and minimum correlation length, represented by the half-length of the major and minor axes of the ellipse.

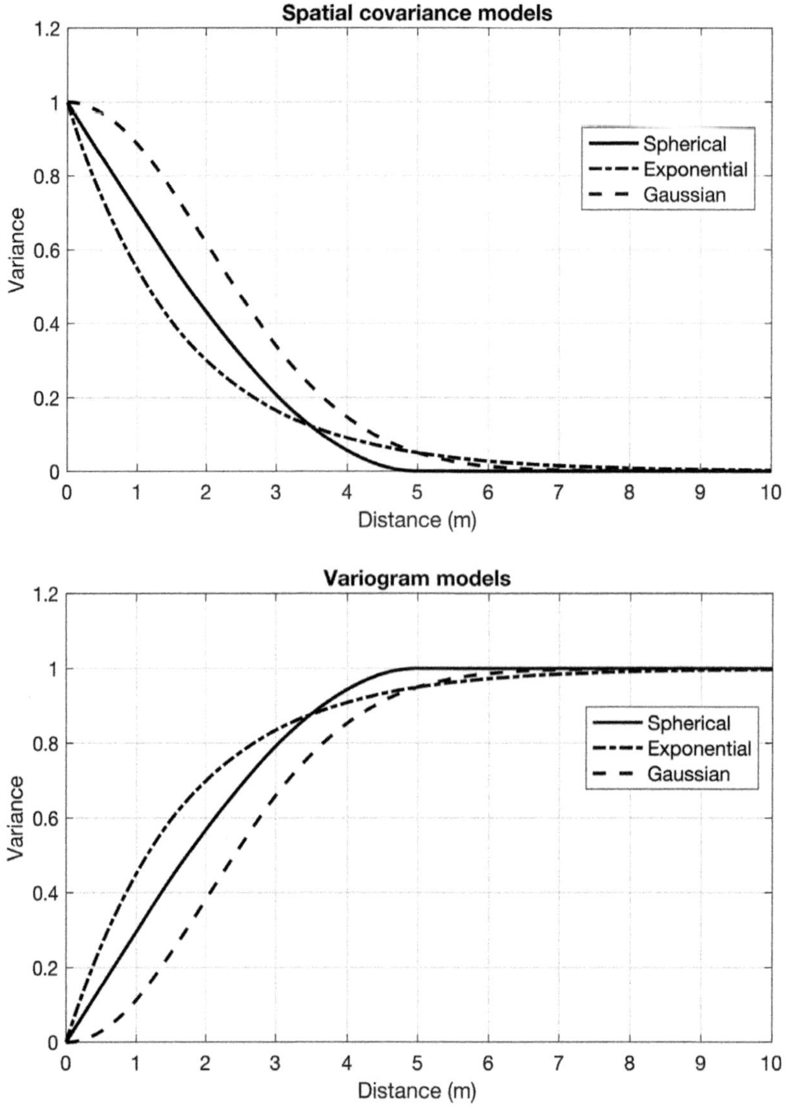

Figure 3.6 Spatial covariance functions and corresponding variogram models with variance equal to 1 and correlation length of 5 m: the solid lines represent the spherical model, the dashed-dotted lines represent the exponential model, and the dashed lines represent the Gaussian model.

The experimental two-dimensional covariance function can be written as:

$$\hat{C}(h,k) = \frac{1}{n_{h,k}} \sum_{i,j=1}^{n_{h,k}} \left(x_{i,j} - \hat{\mu}_X\right)\left(x_{i+h,j+k} - \hat{\mu}_X\right), \tag{3.12}$$

where $n_{h,k}$ is the number of pairs of samples with distance $\sqrt{h^2 + k^2}$. The corresponding experimental two-dimensional correlation function is given by:

$$\hat{\rho}(h,k) = \frac{\hat{C}(h,k)}{\hat{\sigma}_X^2}. \tag{3.13}$$

Anisotropic theoretical models can be built from isotropic functions by adopting a polar coordinate system and assuming that the dependence of the correlation length on the azimuth is expressed by an ellipse. In the two-dimensional space, the polar coordinates are defined by the radial coordinate r, defined as the distance from the origin, and the angular coordinate θ, defined as the counterclockwise angle from the Cartesian horizontal axis. We assume that the correlation length $l(\theta)$ depends on the angular coordinate θ through the following function:

$$l(\theta) = \frac{l_{max} l_{min}}{\sqrt{l_{max}^2 \sin^2(\alpha - \theta) + l_{min}^2 \cos^2(\alpha - \theta)}}. \tag{3.14}$$

Equation (3.14) describes the radius of an ellipse parametrized by the half-length of the major and minor axes (corresponding to the maximum correlation length l_{max} and minimum correlation length l_{min}, respectively) and by the azimuth α. The correlation length $l(\theta)$ is equal to the maximum correlation length l_{max} for $\theta = \alpha$, and to the minimum correlation length l_{min} for $\theta = \alpha + \pi$. The anisotropic correlation model $\rho(h, \theta)$ is then obtained from an isotropic correlation function $\rho(h)$, such as one of the models defined in Eqs. (3.8)–(3.10), by scaling the distance by the azimuthally dependent correlation length $l(\theta)$ as:

$$\rho(h,\theta) = \rho\left(\frac{h}{l(\theta)}\right), \tag{3.15}$$

where $l(\theta)$ is defined by the parametric ellipse in Eq. (3.14). An example of a two-dimensional spatial covariance model with variance equal to 1, maximum correlation length of 10 m, minimum correlation length of 5 m, and azimuth $\alpha = 45°$ is shown in Figure 3.7, for the directions corresponding to $\theta = 45°$, 90°, and 135°.

Spatial correlation and covariance functions and variograms can also be calculated in a three-dimensional space, by adopting a spherical coordinate system and using a three-dimensional ellipsoid defined by the three principal correlation lengths (corresponding to half-lengths of the major, mid, and minor axes) and two angles, namely the azimuth and the dip. In subsurface applications, the shorter correlation length is often observed in the vertical direction, owing to the lateral continuity of geological structures.

The concepts of spatial covariance and correlation functions can also be extended to two random variables to define the cross-covariance and cross-correlation between two properties and relate the spatial variations of one variable to the changes of the other variable. Given a random variable X with stationary mean μ_X and variance σ_X^2, and a random variable Y with stationary mean μ_Y and variance σ_Y^2, then the spatial cross-covariance $C_{X,Y}(h)$ is defined as:

$$C_{X,Y}(h) = E[X(u) - \mu_X][Y(u+h) - \mu_Y]. \tag{3.16}$$

When the distance $h = 0$, then Eq. (3.16) coincides with the definition of covariance, $C_{X,Y}(0) = \sigma_X \sigma_Y$ (Section 1.3.2). The spatial cross-correlation $\rho_{X,Y}(h)$ is then defined as the spatial cross-covariance $C_{X,Y}(h)$ normalized by the product of the standard deviations of the two random variables:

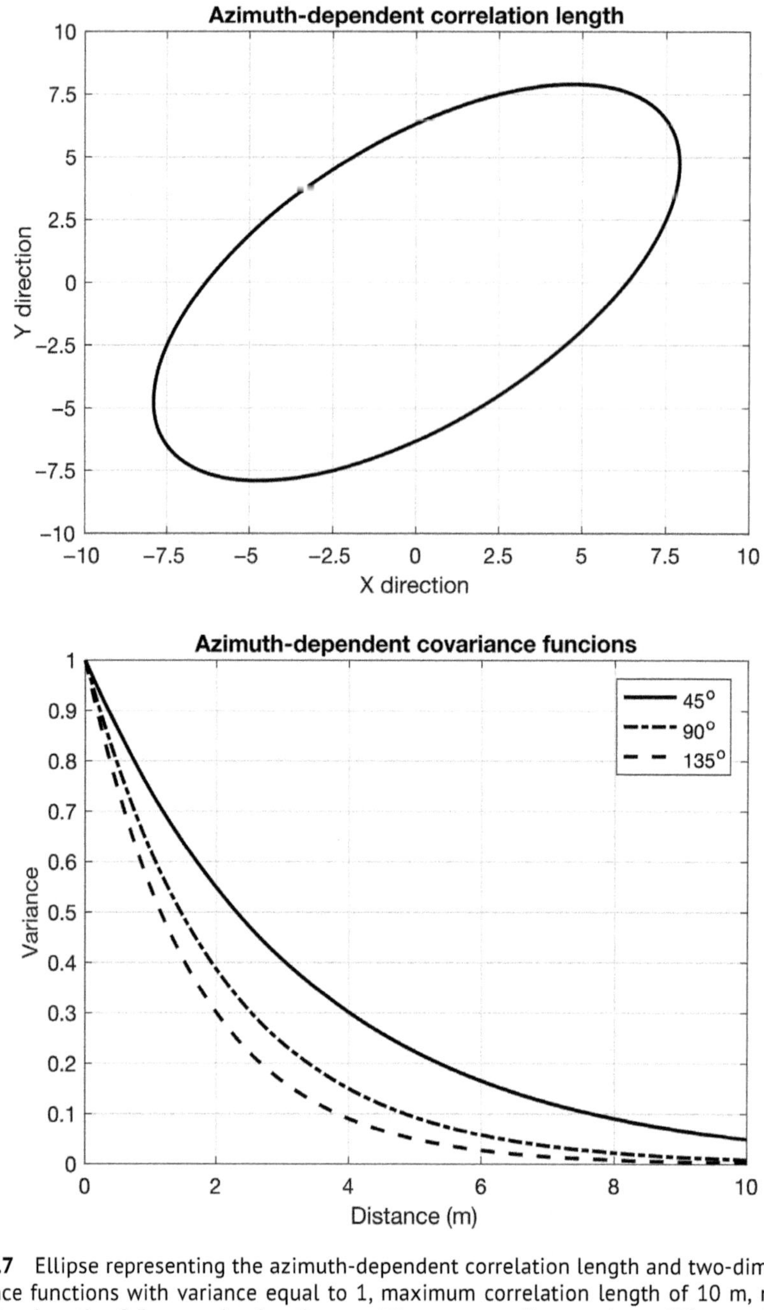

Figure 3.7 Ellipse representing the azimuth-dependent correlation length and two-dimensional covariance functions with variance equal to 1, maximum correlation length of 10 m, minimum correlation length of 5 m, and azimuth $\alpha = 45°$, corresponding to three different directions ($\theta = 45°$, $90°$, and $135°$).

$$\rho_{X,Y}(h) = \frac{C_{X,Y}(h)}{\sigma_X \sigma_Y}. \tag{3.17}$$

The cross-variogram $\gamma_{X,Y}(h)$ is related to the spatial covariance as:

$$\gamma_{X,Y}(h) = C_{X,Y}(0) - C_{X,Y}(h). \tag{3.18}$$

Based on Eqs. (3.6) and (3.18), spatial covariance functions and variograms can be used interchangeably. Spatial correlation and covariance functions and variograms are widely used in spatial interpolation and geostatistical simulation methods to account for the spatial continuity of subsurface properties. Variogram-based geostatistical methods are often called two-point statistics methods, because variograms and spatial covariance functions only account for the spatial correlation between two data points at a time.

3.3 Spatial Interpolation

Spatial interpolation aims to estimate the value of a property at a given location based on the values measured at nearby locations. To understand the challenges of spatial interpolation and the needs of statistical methods, we analyze a simple illustrative problem with three measurements. We aim to estimate the value x_0 of a random variable X at the assigned location $\boldsymbol{u}_0$ with known coordinates, given three measurements, x_1, x_2, and x_3 of the variable X at the locations $\boldsymbol{u}_1$, $\boldsymbol{u}_2$, and $\boldsymbol{u}_3$. For example, we aim to estimate porosity at location $\boldsymbol{u}_0$ given three measurements of porosity at different locations in a two-dimensional map.

A possible solution of the interpolation problem to estimate x_0 is expressed as a linear combination of the measurements of the form:

$$\overline{x}_0 = \sum_{i=1}^{n} w_i x_i, \tag{3.19}$$

where $\overline{x}_0$ is the estimated value, n is the number of available measurements, and the coefficients w_i represent the weights of the linear combination. The solution depends on the criteria and methods used for assigning the weights w_i to the measurements x_i, for $i = 1, \dots, n$. The choice of the weights is related to the spatial distribution of the measurements. To illustrate this concept, we analyze four scenarios with different spatial configurations of the three locations (Figure 3.8).

In the first scenario (Figure 3.8a), the three measurements are distributed at the vertices of an equilateral triangle with location $\boldsymbol{u}_0$ positioned at the centroid of the triangle. In this scenario, the three measurement locations are equally distant from the location $\boldsymbol{u}_0$. In the absence of any information about spatial continuity and anisotropy, the most obvious solution of the interpolation problem is the arithmetic average of the three measurements $\{x_i\}_{i=1,\dots,n}$, with $n = 3$ in our example. The estimate $\overline{x}_0$ is then given by

$$\overline{x}_0 = \frac{1}{n}\sum_{i=1}^{n} x_i, \tag{3.20}$$

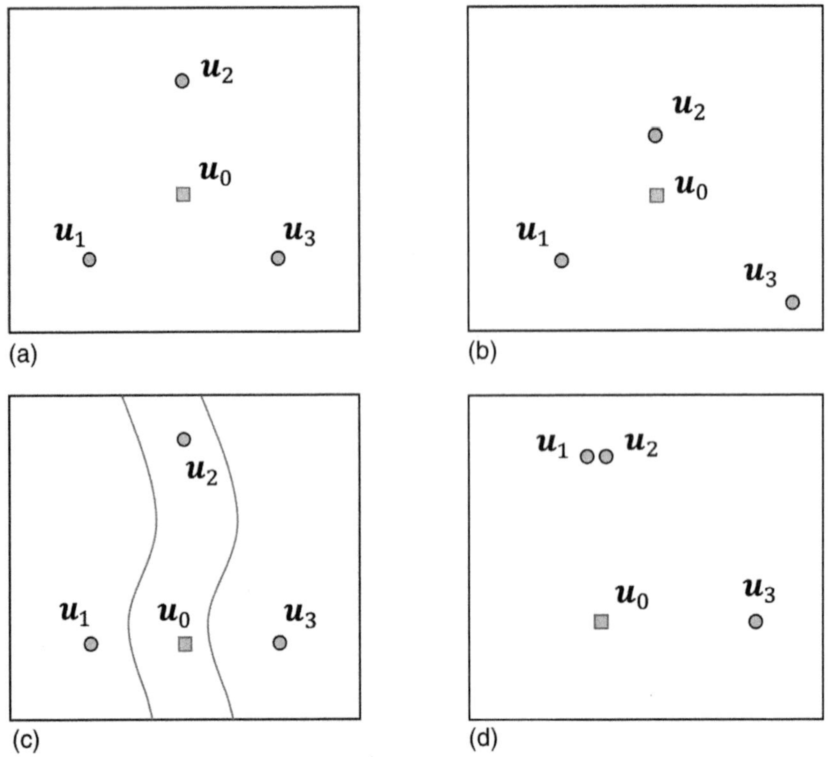

Figure 3.8 Schematic representations of four data configurations. The dots represent the available measurements, whereas the square represents the location $\boldsymbol{u}_0$ of the unknown value x_0. In scenario (a), all the points are equally distant from the location $\boldsymbol{u}_0$. In scenario (b), the three measurements are at three different distances from the location $\boldsymbol{u}_0$. In scenario (c), one of the measurements is located in the same geological formation of the location $\boldsymbol{u}_0$. In scenario (d), all the points are equally distant from the location $\boldsymbol{u}_0$, but two measurements are located in proximity of each other.

where all the weights w_i in Eq. (3.19) are the same and equal to $w_i = 1/n$, for $i = 1, \dots, n$. However, the interpolation in Eq. (3.20) is only valid if all the measurements are equally distant from the location of the prediction.

In the second scenario (Figure 3.8b), we assume that one of the measurements (e.g. location $\boldsymbol{u}_3$) is further from location $\boldsymbol{u}_0$ than the other data, whereas another measurement (e.g. location $\boldsymbol{u}_2$) is closer. The average of the measurements in Eq. (3.20) is not an adequate solution for this scenario, because the value $\overline{x}_0$ at location $\boldsymbol{u}_0$ is expected to be more similar to the value x_2 at location $\boldsymbol{u}_2$ rather than the value x_3 at location $\boldsymbol{u}_3$, because of the proximity of location $\boldsymbol{u}_2$ to location $\boldsymbol{u}_0$. Therefore, a desirable interpolation method should give a larger weight to the value x_2 and a lower weight to the value x_3. A possible solution to this problem is to assign weights based on the distance between the location $\boldsymbol{u}_0$ and the locations of the measurements $\{\boldsymbol{u}_i\}_{i=1,\dots,n}$. For example, we could adopt an average of the measurements weighted by the inverse of the square of the Euclidean distances d_i between the measurement locations $\boldsymbol{u}_i$ and the location $\boldsymbol{u}_0$, for $i = 1, \dots, n$. The estimate $\overline{x}_0$ is then given by:

$$\bar{x}_0 = \sum_{i=1}^{n} w_i x_i = \sum_{i=1}^{n} \frac{\frac{1}{d_i^2}}{\sum_{j=1}^{n} \frac{1}{d_j^2}} x_i. \tag{3.21}$$

If the distances d_i, for $i = 1, \dots, n$, are all equal, then Eq. (3.21) is the same as Eq. (3.20), since $w_i = 1/n$. If the distances of the measurements are different, then the closest data points have a larger weight (greater than $1/n$) in the linear combination and the furthest data points have a lower weight (lower than $1/n$). However, the interpolation in Eq. (3.21) assumes that the estimated value $\bar{x}_0$ only depends on the distance between the locations of the measurements and location $\boldsymbol{u}_0$, but not on their spatial configuration. Under this assumption, two measurements that are equally distant from location $\boldsymbol{u}_0$ have the same weight even if they are located in opposite directions. The interpolation based on the inverse distance in Eq. (3.21) cannot account for spatial information such as anisotropy and continuity. In the third and fourth scenarios, we illustrate two data configurations where standard interpolation methods might not be adequate.

In the third scenario (Figure 3.8c), we assume that prior information related to the spatial continuity of the model is available. The spatial configuration of the measurements includes two location $\boldsymbol{u}_1$ and $\boldsymbol{u}_3$, equally distant from location $\boldsymbol{u}_0$ and a location $\boldsymbol{u}_2$ further from $\boldsymbol{u}_0$ than the other two locations $\boldsymbol{u}_1$ and $\boldsymbol{u}_3$. The available prior information suggests that there is a geological formation (e.g. a channel) in the direction $\overrightarrow{\boldsymbol{u}_0\boldsymbol{u}_2}$. The locations $\boldsymbol{u}_0$ and $\boldsymbol{u}_2$ are within the geological formation, whereas the locations $\boldsymbol{u}_1$ and $\boldsymbol{u}_3$ are outside and belong to a different formation. If we apply the interpolation method based on the inverse distance (Eq. 3.21), the measurement x_2 would have a lower weight than the measurements x_1 and x_3; however, this is not desirable, because of the available prior geological information. For example, if the property of interest is porosity, then we can assume that porosity at location $\boldsymbol{u}_0$ is more similar to the porosity value at location $\boldsymbol{u}_2$ than the values at locations $\boldsymbol{u}_1$ and $\boldsymbol{u}_3$. Therefore, in this scenario, the weights w_i should account for both the distance and the anisotropy owing to the geological continuity in a given direction.

In the fourth scenario (Figure 3.8d), we assume that the locations $\boldsymbol{u}_1$, $\boldsymbol{u}_2$, and $\boldsymbol{u}_3$ of the measurements are equidistant from the location $\boldsymbol{u}_0$, but location $\boldsymbol{u}_1$ is close to location $\boldsymbol{u}_2$. Since subsurface properties are generally spatially correlated, then it is likely that the measurements x_1 and x_2 at locations $\boldsymbol{u}_1$ and $\boldsymbol{u}_2$ have similar values. In other words, the measurements at locations $\boldsymbol{u}_1$ and $\boldsymbol{u}_2$ are redundant owing to their proximity. Therefore, in this scenario, the weights w_i should account for the redundancy of the measurements according to the spatial configuration of the data.

The spatial interpolation problem in the third and fourth scenarios can be solved using kriging, a spatial interpolation method that accounts for the spatial correlation of the variable of interest and the spatial configuration of the measurements, including anisotropy and redundancy of the data.

3.4 Kriging

Kriging is a spatial interpolation method (Matheron 1969) named after Daniel Krige in recognition of his work in geostatistics in mining engineering. Kriging is commonly used in surface and subsurface modeling applications to estimate the unknown value of a variable at a given location based on sparse measurements of the same variable and to create two- and three-dimensional models of the property of interest. There are different versions of kriging, which differ in their statistical assumptions.

3.4.1 Simple Kriging

The basic version of kriging is called simple kriging. For simplicity, we describe the methodology for a two-dimensional spatial interpolation problem, but the methodology can be easily extended to three-dimensional problems. We assume that the property of interest is a continuous random variable X with mean μ_X and variance σ_X^2, assumed to be known and stationary. We also assume that a limited number n of observations $\{x_i\}_{i=1,\dots,n}$ are available at locations $\{\boldsymbol{u}_i\}_{i=1,\dots,n}$ and that the spatial model consists of a two-dimensional isotropic spatial covariance function $C(h)$ (Section 3.2).

In simple kriging, the estimate x_0^{sk} at location $\boldsymbol{u}_0$ is given by the sum of the mean and a linear combination of the residuals $(x_i - \mu_X)$ for $i = 1, \dots, n$, i.e. the difference between the measurements and the mean value. The kriging estimate x_0^{sk} is then given by:

$$x_0^{sk} = \mu_X + \sum_{i=1}^{n} w_i (x_i - \mu_X), \tag{3.22}$$

where the coefficients w_i are the kriging weights. In kriging, we select the weights to minimize the estimation error $x_0 - x_0^{sk}$ in the mean square sense. If the mean μ_X is equal to the sample mean $\mu_X = \frac{1}{n}\sum_{i=1}^{n} x_i$, then the simple kriging estimator is unbiased, i.e. $E\left(x_0 - x_0^{sk}\right) = 0$. The error function is parameterized in terms of the spatial covariance function of the random variable X. In simple kriging, the weights w_i are obtained by solving a system of linear equations, namely the kriging system, that can be written as:

$$\begin{bmatrix} C_{1,1} & C_{1,2} & \cdots & C_{1,n} \\ C_{2,1} & C_{2,2} & \cdots & C_{2,n} \\ \vdots & \vdots & \ddots & \vdots \\ C_{n,1} & C_{n,2} & \cdots & C_{n,n} \end{bmatrix} \begin{bmatrix} w_1 \\ w_2 \\ \vdots \\ w_n \end{bmatrix} = \begin{bmatrix} C_{0,1} \\ C_{0,2} \\ \vdots \\ C_{0,n} \end{bmatrix} \tag{3.23}$$

where the matrix elements $C_{i,j} = C(h_{i,j})$ represent the values of the spatial covariance function at the distance $h_{i,j} = \text{dist}(\boldsymbol{u}_i, \boldsymbol{u}_j)$ between locations $\boldsymbol{u}_i$ and $\boldsymbol{u}_j$, for $i, j = 1, \dots, n$, and the vector elements $C_{0,i} = C(h_{0,i})$ represent the values of the spatial covariance function at the distance $h_{0,i} = \text{dist}(\boldsymbol{u}_0, \boldsymbol{u}_i)$ between locations $\boldsymbol{u}_0$ and $\boldsymbol{u}_i$, for $i = 1, \dots, n$. By construction, the diagonal elements of the matrix are all equal to the variance σ_X^2 of the variable X, i.e. $C_{i,i} = C(h = 0) = \sigma_X^2$ for all $i = 1, \dots, n$. For example, if we assume that four measurements are available ($n = 4$) and the spatial covariance function is known, then the kriging system is given by:

$$\begin{bmatrix} \sigma_X^2 & C_{1,2} & C_{1,3} & C_{1,4} \\ C_{2,1} & \sigma_X^2 & C_{2,3} & C_{2,4} \\ C_{3,1} & C_{3,2} & \sigma_X^2 & C_{3,4} \\ C_{4,1} & C_{4,2} & C_{4,3} & \sigma_X^2 \end{bmatrix} \begin{bmatrix} w_1 \\ w_2 \\ w_3 \\ w_4 \end{bmatrix} = \begin{bmatrix} C_{0,1} \\ C_{0,2} \\ C_{0,3} \\ C_{0,4} \end{bmatrix}. \tag{3.24}$$

The graphical interpretation is shown in Figure 3.9. The dots represent the measurement locations $\{\boldsymbol{u}_i\}_{i=1,\dots,n}$, whereas the square represents the location $\boldsymbol{u}_0$ where we aim to compute the kriging estimate. The values in the kriging matrix in Eq. (3.24) correspond to the values of the spatial covariance function at the distances between the measurement locations $\{\boldsymbol{u}_i\}_{i=1,\dots,n}$. Because all the elements on the diagonal are equal to the variance of the random variable (i.e. the value of the spatial covariance function for the distance $h = 0$) and because of the symmetry of the distance, we only have to compute six values of the spatial covariance model, corresponding to the distances associated with the length of the four sides and two diagonals of the quadrilateral shape formed by the four measurements. The values in the kriging vector in Eq. (3.24) correspond to the values of the spatial covariance function at the distances between the measurement locations $\{\boldsymbol{u}_i\}_{i=1,\dots,n}$ and location $\boldsymbol{u}_0$.

Example 3.2 We compute the kriging estimate x_0^{sk} of P-wave velocity at a given location $\boldsymbol{u}_0$ using the data configuration in Figure 3.9. We assume that four values are available: $x_1 = 3.2$ km/s, $x_2 = 3.8$ km/s, $x_3 = 4.0$ km/s, and $x_4 = 3.0$ km/s. We also assume that the mean of P-wave velocity is the empirical mean of the samples, $\mu_X = 3.5$ km/s, that the variance is $\sigma_X^2 = 0.1$, and that the spatial correlation model is an isotropic exponential covariance function with correlation length 9 m.

The distances between the measurement locations are given in the following matrix:

$$\mathbf{D} = \begin{bmatrix} 0 & 11 & 15 & 10 \\ 11 & 0 & 10 & 15 \\ 15 & 10 & 0 & 11 \\ 10 & 15 & 11 & 0 \end{bmatrix}$$

and the distances between the measurement locations and location $\boldsymbol{u}_0$ are given in the following vector:

$$\boldsymbol{d} = \begin{bmatrix} 9 \\ 6 \\ 6 \\ 6 \end{bmatrix}.$$

All the distances are in meters and are obtained from the spatial configuration in Figure 3.9. By computing the exponential covariance function for the distances in matrix $\mathbf{D}$ and vector $\boldsymbol{d}$, we obtain the following kriging system:

$$\begin{bmatrix} 0.1 & 0.0026 & 0 & 0.0036 \\ 0.0026 & 0.1 & 0.0036 & 0 \\ 0 & 0.0036 & 0.1 & 0.0026 \\ 0.0036 & 0 & 0.0026 & 0.1 \end{bmatrix} \begin{bmatrix} w_1 \\ w_2 \\ w_3 \\ w_4 \end{bmatrix} = \begin{bmatrix} 0.0050 \\ 0.0135 \\ 0.0135 \\ 0.0050 \end{bmatrix}$$

where values less than 0.001 are set to 0, for simplicity. By solving the linear system of equations, we obtain the following weights:

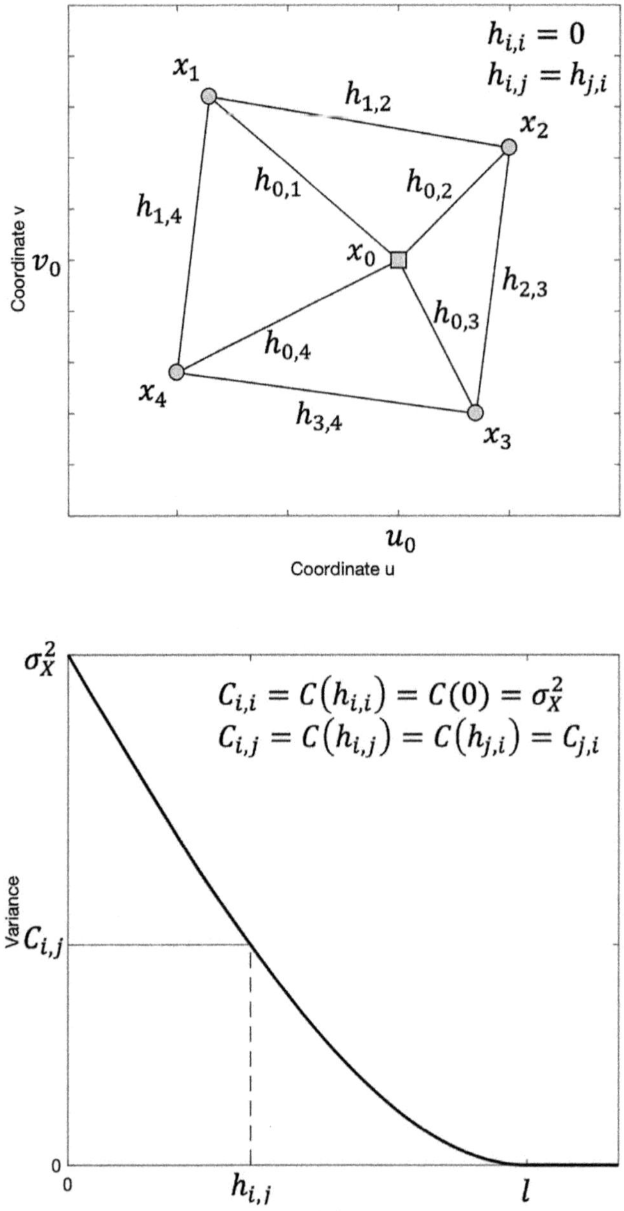

Figure 3.9 Spatial configuration and covariance model used for the kriging system in Eq. (3.24) with $n = 4$ measurements.

$$\begin{bmatrix} w_1 \\ w_2 \\ w_3 \\ w_4 \end{bmatrix} = \begin{bmatrix} 0.0450 \\ 0.1292 \\ 0.1292 \\ 0.0450 \end{bmatrix}.$$

The weights w_2 and w_3 associated with the measurements x_2 and x_3 are greater than the other two weights, as expected based on the proximity of the two measurements to location $\boldsymbol{u}_0$. The kriging estimate x_0^{sk} (Eq. 3.22) is then given by:

$$\begin{aligned} x_0^{sk} &= 3.5 + [0.045 \times (-0.3) + 0.1292 \times 0.3 + 0.1292 \times 0.5 + 0.045 \times (-0.5)] \\ &= 3.5 + 0.0673 = 3.5673 \text{ km/s}. \end{aligned}$$

The so-obtained kriging estimate x_0^{sk} is greater than the prior mean μ_X owing to the influence of the measurements x_2 and x_3 in the kriging system.

The kriging matrix $\mathbf{C} = [C_{i,j}]_{i,j=1,\dots,n}$ in Eq. (3.23) only depends on the spatial correlations of the measurements and the kriging vector $\boldsymbol{C}_0 = [C_{0,i}]_{i=1,\dots,n}^T$ in Eq. (3.23) only depends on the spatial correlations between the measurements and the unknown value to be estimated. Therefore, the kriging system depends on the locations of the data, the location of the unknown value to be estimated, and the spatial covariance function.

In the kriging system, the kriging matrix $\mathbf{C} = [C_{i,j}]_{i,j=1,\dots,n}$ is symmetric by definition, because the distance is symmetric and it is positive definite if the elements of the matrix are computed using a theoretical spatial covariance function (Section 3.2). The vector $\boldsymbol{w} = [w_i]_{i=1,\dots,n}^T$ is then calculated by computing the inverse of the kriging matrix $\mathbf{C}$ and multiplying it by the kriging vector $\boldsymbol{C}_0$. The steps of building and solving the kriging system to compute the kriging weights are repeated at each location where we aim to compute an estimate of the random variable X. Kriging is an exact interpolator since it exactly returns the values of the measurements at the measurement locations. Indeed, at the locations where the measurements are available, we obtain $x_i^{sk} = x_i$, for $i = 1, \dots, n$.

The kriging matrix has size $n \times n$ where n is the number of measurements. For large datasets, assembling the kriging matrix can be computationally demanding. However, the kriging matrix is often sparse (i.e. many of the matrix elements are 0), because for distances larger than the correlation length the value of the spatial covariance function is 0. Therefore, generally, it is not necessary to use all the measurements to build the kriging system. Practical implementations make use of a moving search neighborhood, a method that allows using a limited number of data points by selecting only the measurements inside a region, namely the search neighborhood, centered around the location of the data point to be estimated. When applying the moving search neighborhood method, the kriging system is solved using only the measurements in the search neighborhood; hence the dimension is generally smaller than the full kriging approach including all the data points. The search neighborhood generally has the shape of an ellipse to better capture the anisotropic behavior of the property of interest.

Kriging also provides a quantification of the estimation error, called the kriging variance $\sigma_{X,sk}^2$, given by:

$$\sigma_{X,sk}^2 = \sigma_X^2 - \sum_{i=1}^{n} w_i C_{0,i}. \tag{3.25}$$

The kriging variance depends on the spatial covariance model. The kriging variance is 0 at the locations of the measurements, because the kriging estimates at the measurement locations are equal to the values of the measurements, and it is equal to the variance when the distance from the closest measurement location is larger than the correlation length.

The simple kriging approach presented in this section can be extended to three-dimensional interpolation problems as well as interpolation problems with anisotropic spatial covariance functions and with any finite number of measurements.

Example 3.3 We present an application of simple kriging to a topographic problem. Figure 3.10 shows the elevation map in the Yellowstone National Park. We use this example so that we can compare the results with the exact measurements. We assume that the

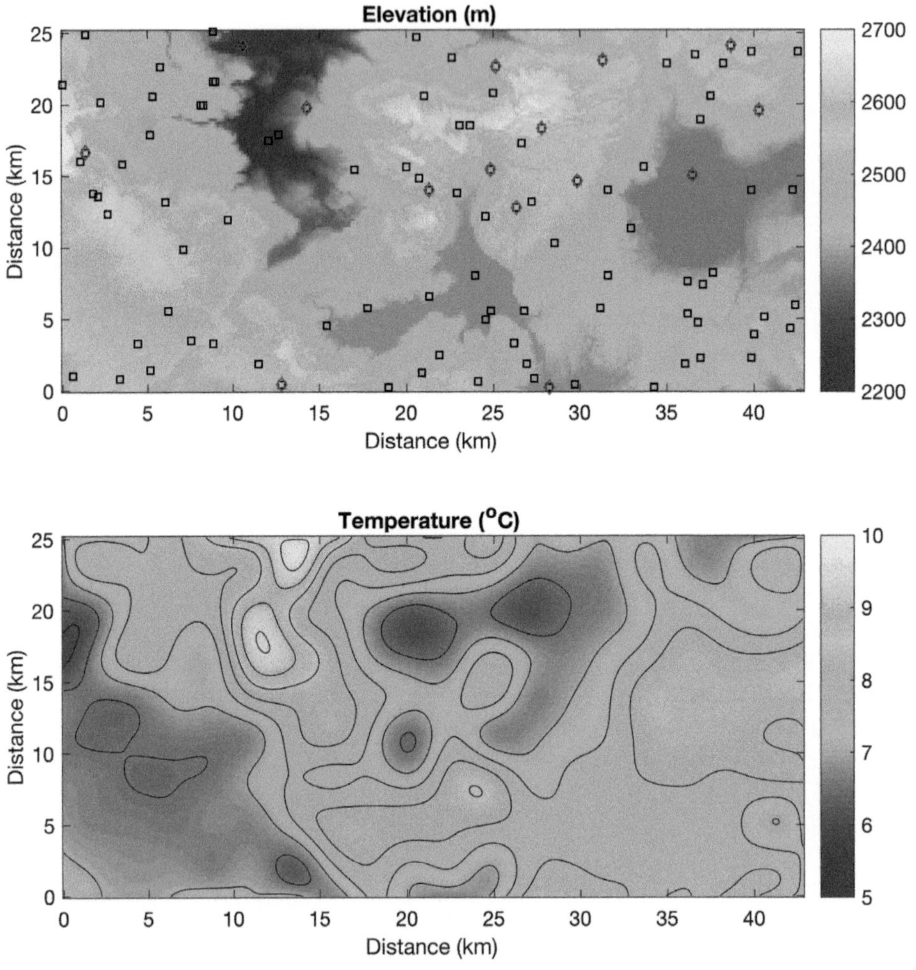

Figure 3.10 Elevation and temperature maps of Yellowstone National Park. The black squares represent the 100 randomly selected locations of the available measurements (full dataset). The red diamonds represent 15 randomly selected locations (sparse dataset). The black lines represent the isotemperature lines.

elevation is unknown except for a set of 100 randomly selected locations (full dataset). We also create a reduced dataset of 15 randomly selected locations (sparse dataset) to investigate the effect of the number of measurements on the kriging estimate. The model grid is approximately 42 km × 25 km and includes 123 × 292 = 35916 regularly spaced locations.

The mean of the elevation is 2476 m and the standard deviation is 94 m. We assume a two-dimensional anisotropic spherical covariance function with a major correlation length of 15 km in the north–south direction and a minor correlation length of 10 km in the east–west direction. We apply simple kriging using the sparse and full datasets of 15 and 100 measurements, respectively. The results are shown in Figure 3.11. In general, both kriging maps are smoother than the true elevation map shown in Figure 3.10. This feature is typical of kriging maps, because kriging is an interpolation method that minimizes the error in the mean square sense. The kriging estimate map obtained using the sparse dataset is smoother than the map obtained using the full dataset, owing to the smaller number of conditioning data. In regions with no measurements, the kriging estimate converges to the mean. The kriging variance maps show the locations of the measurements.

3.4.2 Data Configuration

The spatial covariance model allows accounting for the spatial configuration of the measurements in terms of distance, anisotropy, and redundancy. In the next set of examples, we further investigate the impact of these elements on the kriging estimate. In the first example, we assume that three measurements are available ($n = 3$) according to the data configuration in Figure 3.8a. We also assume that the two-dimensional spatial covariance function is isotropic. In this case, all the measurements are equidistant from each other; therefore, we denote this distance with the constant value h_d and the corresponding spatial covariance value with the value C_d. The measurements are also equidistant from the estimation location, with the distance being h_0 and the corresponding spatial covariance value C_0. The kriging system is given by:

$$\begin{bmatrix} \sigma_X^2 & C_d & C_d \\ C_d & \sigma_X^2 & C_d \\ C_d & C_d & \sigma_X^2 \end{bmatrix} \begin{bmatrix} w_1 \\ w_2 \\ w_3 \end{bmatrix} = \begin{bmatrix} C_0 \\ C_0 \\ C_0 \end{bmatrix}. \tag{3.26}$$

By solving the kriging system, we obtain that all weights are the same and equal to $w_i = C_0/\left(\sigma_X^2 + 2C_d\right)$ for $i = 1, \dots, n$. Then, the simple kriging estimate x_0^{sk} in Eq. (3.22) becomes

$$x_0^{sk} = \mu_X + \frac{C_0}{\sigma_X^2 + 2C_d} \sum_{i=1}^{n} (x_i - \mu_X). \tag{3.27}$$

If the mean is assumed to be equal to the sample mean of the data $\mu_X = \frac{1}{n}\sum_{i=1}^{n} x_i$, then the simple kriging estimate is equal to the mean $x_0^{sk} = \mu_X$.

In the second example, we assume an irregular data distribution of three measurements, as in the data configuration in Figure 3.8b. However, in this example, we ignore the spatial correlation between the data (i.e. $C_{i,j} = 0$, for $i, j = 1, \dots, n$ with $i \neq j$) and the spatial correlation between the data and the unknown value to be estimated (i.e. $C_{0,i} = 0$, for $i = 1, \dots, n$).

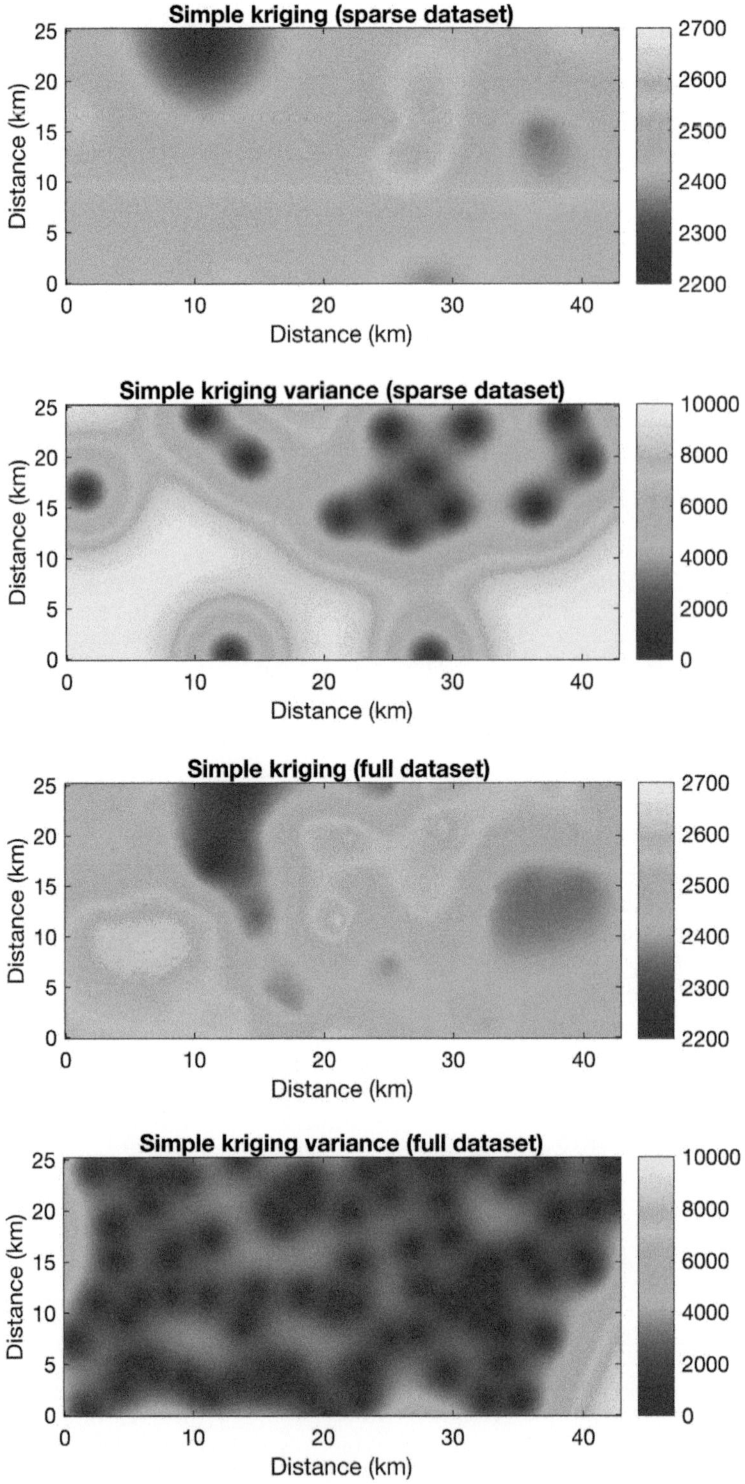

Figure 3.11 Simple kriging estimate maps of elevation and corresponding kriging variance maps. The top two plots show the results for the sparse dataset of 15 measurements; the bottom two plots show the results for the full dataset of 100 measurements.

This assumption is equivalent to assume a spatial covariance function corresponding to white noise, where the spatial covariance is equal to the variance of the variable when the distance is 0, and it is 0 for all the other distances. The kriging system then becomes:

$$\begin{bmatrix} \sigma_X^2 & 0 & 0 \\ 0 & \sigma_X^2 & 0 \\ 0 & 0 & \sigma_X^2 \end{bmatrix} \begin{bmatrix} w_1 \\ w_2 \\ w_3 \end{bmatrix} = \begin{bmatrix} 0 \\ 0 \\ 0 \end{bmatrix}. \tag{3.28}$$

Since $\sigma_X^2 > 0$, we obtain that $w_i = 0$ for $i = 1, ..., n$, and the kriging estimate simply coincides with the mean:

$$x_0^{sk} = \mu_X. \tag{3.29}$$

If we ignore the spatial correlation between the data, but we account for the spatial correlation between the data and the unknown value to be estimated, then the off-diagonal elements of the kriging matrix are still 0, whereas the kriging vector elements are given by the spatial covariance values $C_{0,i}$ for $i = 1, ..., n$. If the data are located at the vertices of an equilateral triangle (Figure 3.8a), then $C_{0,1} = C_{0,2} = C_{0,3}$ as in the first example; otherwise, in a general case (Figure 3.8b), the spatial covariance values $C_{0,i}$ are different from each other. The kriging system is then:

$$\begin{bmatrix} \sigma_X^2 & 0 & 0 \\ 0 & \sigma_X^2 & 0 \\ 0 & 0 & \sigma_X^2 \end{bmatrix} \begin{bmatrix} w_1 \\ w_2 \\ w_3 \end{bmatrix} = \begin{bmatrix} C_{0,1} \\ C_{0,2} \\ C_{0,3} \end{bmatrix}. \tag{3.30}$$

By solving the kriging system, we obtain that the kriging weights are given by $w_i = C_{0,i}/\sigma_X^2$ for $i = 1, \dots, n$, which means that the kriging weights are directly proportional to the spatial covariance between the data and the unknown value to be estimated. Then, the simple kriging estimate becomes:

$$x_0^{sk} = \mu_X + \sum_{i=1}^{n} \frac{C_{0,i}}{\sigma_X^2}(x_i - \mu_X). \tag{3.31}$$

In the third example, we adopt the data configuration in Figure 3.8c, where three measurements are located at the vertices $\boldsymbol{u}_1$, $\boldsymbol{u}_2$, and $\boldsymbol{u}_3$ of an isosceles triangle and the estimation location is at the midpoint of the segment $\overline{\boldsymbol{u}_1\boldsymbol{u}_3}$. We assume that the two-dimensional spatial covariance model is anisotropic with maximum correlation length along the direction $\overrightarrow{\boldsymbol{u}_0\boldsymbol{u}_2}$ and minimum correlation length in the direction $\overrightarrow{\boldsymbol{u}_1\boldsymbol{u}_3}$. We indicate with C^l the covariance function in the direction with the longest correlation length and with C^s the covarance function in the direction with the shortest correlation length. For simplicity, we assume that the distance between the data is larger than the maximum correlation length; hence, the off-diagonal elements of the kriging matrix are 0. The kriging system is then:

$$\begin{bmatrix} \sigma_X^2 & 0 & 0 \\ 0 & \sigma_X^2 & 0 \\ 0 & 0 & \sigma_X^2 \end{bmatrix} \begin{bmatrix} w_1 \\ w_2 \\ w_3 \end{bmatrix} = \begin{bmatrix} C_{0,1}^s \\ C_{0,2}^l \\ C_{0,3}^s \end{bmatrix} \tag{3.32}$$

and the kriging weights are given by: $w_1 = C^s_{0,1}/\sigma^2_X$, $w_2 = C^l_{0,2}/\sigma^2_X$, and $w_3 = C^s_{0,3}/\sigma^2_X$. The simple kriging estimate is then:

$$x_0^{sk} = \mu_X + \frac{C^s_{0,1}}{\sigma^2_X}(x_1 - \mu_X) + \frac{C^l_{0,2}}{\sigma^2_X}(x_2 - \mu_X) + \frac{C^s_{0,3}}{\sigma^2_X}(x_3 - \mu_X). \tag{3.33}$$

Therefore, the measurements are weighted differently according to the directional spatial covariance model and the measurement x_2 at location $\boldsymbol{u}_2$ can have a larger weight than the other measurements even if it is located at a further distance from the estimation location owing to the anisotropic assumption.

In the fourth example, we adopt the data configuration in Figure 3.8d, where the locations $\boldsymbol{u}_1$ and $\boldsymbol{u}_2$ are close, with distance $h_{1,2} \approx 0$ (i.e. $C_{1,2} \approx \sigma^2_X$). For simplicity, we adopt an isotropic spatial covariance model and we assume that the distances $h_{1,3}$ between the data $\boldsymbol{u}_1$ and $\boldsymbol{u}_3$ and $h_{2,3}$ between the data $\boldsymbol{u}_2$ and $\boldsymbol{u}_3$ are larger than the maximum correlation length. The kriging system can then be expressed as:

$$\begin{bmatrix} \sigma^2_X & \sigma^2_X & 0 \\ \sigma^2_X & \sigma^2_X & 0 \\ 0 & 0 & \sigma^2_X \end{bmatrix} \begin{bmatrix} w_1 \\ w_2 \\ w_3 \end{bmatrix} = \begin{bmatrix} C_0 \\ C_0 \\ C_0 \end{bmatrix}. \tag{3.34}$$

The kriging weights are given by: $w_1 = C_0/\left(2\sigma^2_X\right)$, $w_2 = C_0/\left(2\sigma^2_X\right)$, and $w_3 = C_0/\sigma^2_X$, and the simple kriging estimate is:

$$x_0^{sk} = \mu_X + \frac{C_0}{2\sigma^2_X}(x_1 - \mu_X) + \frac{C_0}{2\sigma^2_X}(x_2 - \mu_X) + \frac{C_0}{\sigma^2_X}(x_3 - \mu_X) \tag{3.35}$$

Therefore, the measurements at locations in proximity of each other are assigned a reduced kriging weight, and the kriging system acts as a declustering operator that accounts for data redundancy.

3.4.3 Ordinary Kriging and Universal Kriging

Ordinary kriging is another form of kriging where the mean μ_X of the variable X is assumed to be unknown. In ordinary kriging, the mean μ_X is locally estimated from the available measurements. The local empirical mean $\hat{\mu}_X$ is generally computed using a moving search neighborhood centered at the estimation location. The ordinary kriging estimate x_0^{ok} is then given by:

$$x_0^{ok} = \hat{\mu}_X + \sum_{i=1}^{n} w_i (x_i - \hat{\mu}_X), \tag{3.36}$$

where the ordinary kriging weights w_i are computed by solving the ordinary kriging system. The ordinary kriging estimate is unbiased if the sum of the weights is 1, $\sum_{i=1}^{n} w_i = 1$. This condition can be imposed by including the additional constraint $\sum_{i=1}^{n} w_i = 1$ in the kriging system using the Lagrange multiplier λ, as:

$$\begin{bmatrix} C_{1,1} & C_{1,2} & \dots & C_{1,n} & 1 \\ C_{2,1} & C_{2,2} & \dots & C_{2,n} & 1 \\ \vdots & \vdots & \ddots & \vdots & \vdots \\ C_{n,1} & C_{n,2} & \dots & C_{n,n} & 1 \\ 1 & 1 & \dots & 1 & 0 \end{bmatrix} \begin{bmatrix} w_1 \\ w_2 \\ \vdots \\ w_n \\ \lambda \end{bmatrix} = \begin{bmatrix} C_{0,1} \\ C_{0,2} \\ \vdots \\ C_{0,n} \\ 1 \end{bmatrix}. \tag{3.37}$$

The kriging weights are then computed by solving the linear system in Eq. (3.37) and used in Eq. (3.36) to obtain the ordinary kriging estimate. Because of the constraint $\sum_{i=1}^{n} w_i = 1$, Eq. (3.36) reduces to $x_0^{ok} = \hat{\mu}_X + \sum_{i=1}^{n} w_i x_i - \hat{\mu}_X = \sum_{i=1}^{n} w_i x_i$. The use of a local search neighborhood to estimate the mean in ordinary kriging allows mitigating the problem of the non-stationarity of the mean of the variable, especially for large datasets where the mean locally shows limited spatial variations. However, for large datasets with clustered and sparse measurements, ordinary kriging can generate discontinuities in the results.

Other forms of kriging have been developed based on simple and ordinary kriging approaches. In kriging with locally variable mean, kriging with external drift, and universal kriging, the sample mean is replaced by a function of space (or time) such as a polynomial function. In these forms of kriging, the random variable X is decomposed into the sum of a deterministic function, such as a polynomial function $m(\boldsymbol{u})$ of degree p of the spatial coordinates $\boldsymbol{u}$, and a random variable R with 0 mean and stationary covariance, representing the residual term:

$$X(\boldsymbol{u}) = m(\boldsymbol{u}) + R(\boldsymbol{u}) = \sum_{k=0}^{p} a_k \boldsymbol{u}^k + R(\boldsymbol{u}). \tag{3.38}$$

In kriging with locally variable mean, the deterministic function is assumed to be known, whereas, in universal kriging, it is assumed to be a polynomial function with unknown coefficients. At a given location $\boldsymbol{u}_0$, the kriging estimate $\hat{x}(\boldsymbol{u}_0)$ can then be expressed in the general form:

$$\hat{x}(\boldsymbol{u}_0) = a_0 + a_1 \boldsymbol{u}_0 \dots + a_p \boldsymbol{u}_0^p + \sum_{i=1}^{n} w_i R(\boldsymbol{u}_i), \tag{3.39}$$

where the coefficients w_i represent the kriging weights. If the polynomial coefficients $a_0, \dots, a_p$ are known, the kriging method is equivalent to applying simple kriging to the residual R and summing the kriging estimate to the polynomial function representing the trend of the mean. This approach is particularly useful in subsurface applications in the presence of spatial trends, such as the vertical trends of porosity and P-wave velocity owing to compaction, where the prior trend is often estimated using physical models or fitting data measurements with polynomial functions. In universal kriging, the polynomial coefficients $a_0, \dots, a_p$ are determined as part of an extended kriging system of $n + p + 1$ equations. If the polynomial function is defined externally as a function of auxiliary variables, rather than the spatial coordinates, the term kriging with external drift is generally used. Another form of kriging is Bayesian kriging (Omre 1987), where the model parameters are expressed in terms of probability density functions and are updated conditioned on the measurements.

3.4.4 Cokriging

Cokriging is a multivariate extension of kriging (Matheron 1969) to predict an unknown variable X (primary variable) at a given location $\mathbf{u}_0$ based on a set of measurements $\{x_i\}_{i=1,\dots,n}$ at locations $\{\mathbf{u}_i\}_{i=1,\dots,n}$ and a set of measurements $\{y_j\}_{j=1,\dots,m}$ at locations $\{\mathbf{u}_j\}_{j=1,\dots,m}$ of another variable Y (secondary variable) that is assumed to be correlated to X. Cokriging is often used when there are a limited number of direct measurements of the primary variable but dense and regularly sampled measurements of a secondary variable are available, or when the measurements of the secondary variable are closer to the estimation location. This situation is often encountered in subsurface modeling when direct measurements of the properties of interest are limited but additional geophysical measurements are available at the estimation location. Porosity, for example, can be measured directly only at the borehole locations, but seismic velocities are often available and are generally negatively correlated with porosity. Therefore, in seismic reservoir characterization, dense and regularly sampled geophysical attributes are combined with sparsely sampled direct measurements to estimate the property of interest in the entire three-dimensional model (Doyen 1988, 2007).

We assume that X is a continuous variable with mean μ_X and variance σ_X^2 and Y is a continuous variable with mean μ_Y and variance σ_Y^2. We also assume that the spatial correlation model is described by the spatial covariance functions $C_X(h)$ and $C_Y(h)$ of the variables X and Y, and the spatial cross-covariance function $C_{X,Y}(h)$. The cokriging estimate is given by:

$$x_0^{ck} = \mu_X + \sum_{i=1}^{n} w_i (x_i - \mu_X) + \sum_{j=1}^{m} v_j (y_j - \mu_Y), \tag{3.40}$$

where w_i and v_j are the cokriging weights. Similar to kriging, the cokriging weights are computed to minimize the estimation error $x_0 - x_0^{ck}$ in the mean square sense, by solving the associated cokriging system of equations that depends on the covariance functions $C_X(h)$, $C_Y(h)$, and $C_{X,Y}(h)$. Equation (3.40) can also be extended to multiple conditioning variables.

The general form of cokriging is not commonly used because with sparse datasets the estimation of the covariance functions might be challenging, whereas with dense datasets the kriging system becomes relatively large and computationally demanding. A simplified version of cokriging, namely collocated cokriging (Xu et al. 1992), is used more frequently. Collocated cokriging only uses a single measurement y_0 of the secondary variable Y, located at (or close to) the estimation location $\mathbf{u}_0$. Using a Markov assumption, the cross-covariance $C_{X,Y}(h)$ is expressed in terms of the covariance function $C_X(h)$ of the primary variable X, as:

$$C_{X,Y}(h) = \frac{\sigma_Y}{\sigma_X} \rho_{X,Y} C_X(h), \tag{3.41}$$

where $\rho_{X,Y}$ is the linear correlation coefficient. Therefore, collocated cokriging does not require the three covariance functions $C_X(h)$, $C_Y(h)$, and $C_{X,Y}(h)$ as for full cokriging, but only the covariance function $C_X(h)$, the variances σ_X^2 and σ_Y^2, and the correlation coefficient $\rho_{X,Y}$. Under this assumption, the collocated cokriging estimate x_0^{cck} is:

$$x_0^{cck} = \mu_X + \sum_{i=1}^{n} w_i (x_i - \mu_X) + v(y_0 - \mu_Y), \tag{3.42}$$

where the weights w_i and v are obtained by solving a system of linear equations, namely the collocated cokriging system, of the following form:

$$\begin{bmatrix} \sigma_X^2 & C_X(h_{1,2}) & \dots & C_X(h_{1,n}) & C_{X,Y}(h_{0,1}) \\ C_X(h_{2,1}) & \sigma_X^2 & \dots & C_X(h_{2,n}) & C_{X,Y}(h_{0,2}) \\ \vdots & \vdots & \ddots & \vdots & \vdots \\ C_X(h_{n,1}) & C_X(h_{n,2}) & \dots & \sigma_X^2 & C_{X,Y}(h_{0,n}) \\ C_{X,Y}(h_{0,1}) & C_{X,Y}(h_{0,2}) & \dots & C_{X,Y}(h_{0,n}) & \sigma_Y^2 \end{bmatrix} \begin{bmatrix} w_1 \\ w_2 \\ \vdots \\ w_n \\ v \end{bmatrix} = \begin{bmatrix} C_X(h_{0,1}) \\ C_X(h_{0,2}) \\ \vdots \\ C_X(h_{0,n}) \\ \sigma_{X,Y} \end{bmatrix} \tag{3.43}$$

with $h_{i,j}$ being the distance between locations $\boldsymbol{u}_i$ and $\boldsymbol{u}_j$, for $i, j = 0, \dots, n$. The collocated cokriging variance $\sigma_{X,cck}^2$ is given by:

$$\sigma_{X,cck}^2 = \sigma_X^2 - \sum_{i=1}^{n} w_i C_X(h_{0,i}) - v\sigma_{X,Y}. \tag{3.44}$$

Collocated cokriging can also be reformulated in the Bayesian framework, according to the method named Bayesian updating of kriging (Doyen 2007), where the Bayesian formulation (Eq. 1.25) is applied to a local prior Gaussian distribution $f(x)$ with mean equal to the simple kriging estimate x_0^{sk} (Eq. 3.22) and variance equal to the kriging variance $\sigma_{X,sk}^2$ (Eq. 3.25) to obtain the conditional distribution $f(x|y_0)$ conditioned on the measurement y_0 of the secondary variable. The mean x_0^{buk} of the conditional distribution $f(x|y_0)$ is equivalent to the collocated cokriging estimate in Eq. (3.42) and can be analytically formulated as:

$$x_0^{buk} = \frac{1}{\rho_{X,Y}^2\left(\sigma_{X,sk}^2 - 1\right) + 1}\left(\rho_{X,Y}\sigma_{X,sk}^2 y_0 + \left(1 - \rho_{X,Y}^2\right)x_0^{sk}\right), \tag{3.45}$$

where $\rho_{X,Y}$ is the linear correlation coefficient and the denominator $\rho_{X,Y}^2\left(\sigma_{X,sk}^2 - 1\right) + 1$ is a normalizing constant.

Example 3.4 We show an application of cokriging to the Yellowstone dataset, by introducing a secondary variable, namely the temperature (Figure 3.10), which is assumed to be available at each location in the model. However, the temperature map has a lower resolution than the elevation map. The mean temperature is 7.43°C, the standard deviation is 0.66°C, and the correlation coefficient between elevation and temperature is −0.93. Figure 3.12 shows the cokriging estimate maps of elevation using full cokriging, with temperature as a secondary variable. As for the simple kriging results (Figure 3.11), we calculate the cokriging estimate for a sparse dataset of 15 measurements and for the full dataset of 100 measurements of the primary variable. Because elevation and temperature are highly correlated, the number of measurements of the primary variable does not affect the results as much as for simple kriging and the two cokriging estimate maps are similar. Overall, the two cokriging estimate maps are more accurate than the corresponding simple kriging maps. The results in Figure 3.12 depend on the high correlation between the primary and the secondary variable. Figure 3.13 investigates the effect of correlation in two cases: medium correlation ($\rho = -0.5$) and low correlation ($\rho = -0.1$), using the sparse dataset. The results of the low correlation case resemble the simple kriging estimate map, since the effect of the secondary variable on the cokriging estimate map is almost negligible owing to the low correlation.

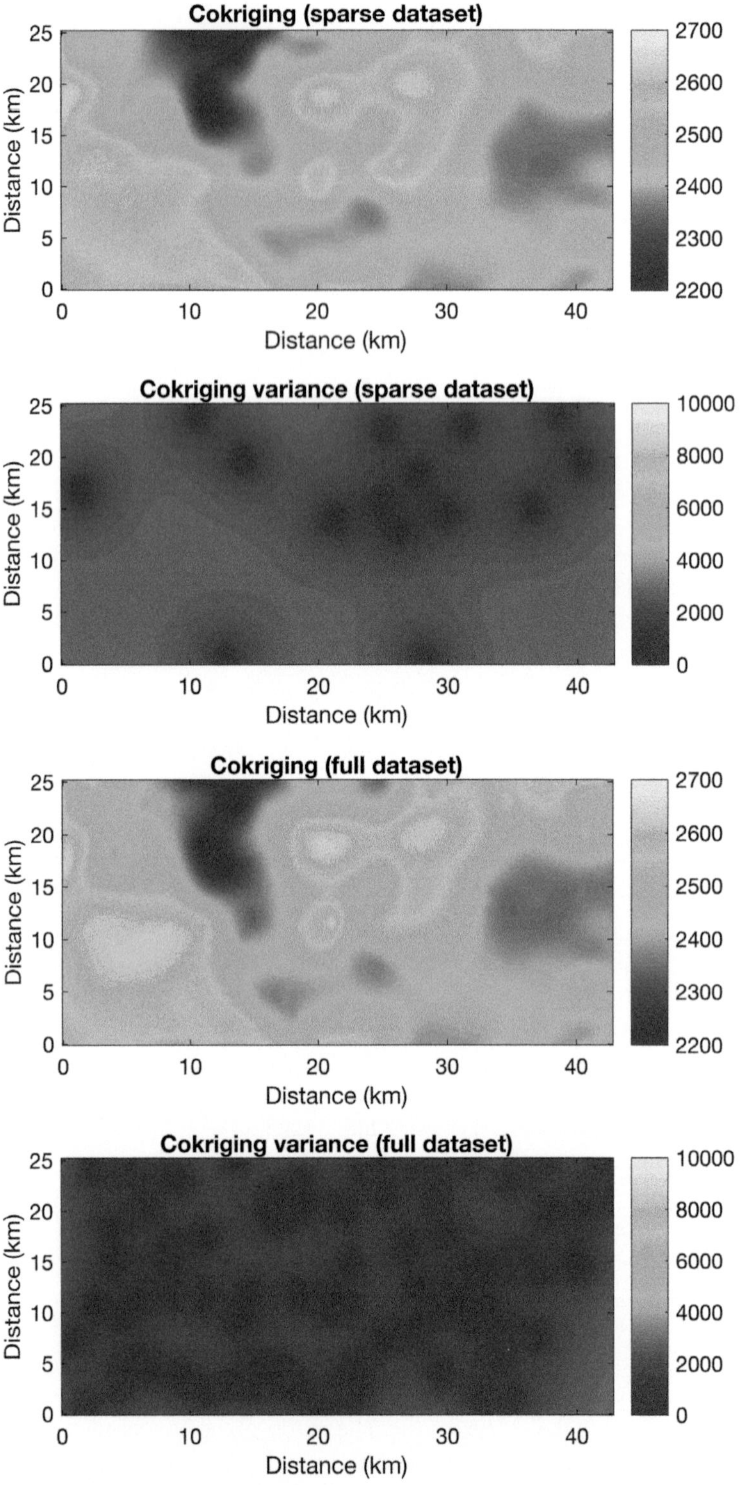

Figure 3.12 Cokriging estimate maps of elevation and corresponding cokriging variance maps. The top two plots show the results for the sparse dataset of 15 measurements; the bottom two plots show the results for the full dataset of 100 measurements.

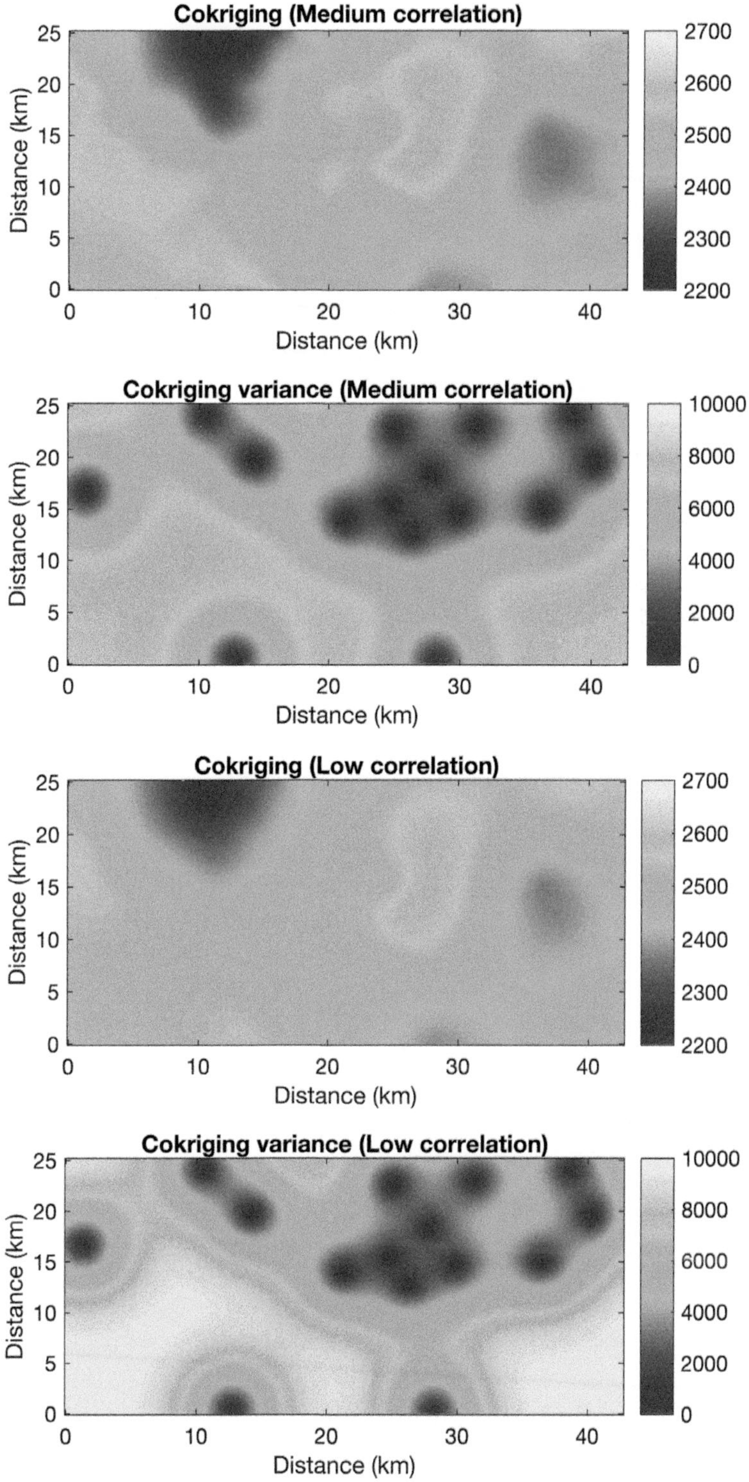

Figure 3.13 Cokriging estimate maps of elevation and corresponding cokriging variance maps. The top two plots show the results assuming medium correlation ($\rho = -0.5$); the bottom two plots show the results assuming low correlation ($\rho = -0.1$). Both results use the sparse dataset of 15 measurements.

There are other forms of kriging, such as factorial kriging for spatial filtering operations, block kriging for averaging problems over areas or volumes, and functional kriging for the interpolation of random functions. Factorial kriging is often used for removing spatially correlated noise, whereas functional kriging is generally applied to predict an operator response based on a limited number of model simulations.

3.5 Sequential Simulations

Kriging and cokriging are estimation methods that provide the estimate that minimizes the error in the mean square sense, for spatial interpolation problems. The examples in Section 3.4 show that kriging generally provides local estimates with smooth spatial variations. Low values are generally overestimated whereas high values are generally underestimated, unless these values are sampled in the measured dataset. From a geological point of view, the local smoothness is not always desirable, because it cannot represent the spatial heterogeneity of subsurface properties, such as spatial variations in porosity and permeability. Simulating heterogeneities in the spatial distribution of geological properties requires spatially correlated sampling algorithms, i.e. statistical methods where values of the random variable are sampled according to the available data and a spatial correlation model.

Geostatistical sequential simulations are stochastic algorithms that generate highly detailed realizations of the properties of interest, by sampling multidimensional random fields with a spatial correlation model to mimic the expected spatial variability. In sequential simulations, the values of the random variable are generated by sequentially visiting the model locations in the two- or three-dimensional space, following a random path. At each location, the value of the random variable is sampled from a conditional probability density function. The simulated value depends on the prior distribution of the random variable, on the direct measurements (if available), and on the previously simulated values in the neighboring locations. This procedure is repeated for all the locations in the model. The use of conditional distributions based on previously simulated values is essential to ensure the spatial continuity of the model. Indeed, if the sampling is performed independently at each location, the resulting realization would not show any spatial continuity.

3.5.1 Sequential Gaussian Simulation

Sequential Gaussian simulation (SGS) is one of the most popular sequential simulation methods for generating spatially correlated realizations of a continuous variable (Deutsch and Journel 1998; Pyrcz and Deutsch 2014). The realizations can be conditioned on direct measurements (often called hard data) or they can be unconditional. The spatial correlation model is defined by a spatial covariance function or variogram (Section 3.2).

We assume that the property of interest is a continuous random variable X distributed according to a Gaussian distribution $\mathcal{N}(X;\mu_X,\sigma_X^2)$ with prior mean μ_X and prior variance σ_X^2. We also assume that a limited number n of observations $\{x_i\}_{i=1,\dots,n}$ are available at locations $\{\boldsymbol{u}_i\}_{i=1,\ldots,n}$ and that the spatial model consists of a spatial covariance function $C(h)$. In SGS, the sample x_0 at a given location $\boldsymbol{u}_0$ is obtained by sampling from a Gaussian distribution $\mathcal{N}\left(X; x_0^{sk}, \sigma_{X,sk}^2\right)$, with conditional mean equal to the kriging estimate x_0^{sk} at location $\boldsymbol{u}_0$ and conditional variance equal to the kriging variance $\sigma_{X,sk}^2$:

$$f(x_0|x_1, \ldots, x_n) \sim \mathcal{N}\left(X; x_0^{sk}, \sigma_{X,sk}^2\right) = \frac{1}{\sqrt{2\pi\sigma_{X,sk}^2}} \exp\left(-\frac{1}{2}\frac{\left(x_0 - x_0^{sk}\right)^2}{\sigma_{X,sk}^2}\right), \tag{3.46}$$

where the parameters of the Gaussian distributions are computed using simple kriging (Eqs. 3.22 and 3.25) or other forms of kriging. The kriging estimates are computed based on the n direct measurements as well as previously simulated values at grid locations that were sampled before the current location. Therefore, the kriging system of equations changes at every grid location and a new kriging system has to be solved at each new location. To sample from the distribution in Eq. (3.46), we can use a random number generator to sample a random number u from a uniform distribution in the interval [0, 1] and calculate the simulated value x_0 as:

$$x_0 = F^{-1}(u), \tag{3.47}$$

where F^{-1} is the inverse of the cumulative distribution function $F(x_0|x_1, \ldots, x_n)$ of the probability density function $f(x_0|x_1, \ldots, x_n)$ in Eq. (3.46). This procedure is repeated at all the locations by adding the previously simulated values to the vector of conditioning data.

The SGS methodology is illustrated in Figure 3.14 for a two-dimensional problem, where we assume that k locations $\{\boldsymbol{u}_i\}_{i=1,\ldots,k}$ have already been simulated or that direct measurements are available at those locations. In Figure 3.14, k is 4. Then, we simulate the next value according to the following steps:

1) We randomly pick a location $\boldsymbol{u}_{k+1}$ where the value of X is unknown (e.g. location $\boldsymbol{u}_5$ in Figure 3.14).
2) We compute the kriging estimate x_{k+1}^{sk} using Eq. (3.22), where the weights are computed by solving the kriging system in Eq. (3.23) (e.g. x_5^{sk} in Figure 3.14).

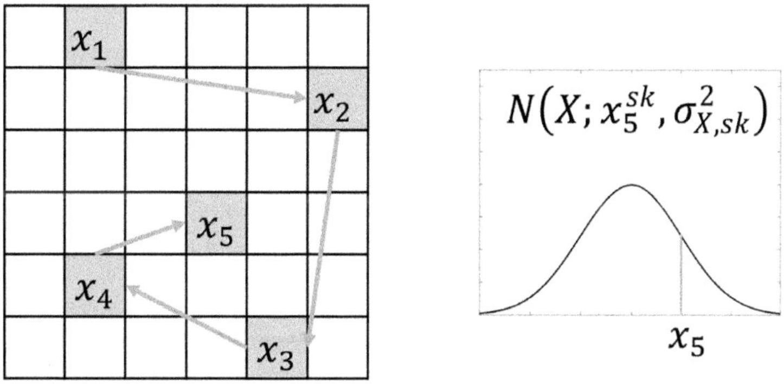

Figure 3.14 Schematic representation of sequential Gaussian simulation for a two-dimensional problem.

3) We compute the kriging variance $\sigma^2_{X,sk}$ using Eq. (3.25).
4) We sample a random value x_{k+1} from the Gaussian distribution $f(x_{k+1}|x_1, \dots, x_k) \sim \mathcal{N}(X; x^{sk}_{k+1}, \sigma^2_{X,sk})$ in Eq. (3.46), and we assign the value x_{k+1} to location $\boldsymbol{u}_{k+1}$ (e.g. x_5 in Figure 3.14).
5) We add the simulated value x_{k+1} to the vector of conditioning data $x_1, \dots, x_k$.
6) We repeat steps 1–5 for all the remaining locations.

Similar to kriging, in SGS, a moving search neighborhood is often implemented to limit the number of conditioning measurements (and previously simulated values). At each measurement location, the kriging estimate coincides with the measurement value and the kriging variance is 0; therefore, the predicted value coincides with the measured observation. If the continuous variable X is not Gaussian, then a normal score transformation must be applied, the simulation is performed in the transformed space and the results are transformed back into the original domain. Multiple realizations are generated by selecting different random paths. For a large number of realizations, their pointwise average converges to the kriging estimate. The SGS realizations approximately reproduce the spatial covariance model used to solve the kriging system of equations, although the experimental covariance functions of the individual realizations do not exactly match the theoretical model. In particular, for highly skewed prior distributions, SGS reproduces the spatial covariance function in the normal score domain but the reproduction of the spatial covariance function in the original domain is not guaranteed. SGS can also be applied using other forms of kriging, by replacing the simple kriging estimate in Eq. (3.46) with the desired formulation.

Example 3.5 Figure 3.15 shows an example of application of SGS using the dataset of the Yellowstone National Park (Figure 3.10). In this example, we generate 100 realizations of the elevation under the same assumptions of the simple kriging results (Figure 3.11), using the sparse dataset of 15 measurements. While kriging provides a smooth map, the realizations attempt to mimic the spatial heterogeneity and natural variability of elevation, according to the statistical assumptions of the spatial correlation model. Figure 3.15 shows three random realizations and the average of the full set of 100 realizations. The average of the SGS realizations in Figure 3.15 is approximately equal to the kriging estimate in Figure 3.11. Figure 3.16 shows the same application using the full dataset of 100 realizations. Since the number of available measurements is larger than in the previous example (Figure 3.15), the variability of the realizations is reduced compared with the variability of the results in Figure 3.15. The so-obtained realizations resemble the true elevation map (Figure 3.10) despite using only 100 measurements for 35916 total locations.

Figure 3.17 shows a comparison of the results obtained using simple kriging and SGS by analyzing the histograms, boxplots, and experimental variograms of the following datasets: the true model, the simple kriging estimate map, one random SGS realization, and the SGS ensemble average of 100 realizations. First, we compare the results obtained using the sparse dataset. The kriging map has the correct mean, but the variability is smaller and most of the values regress toward the mean, whereas the SGS realization shows the correct mean and variance and includes values close to the minimum and maximum values of the true model. The experimental variogram is computed for the 45° direction. The simple kriging

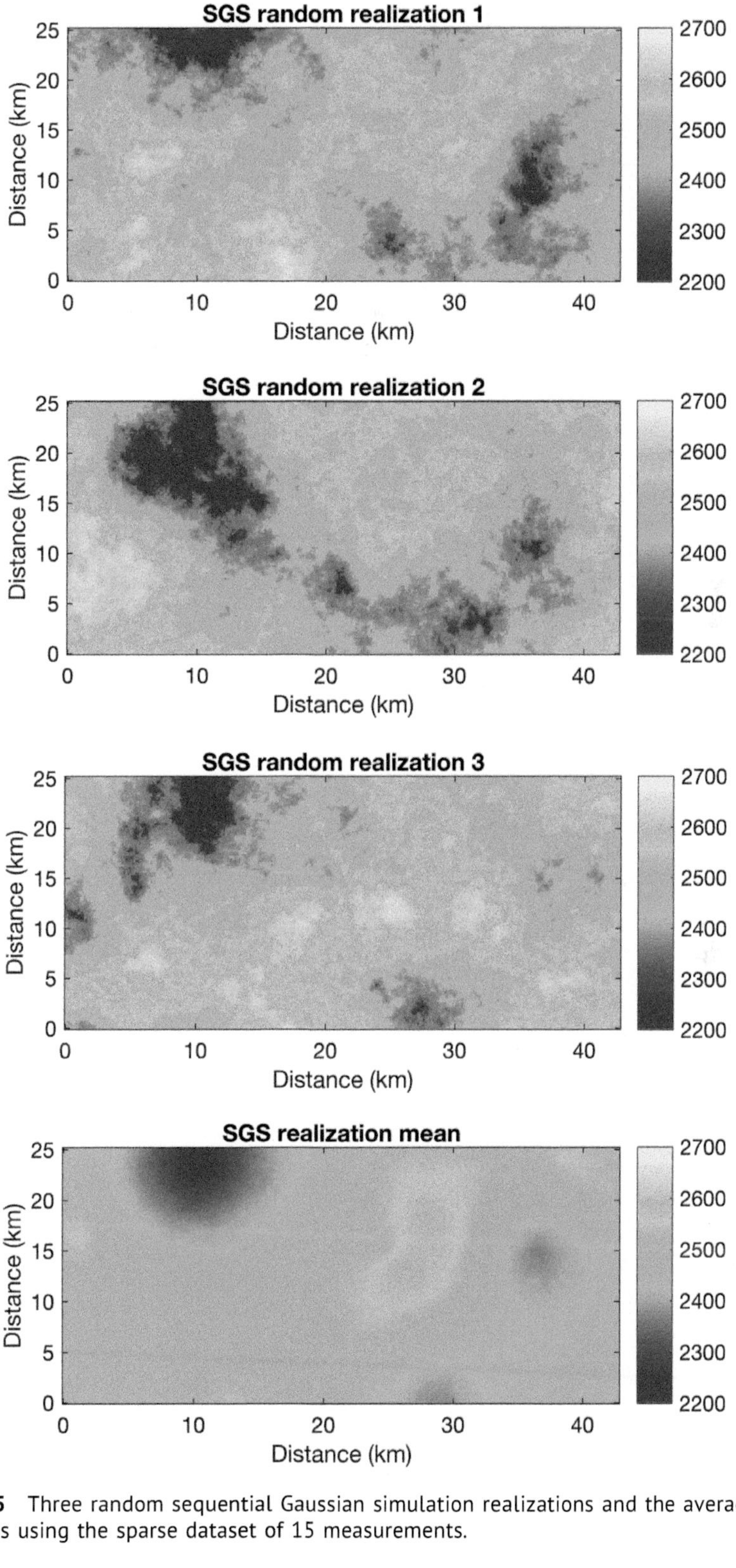

Figure 3.15 Three random sequential Gaussian simulation realizations and the average of 100 realizations using the sparse dataset of 15 measurements.

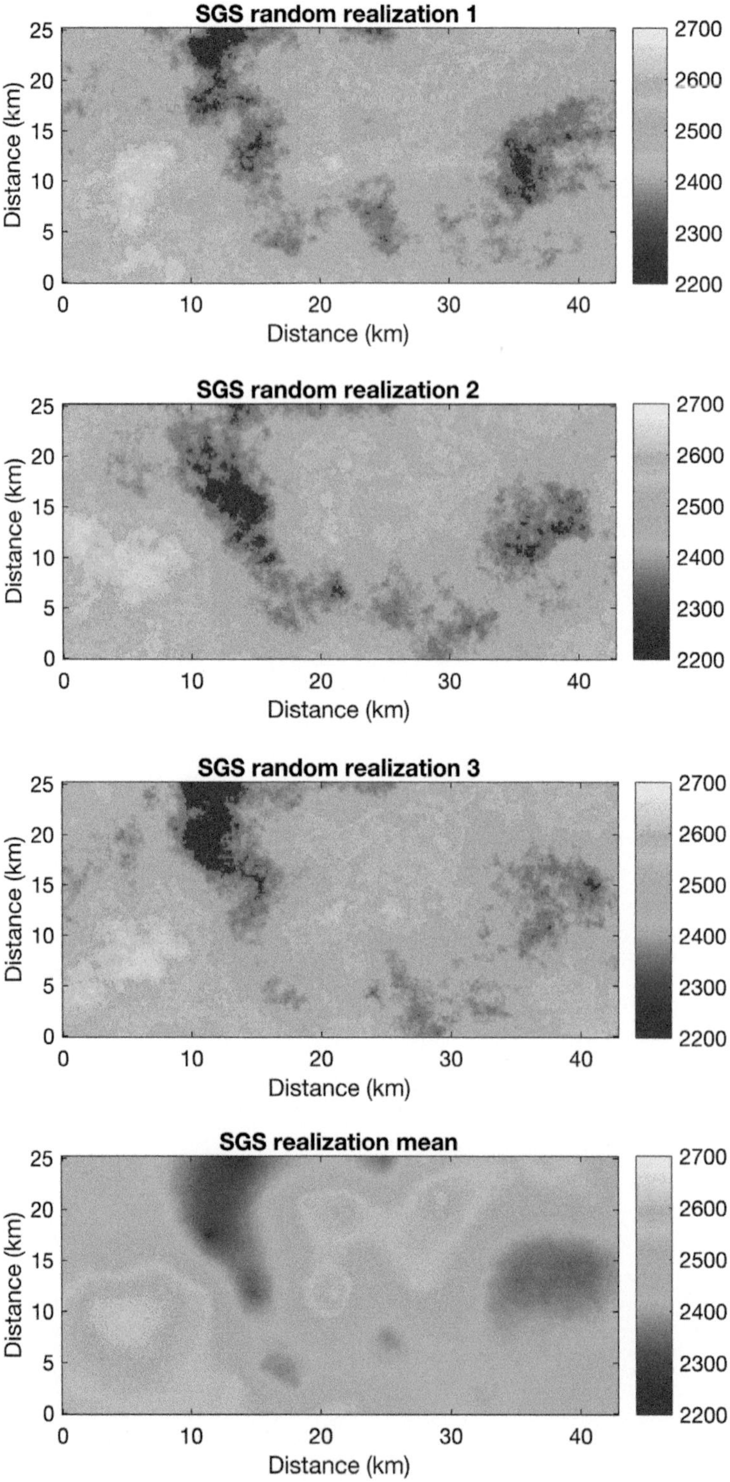

Figure 3.16 Three random sequential Gaussian simulation realizations and the average of 100 realizations using the full dataset of 100 measurements.

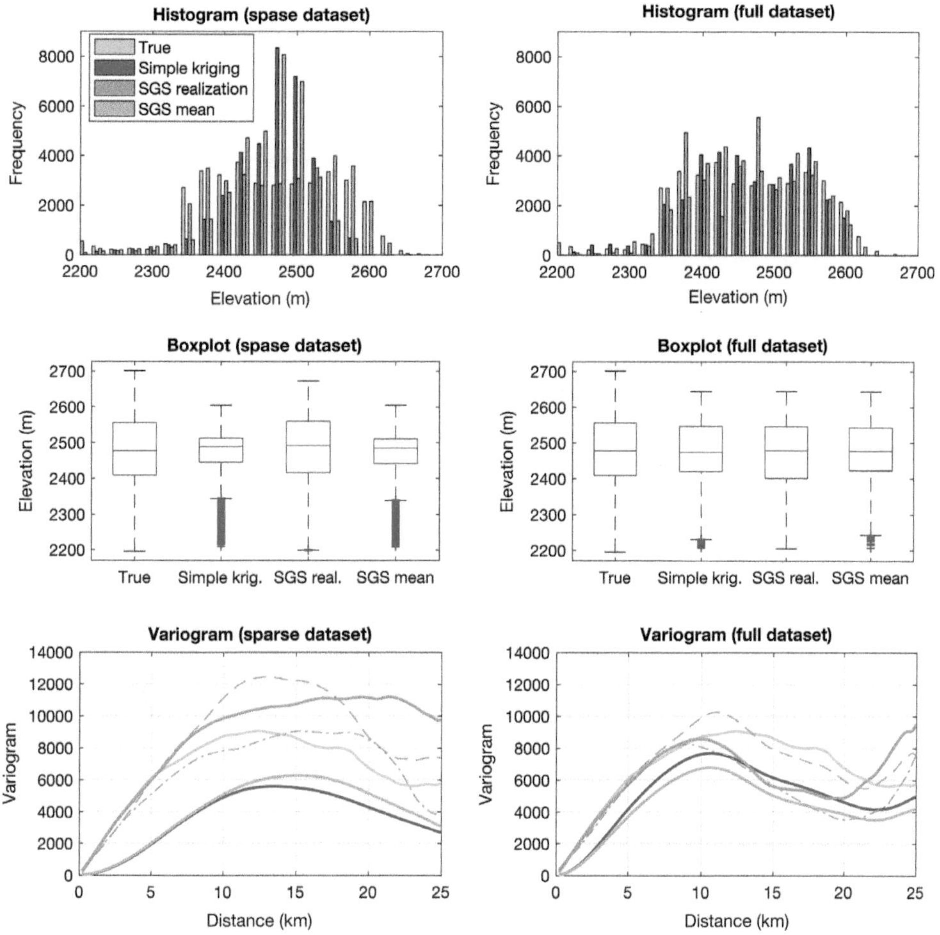

Figure 3.17 Comparison of simple kriging and sequential Gaussian simulation using histograms, boxplots, and variograms of the elevation values (light blue represents the true model, red represents the simple kriging estimate, green represents one random sequential Gaussian simulation realization, yellow represents the average of 100 sequential Gaussian simulation realizations). The dashed and dashed-dotted green lines represent the variograms of two additional realizations for comparison. The left plots show the results for the sparse dataset of 15 measurements; the right plots show the results for the full dataset of 100 measurements.

variogram shows a longer correlation range than the true one owing to the smoother nature of the kriging estimate map, whereas the variograms of three randomly selected realizations approximate but do not exactly match the true one and show some variability between the different realizations. The SGS ensemble average shows the same features as the kriging map. Then, we compare the results for the full dataset. The conclusions are similar to the previous case, but the differences are less evident owing to the larger number of conditioning measurements and the smaller variability of the realizations.

3.5.2 Sequential Gaussian Co-Simulation

In subsurface modeling, it is common to build a low-frequency model of the primary variable from the secondary variable, for example by applying a linear regression, and use the low-frequency model as the locally variable mean in SGS. For example, we can generate multiple realizations of porosity using SGS with a locally variable mean computed by applying a rock physics model (Chapter 2) to the seismic velocities or impedances. SGS can also be generalized to multivariate applications to sequentially simulate multiple correlated variables or by constraining the simulation of the primary variable using a linearly correlated secondary variable (Pyrcz and Deutsch 2014).

SGS can be applied with cokriging or collocated cokriging; this method is often called sequential Gaussian co-simulation (CoSGS). In this case, the local probability density function in Eq. (3.46) is computed using the cokriging or collocated cokriging estimate and variance. For example, in SGS with collocated cokriging, the prediction x_0 at a given location $\boldsymbol{u}_0$ is obtained by sampling from the following Gaussian distribution:

$$f(x_0|x_1, \ldots, x_n, y_0) \sim \mathcal{N}\left(X; x_0^{cck}, \sigma_{X,cck}^2\right) = \frac{1}{\sqrt{2\pi\sigma_{X,cck}^2}} \exp\left(-\frac{1}{2}\frac{\left(x_0 - x_0^{cck}\right)^2}{\sigma_{X,cck}^2}\right), \tag{3.48}$$

where y_0 is the value of the secondary variable at location $\boldsymbol{u}_0$ and the parameters x_0^{cck} and $\sigma_{X,cck}^2$ of the Gaussian distributions are computed using the collocated cokriging formulation (Eqs. 3.42 and 3.44). The local probability density function is narrower than in the kriging case as the cokriging variance is smaller than the corresponding kriging variance. The CoSGS realizations approximately reproduce the correlation between the primary and the secondary variables. However, for highly skewed distributions of the primary variable, the normal score and back-transformations might alter the linear relation between the primary and the secondary variables, owing to the non-linearity of the transformations.

Example 3.6 Figure 3.18 shows an application of CoSGS with cokriging to the dataset of the Yellowstone National Park (Figure 3.10) using the sparse dataset of 15 elevation measurements as hard data and the temperature map as a secondary variable. Compared with the SGS results in Figure 3.15, the realizations are more constrained owing to the high correlation between elevation and temperature. The average of the 100 simulations is almost identical to the cokriging estimate map obtained with the sparse dataset in Figure 3.12.

Further developments in SGS, such as SGS with local Bayesian updating (Doyen 2007) and SGS with Markov–Bayes soft indicator kriging (Pyrcz and Deutsch 2014), extend the application to multivariate problems with non-linear relations between the primary and secondary variables. For example, in SGS with local Bayesian updating, we compute the local probability density function $f(x_0|x_1, \ldots, x_n)$ using simple kriging (Eq. 3.46), and then use it as a prior distribution in a Bayesian formulation to compute the local posterior distribution as the product of the prior distribution and the likelihood function of the secondary variable conditioned on the primary variable $f(y_0|x_0)$. The likelihood function is computed from the joint distribution of the primary and secondary variables. In SGS with

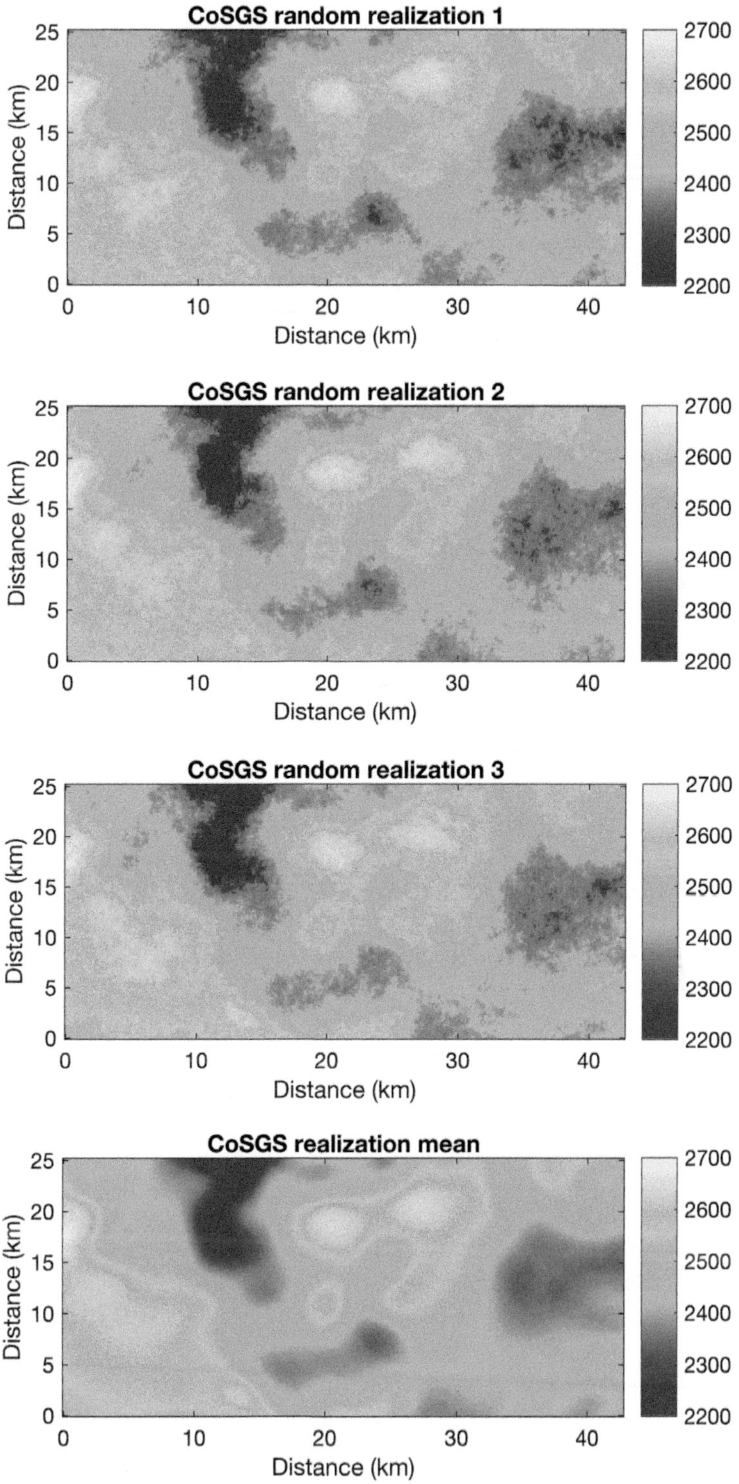

Figure 3.18 Three random sequential Gaussian co-simulation realizations and the average of 100 realizations using the sparse dataset of 15 measurements.

cokriging, the joint distribution is assumed to be a multivariate Gaussian distribution, whereas in SGS with local Bayesian updating, it can be any non-parametric probability density function (Doyen 2007).

An extension of SGS to Gaussian mixture models (Section 1.4.5), namely sequential Gaussian mixture simulation (SGMS), has been proposed in Grana et al. (2012b). The SGMS method allows generating realizations of a Gaussian mixture random field of a continuous property, by extending the sequential simulation method to the mixed discrete–continuous problem. Similar to the SGS case, in SGMS, the means and the variances of the conditional distribution at a given point correspond to the kriging estimates and variances for each component of the mixture. The conditional weights of the local conditional Gaussian mixture correspond to the conditional probability of the underlying discrete property (Grana et al. 2012b). The main advantage of the SGMS is that it simultaneously generates realizations of the discrete and continuous properties.

A similar approach to SGS is the direct sequential simulation (DSS, Soares 2001) method that does not require a normal score transformation for non-Gaussian variables. The normal score and back-transformations are non-linear and might affect the reproduction of the prior distribution, spatial covariance function model, and linear correlation between primary and secondary variables. In the DSS approach, the random values are generated from a local cumulative density function $F(x_0|x_1, \dots, x_n)$ obtained from the prior (global) cumulative density function $F(x)$ by selecting, for each value x_0, a subinterval of the domain of the random variable X centered around the local simple kriging estimate x_0^{sk} according to the simple kriging variance $\sigma^2_{X,sk}$ (Soares 2001). By locally sampling from the cumulative density function, the reproduction of the prior distribution is guaranteed, even for non-Gaussian distributions. The DSS approach relies on the fact that in order to ensure the reproduction of the spatial covariance function in the model realizations, any type of local conditional distribution can be used to simulate values, as long as its mean and variance are equal to the kriging mean and variance. DSS can accommodate any numerical technique for assessing the local conditional distribution, which has made the direct sequential approach a popular and flexible method for variogram-based or multiple-point statistics (MPS) simulations (Section 4.5).

3.6 Other Simulation Methods

Several other geostatistical methods are available to generate spatially correlated realizations of random fields in a multidimensional domain, including the turning bands method (Matheron 1973), the LU decomposition simulation (Alabert 1987; Davis 1987), the moving average (MA; Oliver 1995), the probability field simulation (PFS; Srivastava 1992), and the fast Fourier transform – moving average (FFT-MA; Le Ravalec et al. 2000).

In the turning bands method (Matheron 1973), instead of simulating realizations of two- or three-dimensional random fields, we perform one-dimensional simulations in different directions using directionally variable covariance functions; then, at each location of the model, the realization is obtained as a weighted sum of the corresponding one-dimensional simulations.

For applications in small dimensional domains, a simple method for simulating realizations of Gaussian random fields is to decompose the spatial covariance matrix $\mathbf{C} = \mathbf{LU}$ as the product of a lower triangular matrix $\mathbf{L}$ and an upper triangular matrix $\mathbf{U}$ (LU decomposition). For a positive definite covariance matrix $\mathbf{C}$, the upper triangular matrix can be written as the transpose of $\mathbf{L}$, and the decomposition becomes $\mathbf{C} = \mathbf{LL}^T$ (Cholesky decomposition). For a positive definite matrix, the Cholesky decomposition always exists and it is unique. Then, we can generate a spatially correlated realization $\boldsymbol{y}$ of a Gaussian random field with mean $\boldsymbol{\mu}_y$ and spatial covariance function $\mathbf{C}_y = \mathbf{L}_y\mathbf{L}_y^T$ as:

$$\boldsymbol{y} = \boldsymbol{\mu}_y + \mathbf{L}_y\boldsymbol{z}, \tag{3.49}$$

where $\boldsymbol{z} \sim \mathcal{N}(\boldsymbol{z}; \mathbf{0}, \mathbf{I})$ represents a realization of a spatially uncorrelated random variable distributed according to a Gaussian distribution with $\mathbf{0}$ mean and covariance matrix equal to the identity $\mathbf{I}$ (Alabert 1987; Davis 1987). The uncorrelated Gaussian realization can be generated by independently sampling from a standard Gaussian distribution at each location of the model. Because $\boldsymbol{z}$ has $\mathbf{0}$ mean, the expected value of the random variable $\boldsymbol{y}$ is equal to $\boldsymbol{\mu}_y$, as $E[\boldsymbol{y}] = E[\boldsymbol{\mu}_y + \mathbf{L}_y\boldsymbol{z}] = \boldsymbol{\mu}_y + \mathbf{L}_yE[\boldsymbol{z}] = \boldsymbol{\mu}_y$. The expected covariance matrix of the random variable $\boldsymbol{y}$ is equal to $\mathbf{C}_y$, as $E\left[(\boldsymbol{y}-\boldsymbol{\mu}_y)(\boldsymbol{y}-\boldsymbol{\mu}_y)^T\right] = E\left[\mathbf{L}_y\boldsymbol{z}(\mathbf{L}_y\boldsymbol{z})^T\right] = \mathbf{L}_yE[\boldsymbol{z}\boldsymbol{z}^T]\mathbf{L}_y^T = \mathbf{L}_y\mathbf{I}\mathbf{L}_y^T = \mathbf{C}_y$. The Cholesky decomposition method is very efficient for applications to small models, but it becomes unfeasible with large models, since the simulation requires the decomposition of large spatial covariance matrices.

Similarly, the MA method can be used to generate realizations of Gaussian random fields with a given spatial covariance model, by expressing the spatial covariance as a convolution (Oliver 1995). In the one-dimensional case, if we can express the spatial covariance function $C(h)$ as a convolution of an even function $g(h) = g(-h)$ with itself as $C(h) = g(h) * g(h)$ (where $*$ is the convolution operator), then we can generate a one-dimensional spatially correlated realization $y(x)$ with mean $m(x)$ and spatial covariance function $C(h)$ as the convolution of the function $g(x)$ with an uncorrelated random variable $z(x)$, Gaussian distributed with 0 mean and variance 1, as:

$$y(x) = m(x) + g(x) * z(x) = m(x) + \int g(x-t)z(t)dt, \tag{3.50}$$

where the integral is numerically computed as a sum in discretized model grids.

The Cholesky decomposition and the MA methods convolve a standard Gaussian realization with a decomposition of the covariance model; however, the MA method does not require the covariance matrix but uses the analytical decomposition of the covariance function. The main limitation of the MA method is that the calculation of the function $g(h)$ for two- and three-dimensional covariance models is not analytically tractable (Oliver 1995), making it unfeasible in many practical applications.

The MA method can be combined with the FFT method to efficiently compute the convolution in the frequency domain. The FFT has been widely used in geophysics to generate stochastic realizations with specified covariance structures for studying wave propagation in random media (Frankel and Clayton 1986; Ikelle et al. 1993; Mukerji et al. 1995). Spectral simulation based on FFT can be used to generate unconditional realizations of random fields with a given spatial covariance model (Pardo-Iguzquiza and Chica-Olmo 1993). Local data conditioning is typically obtained by adding a kriging residual. Conditional simulations using FFT can be obtained using an iterative approach, for example simulated annealing, to ensure the reproduction of the frequency spectrum (i.e. the covariance model) and approximately match the data values (Yao 1998). Similarly, the FFT-MA simulation method (Le Ravalec et al. 2000) provides a very efficient technique for generating unconditional Gaussian realizations in large grids. The MA method (Eq. 3.50) generates a spatially correlated realization of a Gaussian random field by calculating the convolution of a standard Gaussian variable with a filter operator. The shape of the filter defines the spatial correlation structure of the generated random field. In the FFT-MA method, the convolution in Eq. (3.50) is performed in the frequency domain, using the FFT approach. In the MA approach, the spatial covariance function is expressed as a convolution $C = g * g$; therefore, the Fourier transform $\mathcal{F}(C)$ is the product of the Fourier transforms $\mathcal{F}(C) = \mathcal{F}(g)\,\mathcal{F}(g) = (\mathcal{F}(g))^2$, where we assume that g is a real even function. Equation (3.50) can then be rewritten as:

$$y(x) = m(x) + \mathcal{F}^{-1}(\mathcal{F}(g)\,\mathcal{F}(z)) = m + \mathcal{F}^{-1}\left(\mathcal{F}(C)^{\frac{1}{2}}\,\mathcal{F}(z)\right) \tag{3.51}$$

where $\mathcal{F}^{-1}$ is the inverse Fourier transform.

In a discretized multidimensional domain, according to the FFT-MA method, we first calculate the spatial covariance matrix **C**; we compute its discrete Fourier transform $\mathcal{F}(\mathbf{C})$ and the Fourier transform of the operator filter $\mathcal{F}(\mathbf{C})^{\frac{1}{2}}$; we generate a spatially uncorrelated realization $\mathbf{z}$ of a standard Gaussian random variable $\mathbf{z} \sim \mathcal{N}(\mathbf{z}; \mathbf{0}, \mathbf{I})$; we compute its discrete Fourier transform $\mathcal{F}(\mathbf{z})$; we calculate the product of $\mathcal{F}(\mathbf{C})^{\frac{1}{2}}$ and $\mathcal{F}(\mathbf{z})$; we compute the inverse discrete Fourier transform $\mathcal{F}^{-1}\left(\mathcal{F}(\mathbf{C})^{\frac{1}{2}}\,\mathcal{F}(\mathbf{z})\right)$; and we add the prior mean **m**. In practical applications, the simulations are performed on grids larger than the actual model and the simulated points outside of the original model are discarded to avoid the periodicity effects associated with the Fourier transform (Le Ravalec et al. 2000).

The PFS method is another very efficient technique for generating conditional Gaussian realizations (Srivastava 1992). First, we generate a spatially correlated realization of a Gaussian random field $\mathbf{z} \sim \mathcal{N}(\mathbf{z}; \mathbf{0}, \mathbf{I})$, for example using the FFT-MA method; then, at each location, we multiply the correlated realization by a local standard deviation $\boldsymbol{\sigma}_y$ and we add the local mean $\boldsymbol{\mu}_y$ to obtain the final realization:

$$\mathbf{y} = \boldsymbol{\mu}_y + \boldsymbol{\sigma}_y \mathbf{z}. \tag{3.52}$$

The local mean $\boldsymbol{\mu}_y$ and local standard deviation $\boldsymbol{\sigma}_y$ can be computed using a kriging approach. Data conditioning is automatically achieved by specifying standard deviation equal to 0 at the measurement locations. PFS is widely applied due to its conceptual simplicity and ability to handle local conditional distributions computed with different algorithms; however, PFS can produce artifacts such as local extrema in the vicinity of conditioning data points (Pyrcz and Deutsch 2014).

Two-point statistics methods rely on spatial covariance functions that can only account for the spatial correlation between two data points at a time. For this reason, two-point statistics methods are not able to reproduce complex patterns, such as geometrical features and connectivity of geobodies. MPS methods account for the spatial correlation of more than two data points at a time, by extracting the correlation information from a conceptual model, called training image (Mariethoz and Caers 2014). MPS methods are sequential simulation approaches, where the model locations are sequentially visited according to a random path, and the previously simulated values are used as conditioning data. In several MPS algorithms, the higher-order statistics extracted from the training image are usually stored in a database, which allows the computation of the conditional probabilities for the simulation. MPS methods are described in detail in Section 4.5, since they are more commonly used for discrete properties, such as facies and rock types. However, some of the available MPS methods can also be applied to continuous variables and to multivariate problems. For example, the direct sampling (DS) method (Mariethoz et al. 2010) is a sequential simulation algorithm that directly samples from a training image conditioned on the available measurements. The DS algorithm is applicable to discrete and continuous properties and it can be used for conditional simulations with non-linear relations between variables. An example of application is shown in Section 4.5, for the simulation of reservoir facies in a fluvio-deltaic reservoir with multiple channels. Similarly, the cross-correlation simulation (CCS) method (Tahmasebi et al. 2012) is a sequential simulation algorithm based on a cross-correlation function describing the similarity of a pattern of multiple data points and the patterns of the training image, and it can be used with both discrete and continuous training images.

In contrast to the sequential approach, iterative simulation methods first assign to the model locations initial values sampled from the prior distribution and then perturb the values (except for the hard data points) using a predefined perturbation function, until the spatial covariance model, and other possible statistical parameters such as the correlation with a secondary variable, are honored within a certain tolerance. Markov chain Monte Carlo methods (Section 5.6.1), genetic algorithms, simulated annealing, and other optimization-based approaches (Section 5.7) belong to the class of iterative simulation algorithms (Deutsch 1992; Peredo and Ortiz 2011; Pyrcz and Deutsch 2014). Typically, iterative methods are more computationally intensive than sequential simulation methods.

3.7 Application

Geostatistical methods are often applied in reservoir characterization studies to simulate reservoir properties, such as porosity and permeability, conditioned on direct measurements at the borehole locations and/or secondary measurements including geophysical data. In this section, we show two applications of the sequential simulation method conditioned on direct and indirect measurements.

In the first example, the petrophysical property of interest is porosity and four measurements are available. We assume that the prior distribution is a Gaussian mixture model with two components, corresponding to low- and high-porosity rocks. The prior parameters of the model are the means and variances of the two Gaussian distributions and the weights of the linear combination. We assume that the two Gaussian distributions have the same variance but different means and weights: the first component (associated with low-porosity rocks) has mean 0.05 and weight 0.4, whereas the second component (associated with high-porosity rocks) has mean 0.3 and weight 0.6. We assume that porosity follows an isotropic spherical covariance function, with the same parameters in both components. The locations of the four measurements are shown in Figure 3.19 and their values show that two samples are from low-porosity rocks and two samples are from high-porosity rocks. Using the SGMS (Grana et al. 2012b), we generate 100 realizations conditioned on the four hard data points. Three random realizations and the average of the ensemble of 100 realizations are shown in Figure 3.19.

In the second example, we apply the SGMS approach to a layer map, corresponding to the top horizon of the reservoir, extracted from a three-dimensional geophysical dataset of a hydrocarbon field in the Norwegian Sea (Figure 3.20), to generate realizations of porosity

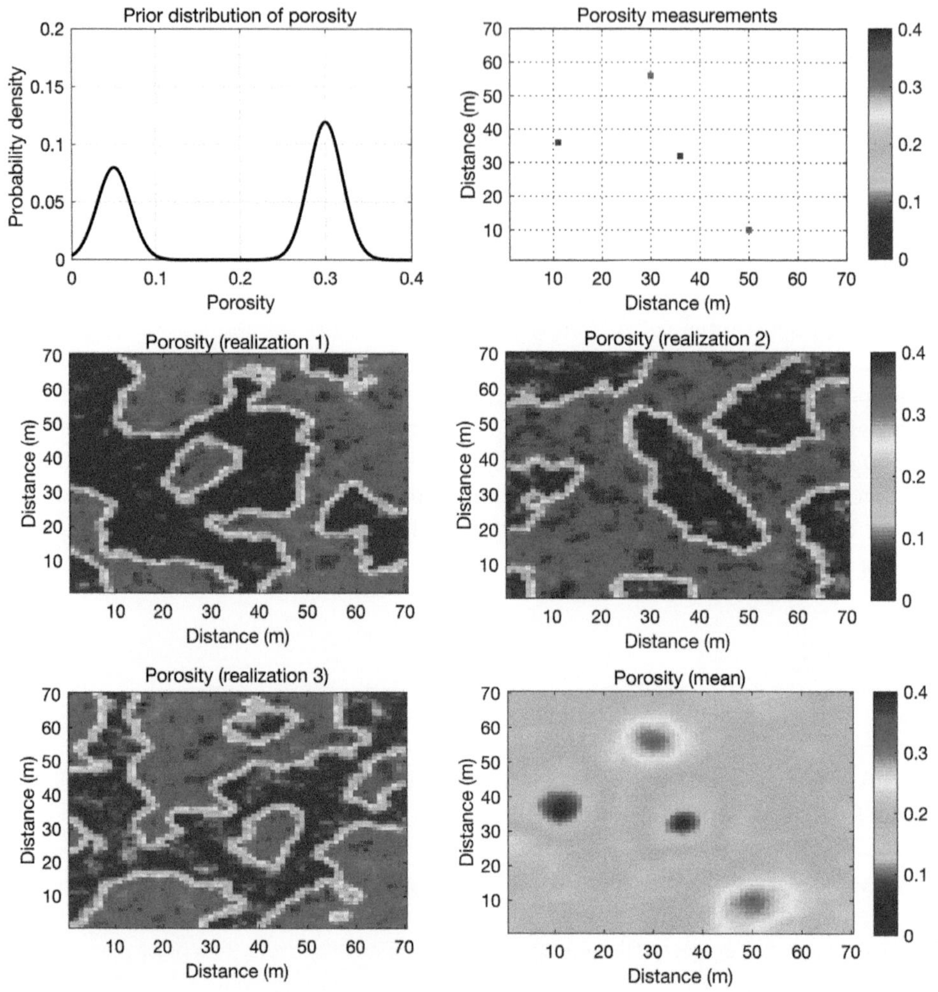

Figure 3.19 Application of geostatistical sequential simulations to reservoir porosity modeling: prior distribution of porosity, spatial configuration of porosity measurements, three random realizations obtained using sequential Gaussian mixture simulation, and the average map of 100 realizations.

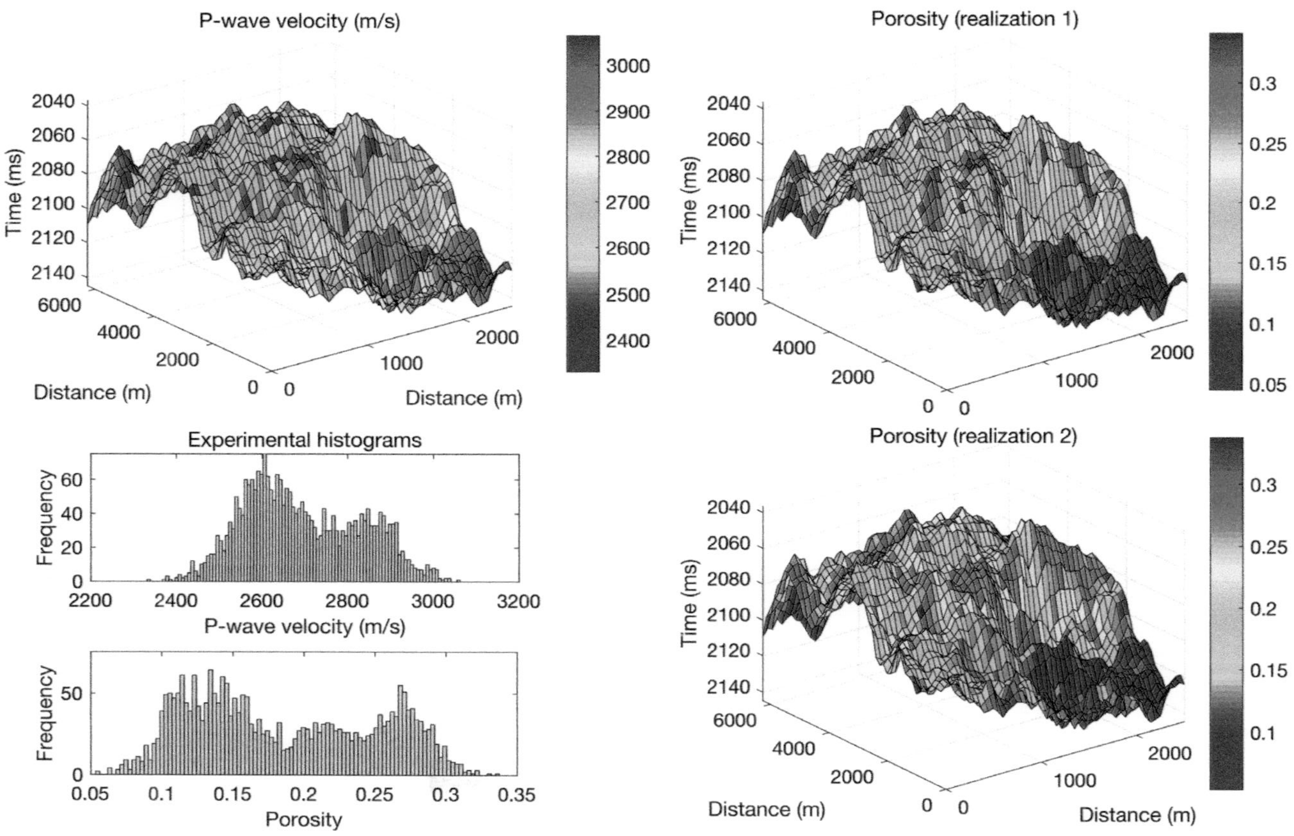

Figure 3.20 Application of geostatistical sequential simulations to reservoir porosity modeling conditioned on geophysical data: map of measured P-wave velocity, two random realizations obtained using different prior distributions, and experimental histograms of measured P-wave velocity and simulated porosity.

conditioned on P-wave velocity. The prior distribution of porosity is a Gaussian mixture model with three components corresponding to different rock types, with means equal to 0.12, 0.22, and 0.27 for the low-porosity, mid-porosity, and high-porosity rock types, respectively. We assume that porosity and P-wave velocity have linear correlation $\rho = -0.8$ and that they both follow an isotropic exponential covariance function with the same correlation length. Figure 3.20 shows two realizations of porosity obtained using SGMS with two different prior models: in the first realization, the prior weights of the three rock types are the same and equal to $0.\overline{3}$, whereas, in the second realization, the high-porosity rock type has a higher prior weight of 0.4 and the other two rock types have weights equal to 0.3. Both realizations preserve the multimodal histograms and the correlation with the measured data; however, the second realization shows a larger presence of high-porosity values according to the corresponding prior model.

4

Geostatistics for Discrete Properties

The spatial distribution of rock and fluid properties in the subsurface often depends on the geological structure, stratigraphy, and sedimentology resulting from earth processes. Based on geological information, we can generally classify sequences of subsurface rocks in different groups, called facies. In general, the term facies refers to a geobody with rocks with similar geological features. For example, sedimentary facies are geobodies that are distinct from adjacent rocks that resulted from different depositional environments (Boggs 2001). In many geophysical studies, the term facies is broadly defined and indicates rock classifications based on common lithological or petrophysical features, such as mineral volumes or porosity. In this book, we use the term facies to indicate groups of rocks with distinct rock and/or fluid properties. From a mathematical point of view, facies are represented using categorial variables, i.e. discrete variables without a preferential ordering. For example, in a sequence of sand and shale, we can label sand facies as 1 and shale facies as 0, or vice versa. Geostatistical methods such as kriging and sequential simulations (Chapter 3) can be extended to discrete properties (Goovaerts 1997; Deutsch and Journel 1998; Deutsch 2002; Dubrule 2003; Mariethoz and Caers 2014; Pyrcz and Deutsch 2014).

The MATLAB codes for the examples included in this chapter are provided in the SeReM package and described in the Appendix, Section A.2.

4.1 Indicator Kriging

In the discrete case, the probability distribution of a random variable π (e.g. facies) with outcomes $k \in \{1, \dots, F\}$ is represented by a probability mass function (Section 1.3.1) with probabilities $p(\pi = k) = p_k$, for $k = 1, \dots, F$. To extend geostatistical methods to the discrete case, we first introduce the concept of an indicator variable, i.e. a binary variable that indicates the occurrence of an event, such as facies occurrence.

We initially assume that only two facies are present, for example, sand and shale. Therefore, facies π can be represented by a binary variable (equal to 1 in sand and 0 in shale) and distributed according to a Bernoulli distribution (Section 1.4.1) with probability $p = P(\pi = 1)$. Hence, in the example of sand and shale, the probability of finding sand is p_{sand} and the

Seismic Reservoir Modeling: Theory, Examples, and Algorithms, First Edition. Dario Grana, Tapan Mukerji, and Philippe Doyen.

probability of finding shale is $1 - p_{sand}$. We then define a binary indicator variable I_u at a given location $\boldsymbol{u}$ as:

$$I_u = \begin{cases} 1 & \text{if sand occurs at location } \boldsymbol{u} \\ 0 & \text{otherwise.} \end{cases} \tag{4.1}$$

The mean μ_I of the indicator variable I_u is the probability p_{sand} of the sand facies that corresponds to the overall sand proportion (i.e. $\mu_I = p_{sand}$), whereas the variance σ_I^2 of the indicator variable I_u is $\sigma_I^2 = p_{sand}(1 - p_{sand})$, i.e. the product of the sand and shale proportions. We can generalize the indicator variable approach to a number of facies F greater than 2, by introducing an indicator variable for each facies category k as:

$$I_u(k) = \begin{cases} 1 & \text{if } k \text{ occurs at location } \boldsymbol{u} \\ 0 & \text{otherwise,} \end{cases} \tag{4.2}$$

for $k = 1, \dots, F$. The mean $\mu_{I(k)}$ of the indicator variable $I_u(k)$ is the probability p_k of the facies k that corresponds to the proportion of the facies k (i.e. $\mu_{I(k)} = p_k$), whereas the variance $\sigma_{I(k)}^2$ is $\sigma_{I(k)}^2 = p_k(1 - p_k)$.

As in the continuous case (Section 3.2), we can model the spatial continuity of the facies distribution using variograms or spatial covariance functions (Goovaerts 1997; Deutsch and Journel 1998; Pyrcz and Deutsch 2014). The indicator spatial covariance $C(h)$ measures the probability of observing the same facies at two locations at a distance h apart. The experimental indicator spatial covariance function can be computed as in the continuous case, using the indicator variables in Eq. (4.2). In the one-dimensional case, given a set of n measured facies $\{\pi_i\}_{i=1,\dots,n}$ at different locations in space $\{u_i\}_{i=1,\dots,n}$, the experimental indicator spatial covariance $\hat{C}(h, k)$ for each facies $k = 1, \dots, F$ can then be computed as:

$$\hat{C}(h, k) = \frac{1}{n_h} \sum_{j=1}^{n_h} \left(I_j(k) - p_k\right)\left(I_{j+h}(k) - p_k\right), \tag{4.3}$$

where n_h is the number of pairs of samples, $\left\{\left(I_j(k), I_{j+h}(k)\right)\right\}_{j=1,\dots,n_h}$ represent the pairs of indicator variables of the samples $\left\{\left(\pi_j, \pi_{j+h}\right)\right\}_{j=1,\dots,n_h}$ measured at locations $\left\{\left(u_j, u_{j+h}\right)\right\}_{j=1,\dots,n_h}$ separated by a distance h, and p_k is the prior probability of the facies k. The experimental indicator covariance function can be approximated using parametric models (Section 3.2). However, each indicator variable requires its own covariance function. Furthermore, if the number of facies F is greater than 2, a spatial cross-covariance function is required for each pair of distinct indicator variables.

Indicator kriging is an interpolation technique that uses indicator variables to predict the probability of occurrence of an outcome at a given location based on direct measurements at other locations. We assume that a limited number n of observations $\{\pi_j\}_{j=1,\dots,n}$ of the discrete variable π is available at locations $\{\boldsymbol{u}_j\}_{j=1,\dots,n}$ and, for each measurement, we define the indicator variables $I_j(k)$, as in Eq. (4.2), for $j = 1, \dots, n$ and for $k = 1, \dots, F$. We also assume that the spatial model consists of a set of two-dimensional isotropic indicator spatial covariance functions, one for each facies.

In indicator kriging, the indicator kriging probability $p_0^{ik}(k)$ at location $\boldsymbol{u}_0$ of the facies outcome k is obtained by adding a linear combination of the indicator variables $I_j(k)$ to the prior mean $\mu_{I(k)} = p_k$ as:

$$p_0^{ik}(k) = p_k + \sum_{j=1}^{n} w_j \left(I_j(k) - p_k \right), \tag{4.4}$$

where the coefficients w_j represent the kriging weights and the values $I_j(k)$ represent the indicator variables at the measurement locations $\boldsymbol{u}_j$ of the facies $\pi = k$, for $j = 1, \dots, n$ and for $k = 1, \dots, F$. As in kriging of continuous variables (Section 3.4), the weights w_j are computed by solving a kriging system (Eq. 3.23) constructed from the indicator spatial covariance function, to minimize the mean square prediction error. The indicator kriging probability in Eq. (4.4) is the expected value of the indicator variable (Eq. 4.2). In reservoir modeling, the indicator kriging probability is used to predict the conditional probability of facies based on available direct measurements.

In many practical applications, it is common to compute the facies estimate $\hat{\pi}_0$ at location $\boldsymbol{u}_0$ as the value $k \in \{1, \dots, F\}$ that maximizes the indicator kriging probability $p_0^{ik}(k)$:

$$\hat{\pi}_0 = \operatorname{argmax}_{k=1,\dots,F} \, p_0^{ik}(k). \tag{4.5}$$

Example 4.1 We compute the indicator kriging probability of facies and the associated prediction at a given location $\boldsymbol{u}_0$ using the data configuration in Figure 3.9. We assume that there are only two facies: sand (facies 1) and shale (facies 0). Four measurements are available: $\pi_1 = 0$, $\pi_2 = 1$, $\pi_3 = 1$, and $\pi_4 = 0$. We recall that in the case of two facies only, the facies values correspond to the indicator variable of facies 1 (sand in our example). We assume that the prior sand proportion is $p_{sand} = 0.6$, which corresponds to the mean of the indicator variable of sand. The variance of the indicator variable is then $\sigma^2_{sand} = 0.6 \times 0.4 = 0.24$. For simplicity, we assume the same spatial covariance model as in Example 3.1, i.e. an isotropic exponential covariance function with a correlation length of 9 m. As shown in Example 3.1 the solution of the kriging system (Eq. 3.23) is given by:

$$\begin{bmatrix} w_1 \\ w_2 \\ w_3 \\ w_4 \end{bmatrix} = \begin{bmatrix} 0.0450 \\ 0.1292 \\ 0.1292 \\ 0.0450 \end{bmatrix}.$$

The sand samples are closer to the location $\boldsymbol{u}_0$, hence they have higher weights than the shale samples. The indicator kriging probability $p_0^{ik}(k = 1)$ of sand at location $\boldsymbol{u}_0$ (Eq. 4.4) is then given by:

$$\begin{aligned} p_0^{ik}(k = 1) &= 0.6 + [0.045 \times (-0.6) + 0.1292 \times 0.4 + 0.1292 \times 0.4 + 0.0450 \times (-0.6)] \\ &= 0.6 + 0.0494 = 0.6494. \end{aligned}$$

According to this result, the probability of sand is 0.6494 and the probability of shale is 0.3506. The predicted facies is then $\hat{\pi}_0 = 1$, hence sand. If the prior sand proportion is $p_{sand} = 0.5$, the indicator kriging probability becomes $p_0^{ik}(k = 1) = 0.5842$ and the predicted facies is still sand, whereas, if the prior sand proportion is $p_{sand} = 0.2$, the indicator kriging probability becomes $p_0^{ik}(k = 1) = 0.3887$, and the predicted facies is shale.

Example 4.2 We show an example of application of indicator kriging using the topographic problem introduced in Example 3.2 (Figure 3.10) based on the dataset of the Yellowstone National Park. We assume that the elevation data are classified in two groups: valleys and peaks. All the locations with elevation greater than 2500 m are classified as peaks; all the other locations are valleys (Figure 4.1). We assume that the valley–peak classification is unknown except for a set of 100 randomly selected locations (full dataset). We also create a reduced dataset of 15 randomly selected locations (sparse dataset).

The proportions of valleys and peaks are 0.575 and 0.425, respectively. We assume a two-dimensional anisotropic indicator covariance function with a major correlation length of 15 km in the north–south direction and a minor correlation length of 10 km in the east–west direction. We apply indicator kriging using the sparse and full datasets of 15 and 100 measurements, respectively. Figure 4.2 shows the kriging classification and the corresponding maps of indicator kriging probability of peaks.

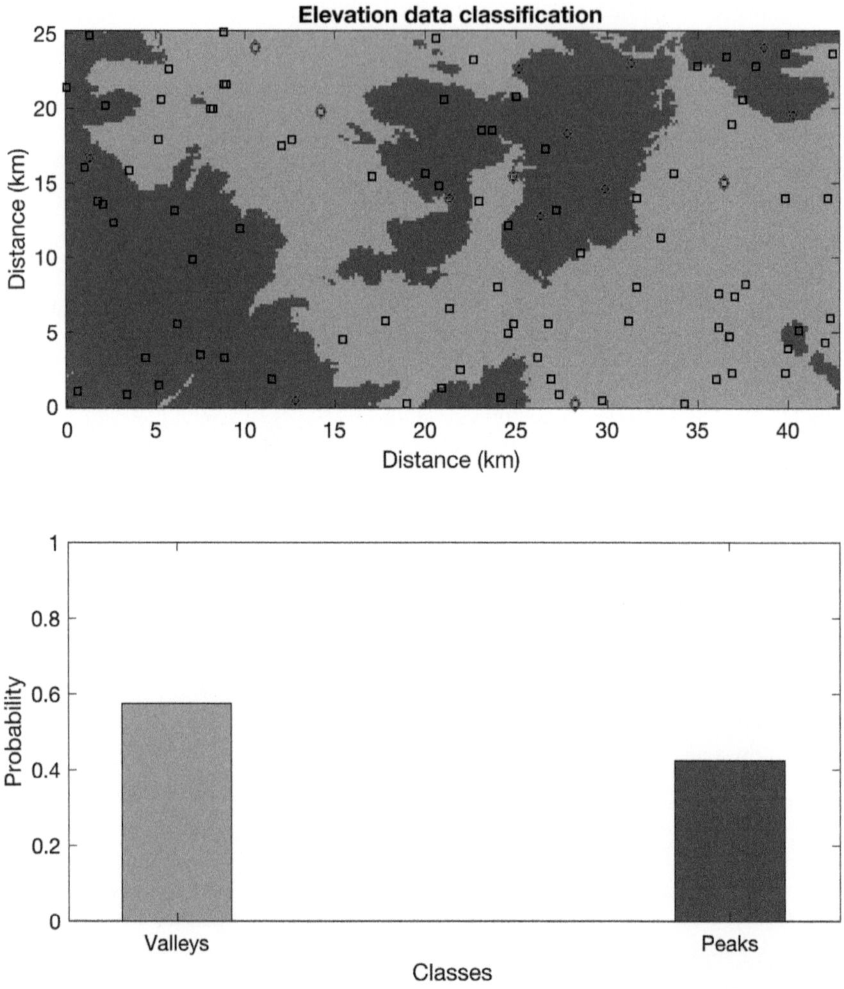

Figure 4.1 Peak–valley classification map of elevation data in the Yellowstone National Park: green indicates the valleys and brown indicates the peaks. The black squares represent the 100 locations of the available measurements (full dataset). The red diamonds represent 15 randomly selected locations (sparse dataset). The bar chart shows the proportions of valleys and peaks.

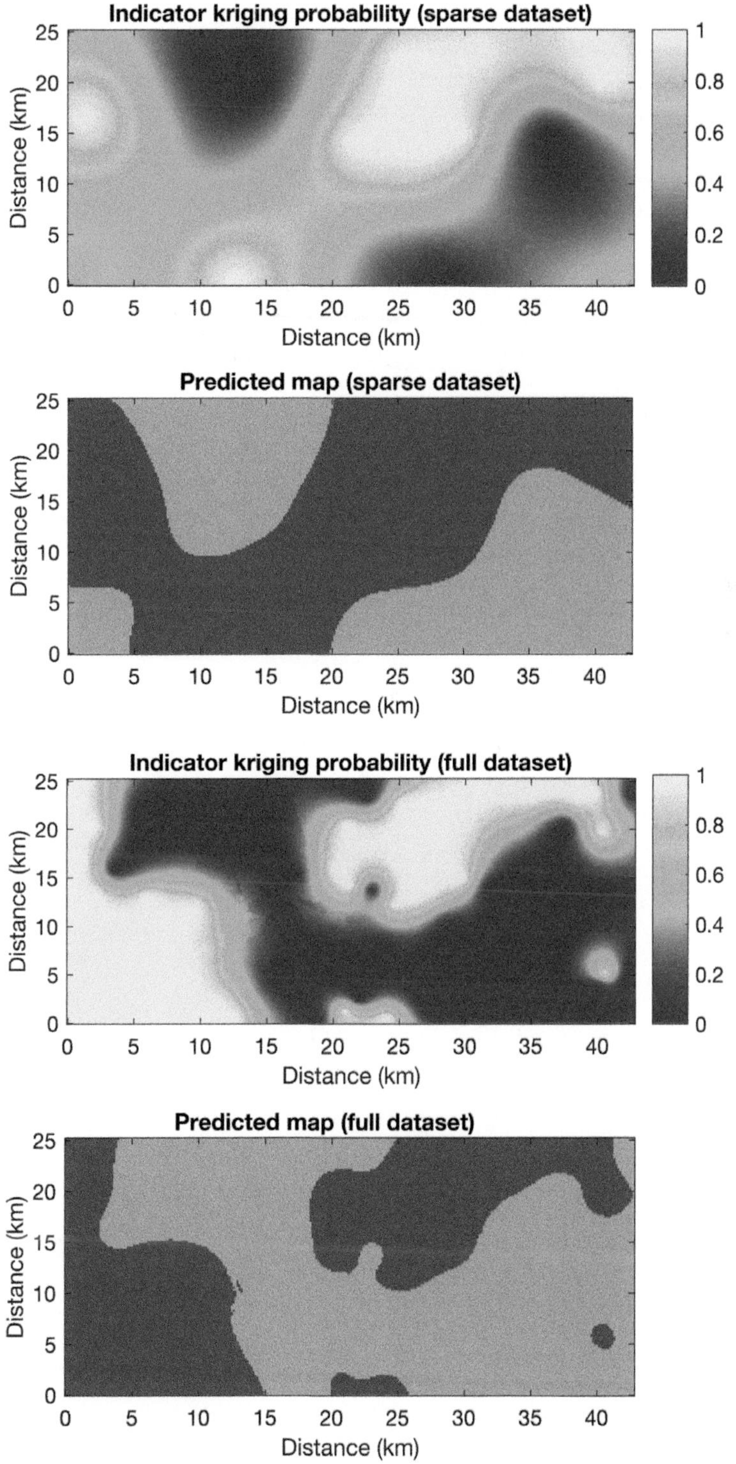

Figure 4.2 Maps of indicator kriging probability of peaks and corresponding predicted maps of valleys and peaks. The top two plots show the results for the sparse dataset of 15 measurements; the bottom two plots show the results for the full dataset of 100 measurements.

Indicator cokriging is a multivariate extension of indicator kriging to predict an unknown variable π (primary variable) at a given location $\boldsymbol{u}_0$ based on a set of measurements $\{\pi_i\}_{i=1,\dots,n}$ at locations $\{\boldsymbol{u}_i\}_{i=1,\dots,n}$ and a set of measurements $\{y_j\}_{j=1,\dots,m}$ at locations $\{\boldsymbol{u}_j\}_{j=1,\dots,m}$ of another variable Y (secondary variable) that is assumed to be correlated to π. In many applications, the secondary variable is continuous; in this case, indicator cokriging can be implemented by using multiple thresholds for the secondary variable Y and the corresponding indicator variables (Pyrcz and Deutsch 2014).

4.2 Sequential Indicator Simulation

Sequential indicator simulation (SIS; Journel and Gomez-Hernandez 1993) is a generalization of sequential Gaussian simulation (SGS) (Section 3.5.1) based on indicator variables and it is commonly applied to simulate realizations of discrete variables, such as facies or rock types. Instead of sampling from a Gaussian probability density function of a continuous variable (Eq. 3.46), we sample from a probability mass function of the facies. The facies probability at a given location, conditioned on the values of the previously simulated locations, is calculated by indicator kriging.

As in SGS, in SIS we visit the spatial locations sequentially, according to a random path. At each location, we calculate the facies probability conditioned on the previously simulated data using indicator kriging and we randomly draw a facies value $k \in \{1, \dots, F\}$ according to the indicator kriging probability. The simulated facies is then used as an additional conditioning value for the simulation at the following locations in the random path.

The SIS methodology is illustrated in Figure 4.3, where we assume that h locations $\{\boldsymbol{u}_j\}_{j=1,\dots,h}$ have already been simulated or that hard data are available at these locations. In Figure 4.3, h is 4 and the number of facies F is 3. Then, we simulate the next value according to the following steps:

1) We randomly pick a location $\boldsymbol{u}_{h+1}$ where the value of π is unknown (e.g. location $\boldsymbol{u}_5$ in Figure 4.3).
2) We compute the indicator kriging probability $p_{h+1}^{ik}(k)$ using Eq. (4.4) where the weights are computed by solving the kriging system in Eq. (3.23) (e.g. $p_5^{ik}(k)$ in Figure 4.3).
3) We sample a random value $\pi_{h+1} \in \{1, \dots, F\}$ from the probability mass function $p_{h+1}^{ik}(k)$ in Eq. (4.4), and we assign the value π_{h+1} to location $\boldsymbol{u}_{h+1}$ (e.g. $\pi_5 = 3$ in Figure 4.3).

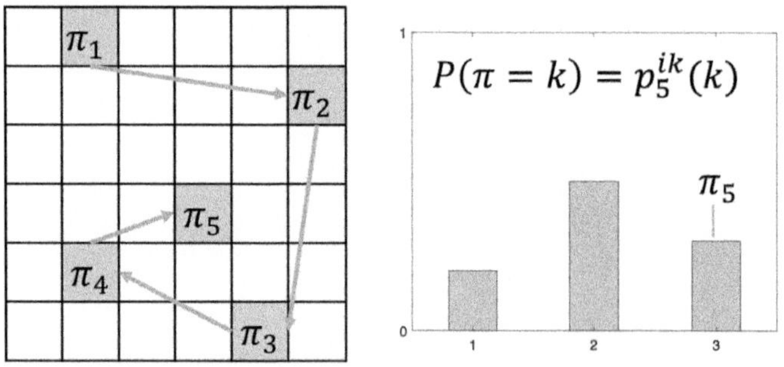

Figure 4.3 Schematic representation of sequential indicator simulation for a two-dimensional problem, with $F = 3$ facies.

4) We compute the indicator variables $I_{h+1}(k)$ for $k = 1, \dots, F$, as follows: $I_{h+1}(\pi_{h+1}) = 1$ and $I_{h+1}(k) = 0$ for $k = 1, \dots, F$ with $k \neq \pi_{h+1}$ (e.g. $I_5(1) = 0$; $I_5(2) = 0$; $I_5(3) = 1$, in Figure 4.3).
5) We add the indicator variables $I_{h+1}(k)$ for $k = 1, \dots, F$ to the set of conditioning data.
6) We repeat steps 1–5 for all the remaining locations.

Example 4.3 Figure 4.4 shows an example of application of SIS using the dataset of the Yellowstone National Park (Figure 4.1). In this example, we generate 100 realizations of the two classes, valleys and peaks, using the full datasets of 100 measurements. Figure 4.4 shows three random realizations and the average of the full set of 100 realizations. The average of the SIS realizations (Figure 4.4) is approximately equal to the indicator kriging probability in Figure 4.2.

Sequential indicator co-simulation (CoSIS) uses secondary data to generate conditional realizations. Because secondary data generally include continuous variables, the conditional simulation requires either the discretization of the continuous variable and the transformation into indicator variables, or a transformation of the continuous variable to conditional probabilities of facies that are then treated as soft indicators. CoSIS methods based on cokriging require the transformation of the secondary variable and the modeling of cross-covariance functions (Pyrcz and Deutsch 2014). Spatially dependent conditional probabilities of the primary (discrete) variable conditioned on the secondary (continuous) variable can also be used as conditioning data. In facies simulation, the conditional probabilities can be interpreted as spatially variable proportions predicted from geophysical measurements or estimated reservoir properties. For example, spatially dependent facies proportions can be computed using Bayesian classification (Section 6.1) and used in CoSIS to generate realizations of facies conditioned on the facies conditional probabilities (Doyen 2007).

Another possible approach to include secondary data is SIS with local Bayesian updating (Doyen 2007) that directly combines SIS with Bayesian facies classification. In SIS with Bayesian updating, we first compute the local indicator kriging probability $p_0^{ik}(k)$ for $k = 1, \dots, F$ using Eq. (4.4); we calculate the likelihood $f(y_0|\pi = k)$ of the secondary variable y_0, such as seismically derived petrophysical or elastic properties, for each facies $k = 1, \dots, F$; then we compute the conditional probability mass function of facies $P(\pi = k|y_0) \propto p_0^{ik}(k) f(y_0|\pi = k)$ using Bayes' theorem, for all $k = 1, \dots, F$; and we simulate a random value by drawing from the conditional probability $P(\pi|y_0)$. The advantage of SIS with Bayesian updating is that it does not require the discretization of the secondary variable and it can be generalized to multiple secondary variables using multivariate distributions that can be modeled using Gaussian or non-parametric PDFs (Doyen 2007).

Example 4.4 An example of application of CoSIS using cokriging is shown in Figure 4.5. The spatially dependent proportions are computed from the temperature data shown in Figure 3.10 using a Bayesian approach that predicts the conditional probability of valleys and peaks conditioned on the temperature measurements. The use of secondary data allows for a better delineation of the morphology of the peaks, despite the use of only 15 direct measurements as hard data. The average of the 100 simulations is approximately equal to the indicator cokriging probability map.

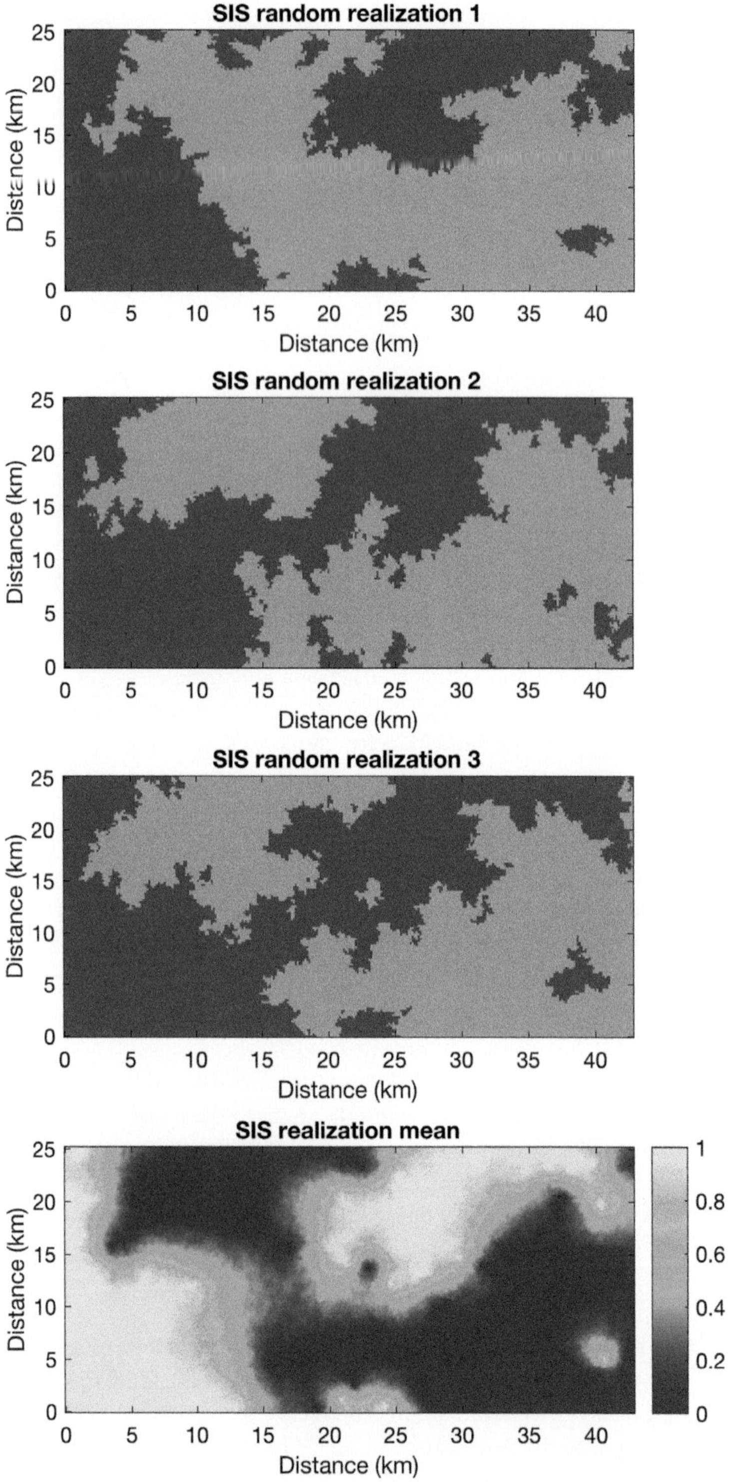

Figure 4.4 Three random sequential indicator simulation realizations and the average of 100 realizations using the full dataset of 100 measurements.

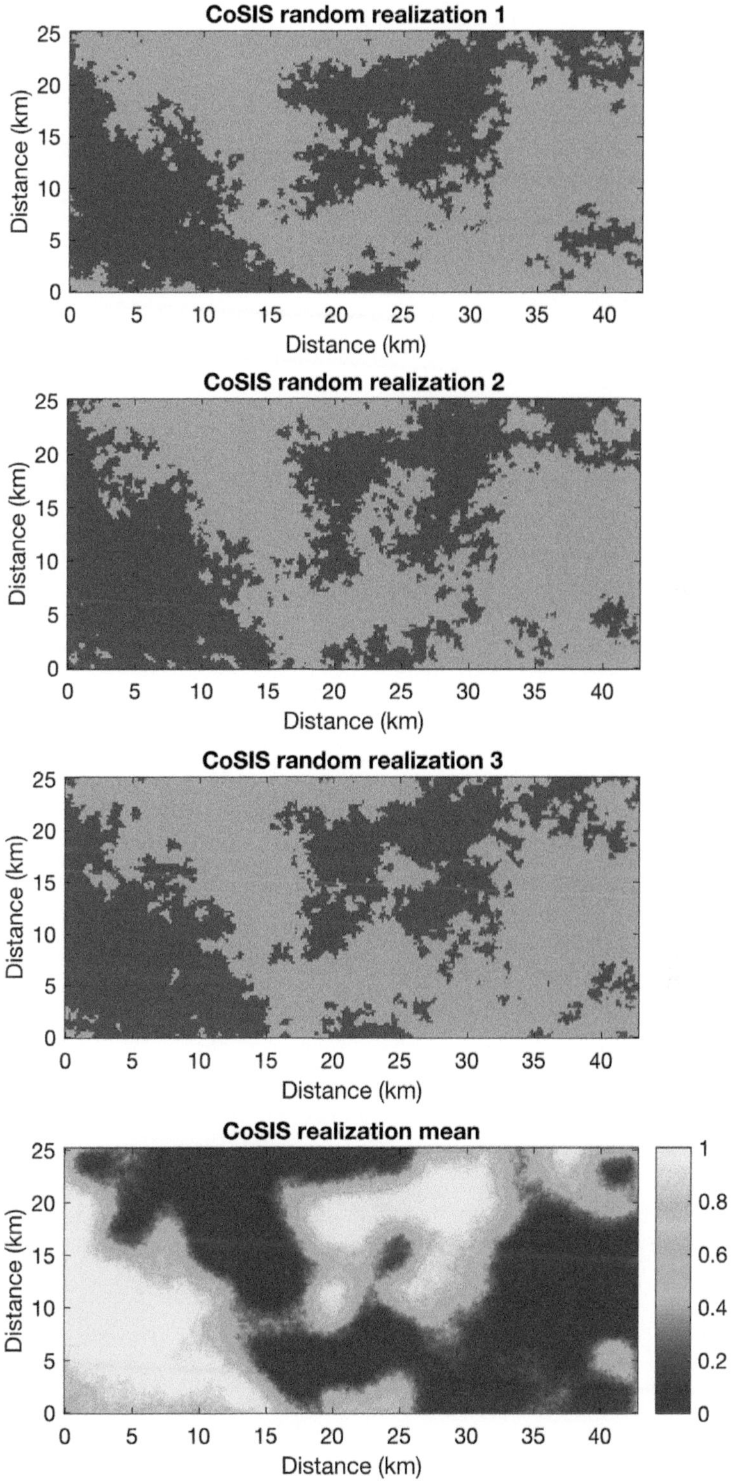

Figure 4.5 Three random sequential indicator co-simulation realizations and the average of 100 realizations using the sparse dataset of 15 measurements.

4.3 Truncated Gaussian Simulation

Several other methods have been developed in geostatistical literature to simulate realizations of discrete properties (Dubrule 2003; Doyen 2007; Lantuéjoul 2013; Pyrcz and Deutsch 2014). Truncated Gaussian simulation (TGS) generates multiple realizations of discrete properties such as facies or rock types (Xu and Journel 1993; Lantuéjoul 2013). In TGS, we first generate realizations of a spatially correlated continuous Gaussian random field, using, for example, SGS (Section 3.5.1) or fast Fourier transform – moving average (FFT-MA) (Section 3.6); then we map the values into the domain [0, 1] using the Gaussian cumulative distribution function (CDF); and we apply a set of thresholds to generate realizations of the discrete property. The spatial correlation is imposed by a spatial covariance function in the continuous domain, whereas the threshold values depend on the facies proportions and are computed from the facies CDF.

It should be noted that, after truncation, the discrete property does not have the same spatial covariance function as the original Gaussian random field. If $\gamma(h)$ denotes the variogram of the continuous Gaussian random field, and $\gamma_t(h)$ the variogram of the discrete field obtained by applying the threshold t, then the two variograms are related according to the following relation (Lantuéjoul 2013):

$$\gamma_t(h) = \frac{1}{\pi}\int_0^{\arcsin\sqrt{\gamma(h)/2}} \exp\left[-\frac{t^2}{2}\left(1 + \tan^2\theta\right)\right] d\theta. \tag{4.6}$$

For example, we assume only two facies, sand and shale, with proportions 0.7 and 0.3, respectively. We then generate a realization of a continuous property $\boldsymbol{x}$, for example using SGS, and map the values into the interval [0, 1]. Then, we classify the locations with values less than or equal to 0.7 (i.e. the sand proportion) as sand and the values greater than 0.7 as shale. The same approach can be applied to a number of facies greater than 2 by introducing multiple threshold values based on the facies CDF. In this case, we can partially control the spatial distribution of facies and their transitions through the order in the association between facies and the numerical value of the categorical variable. For example, in the case of 3 facies, namely sand, shaley sand, and shale, we can associate the value $k = 1$ to sand, $k = 2$ to shaley sand, and $k = 3$ to shale, so that shale and sand are not adjacent in the realization. By construction, individual realizations reproduce the prior proportions. In TGS, we can also integrate spatially variable proportions by locally varying the threshold values. The main advantage of TGS, compared with SIS, is that it does not require the calibration of an indicator spatial covariance function; however, it requires a spatial covariance function for the underlying Gaussian random field that is not observed in the actual data. One of the limitations of TGS is that the resulting spatial covariance model is the same for all the facies.

Example 4.5 Figure 4.6 shows an example of application of TGS to the Yellowstone dataset shown in Figure 4.1. In this example, we generate 100 realizations using truncated realizations of a continuous Gaussian random field, conditioned on 100 measurements, and we transform the continuous realizations into binary realizations according to the true facies proportions shown in the bar chart in Figure 4.1. Figure 4.6 shows three random realizations and the average of the full set of 100 realizations. The results are similar to those obtained by applying SIS (Figure 4.4).

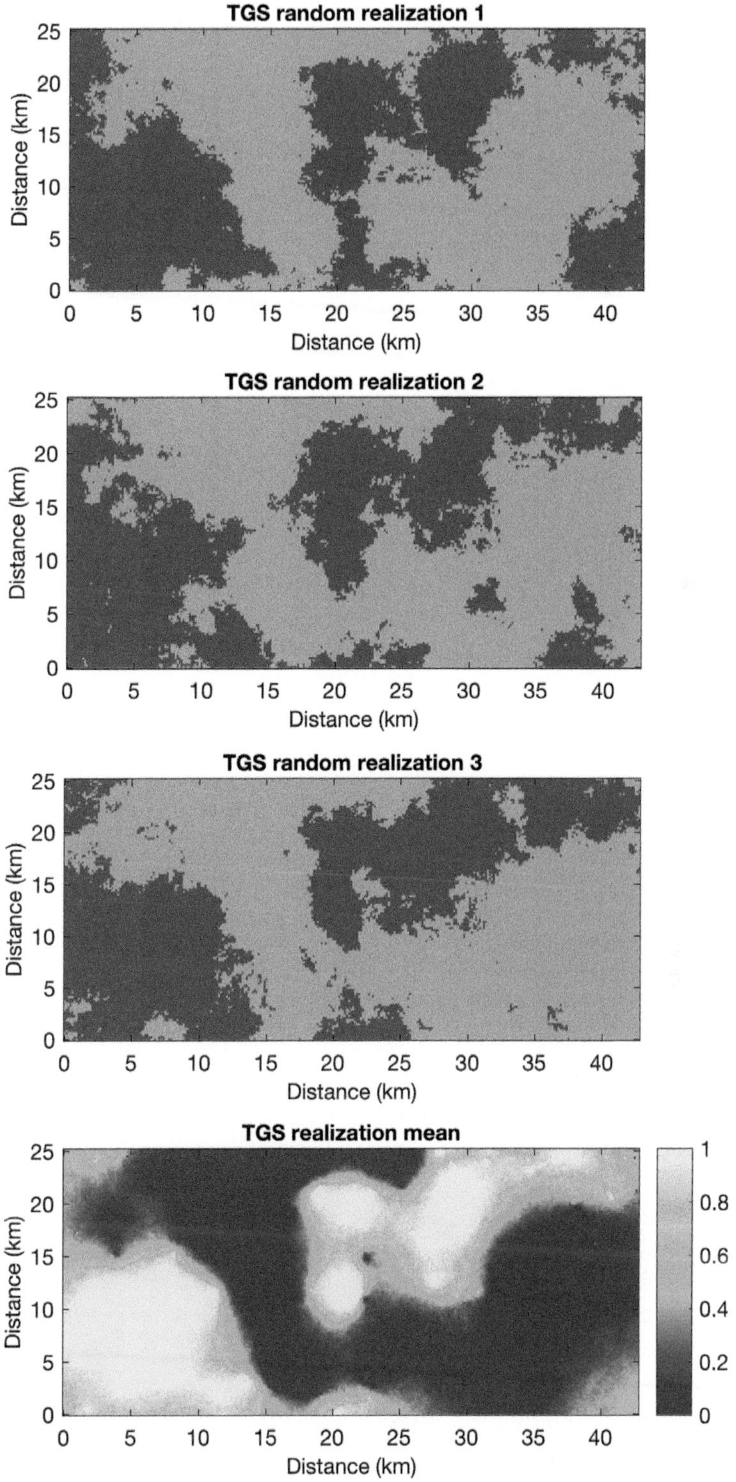

Figure 4.6 Three random truncated Gaussian simulation realizations and the average of 100 realizations using the full dataset of 100 measurements.

The pluri-Gaussian simulation (PGS) method (Armstrong et al. 2011) is an extension of TGS that allows generating realizations with different spatial covariance models in each facies and also provides more control of the spatial relationships between different facies. PGS is based on two or more realizations of continuous Gaussian random fields with different spatial covariance functions. Each realization is then mapped into the domain [0, 1] using the Gaussian CDF. The threshold values are defined in a multidimensional space with dimensions equal to the number of Gaussian random field realizations. For example, if we aim to model three facies and we assume that one facies has a different spatial covariance than the other two facies, then we can simulate two Gaussian random field realizations, $\boldsymbol{x}_1$ and $\boldsymbol{x}_2$, with different spatial covariance functions (e.g. a different azimuth), transform them into the realizations $\boldsymbol{y}_1$ and $\boldsymbol{y}_2$ whose values, at each location, belong to the domain [0, 1], and define the thresholds in the bivariate domain $[0, 1] \times [0, 1]$ based on a set of inequalities. For instance, we define the thresholds $t \in [0, 1]$ and $s \in [0, 1]$, and assign all the locations with $\boldsymbol{y}_1 > t$ and $\boldsymbol{y}_2 < s$ to facies 1. Similarly, we can define inequalities for the other two facies; for instance, we can assign all the locations with $\boldsymbol{y}_1 \leq t$ and $\boldsymbol{y}_2 < s$ to facies 2 and all the locations with $\boldsymbol{y}_2 \geq s$ to facies 3.

Conditional simulations where the realizations are conditioned on hard data are more computationally expensive with TGS and PGS than conditional simulations with SIS, because the observed data are the discrete facies values whereas the underlying continuous variables are not directly observed and they tipically do not have any physical or geological meaning. Furthermore, the spatial correlation of the observed discrete data is not the same as that of the continuous Gaussian variable. Armstrong et al. (2011) and Lantuéjoul (2013) describe algorithms for conditional TGS and PGS simulations involving the Gibbs sampling algorithm (Section 5.6.1).

4.4 Markov Chain Models

An alternative method for the simulation of discrete properties is based on stationary first-order Markov chains (Krumbein and Dacey 1969; Elfeki and Dekking 2001; Eidsvik et al. 2004b). In statistics, a Markov chain is a stochastic process in which the probability distribution of the next state of the process is conditioned on the previous states, rather than the entire sequence of states. The order of the Markov chain indicates the number of previous states in the conditional distribution. Categorical Markov chains represent a special case defined in a discrete-valued state space, where each state belongs to a finite number of classes.

For facies simulation, the state is the facies value at a given location; therefore, in a first-order Markov chain, at each location, the probability of a given facies depends only on the facies at the previous adjacent location in the direction of interest. The simulation is performed sequentially following a given direction. Markov random fields extend the idea of Markov chains to two and three dimensions. Here, we limit our description to one-dimensional Markov chains in the vertical direction. The probabilities of transitioning from a given facies to another one, i.e. the transition probabilities, are generally represented by the transition matrix **T**, where the rows represent the previous adjacent location and the columns represent the current location. For example, for a facies vertical profile with F possible facies, the element $T_{i,j}$ of the transition matrix represents the probability of the transition from facies i located above the interface to facies j located below the interface (downward transition). In this case the transition matrix is a $(F \times F)$ matrix of the form:

$$\mathbf{T} = \begin{bmatrix} T_{1,1} & T_{1,2} & \cdots & T_{1,F} \\ T_{2,1} & T_{2,2} & \cdots & T_{2,F} \\ \vdots & \vdots & \ddots & \vdots \\ T_{F,1} & T_{F,2} & \cdots & T_{F,F} \end{bmatrix}, \tag{4.7}$$

where $T_{i,j} = P(\pi_u = j | \pi_{u-1} = i)$ is the probability of facies π at location u being equal to j, conditioned on the facies π at the location $u - 1$ above being equal to i, for $i, j = 1, \dots, F$. The sum of the rows of the transition matrix must be 1, i.e. $\sum_{j=1}^{F} P(\pi_u = j | \pi_v = i) = 1$, for all $i = 1, \dots, F$. The elements of the transition matrix are often estimated from available data, such as facies and stratigraphic profiles and well logs. The probability distribution of the facies vector $\boldsymbol{\pi}$ of length M (i.e. the facies vertical profile with M samples) can then be written as:

$$P(\boldsymbol{\pi}) = P(\pi_1) \prod_{k=2}^{M} P(\pi_k | \pi_{k-1}), \tag{4.8}$$

where $P(\pi_1)$ is assumed to be equal to the stationary distribution of the facies, i.e. the facies proportions.

To simulate a stochastic realization of the facies profile according to a Markov chain, we sample the first value according to the prior proportions, then we sequentially sample the following values according to the transition probabilities conditioned on the previously simulated sample. In other words, for each of the following samples along the profile, we draw a value from the probability mass function obtained from the row of the transition matrix corresponding to the facies value simulated at the previous sample. In the Markov chain simulation of facies vertical profiles, the transition probabilities in the transition matrix control the facies proportions, the average thickness of the facies, and the frequencies of the transitions.

Example 4.6 In this example, we simulate multiple facies vertical profiles of sand and shale using a stationary first-order Markov chain, assuming three different transition matrices:

$$\mathbf{T}_1 = \begin{bmatrix} 0.5 & 0.5 \\ 0.5 & 0.5 \end{bmatrix}, \mathbf{T}_2 = \begin{bmatrix} 0.9 & 0.1 \\ 0.1 & 0.9 \end{bmatrix}, \mathbf{T}_3 = \begin{bmatrix} 0.1 & 0.9 \\ 0.1 & 0.9 \end{bmatrix}. \tag{4.9}$$

In the transition matrices in Eq. (4.9), shale corresponds to index 1 and sand to index 2, in each row and column. The transition matrix $\mathbf{T}_1$ shows equal sand and shale proportions with the same probability of transitioning from one facies to itself or to the other one

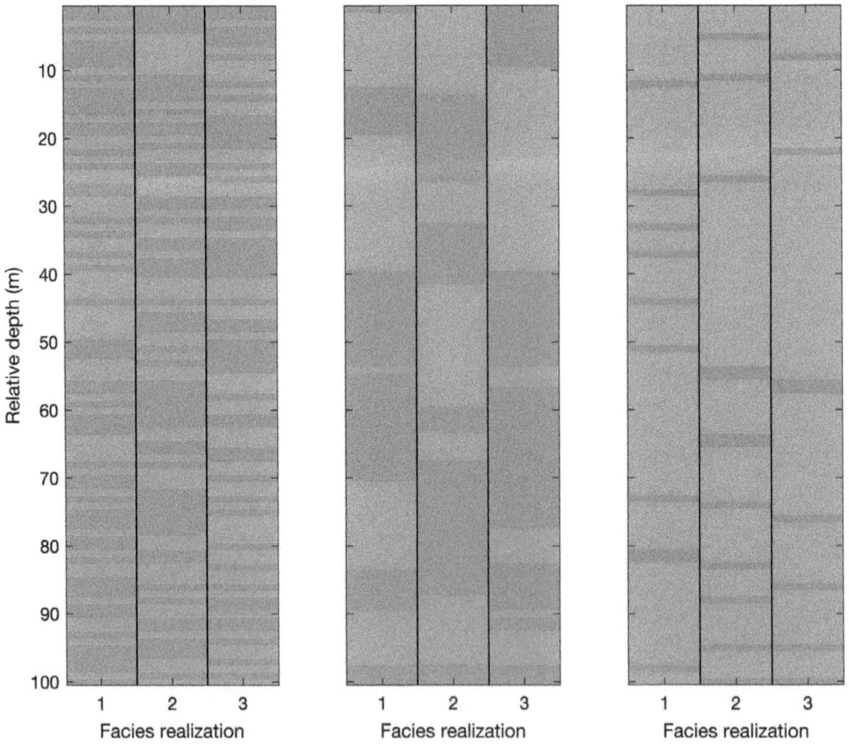

Figure 4.7 Facies simulations obtained using a stationary first-order Markov chain with the transition matrices given in Eq. (4.9). Green represents shale and yellow represents sand.

($P = 0.5$); the transition matrix $\mathbf{T}_2$ shows equal sand and shale proportions with low probability of transitioning from sand to shale and from shale to sand ($P = 0.1$) and high probability of staying in the same facies ($P = 0.9$); and the transition matrix $\mathbf{T}_3$ shows predominant sand proportions with high probability of transitioning to sand ($P = 0.9$) from both shale and sand, and low probability of transitioning to shale ($P = 0.1$) from both sand and shale. For each of the models, we show three random realizations in Figure 4.7. The first plot shows profiles of sand and shale with equal proportions and frequent transitions; the second plot shows profiles with equal proportions and predominantly thick layers; whereas the third plot shows profiles with a predominance of thick sand layers alternating with thin shale layers.

The use of conditioning data associated with a secondary variable is relatively challenging in this formulation, especially in subsurface modeling, where the data often show non-stationary trends that are not compatible with the stationary assumption of the Markov chain. In these cases, Markov chain models can be used to sample unconditional simulations to generate prior realizations in stochastic algorithms, such as Markov chain Monte Carlo methods (Sections 6.2 and 6.3), to sample the posterior probability of facies conditioned on the measured data (Eidsvik et al. 2004a). Recent developments for the application of Markov chain models to seismic reservoir characterization problems have been presented in Rimstad et al. (2012), Lindberg and Omre (2014), Grana et al. (2017a), and Fjeldstad and Grana (2018). The extension to the lateral directions is generally challenging and it is an ongoing research topic. Markov random fields (Tjelmeland and Besag 1998) have been proposed to generate realizations of discrete properties according to high-order spatial statistics models, based on Metropolis–Hastings algorithms, but their convergence is generally slow. Markov mesh models (Daly 2005; Stien and Kolbjørnsen 2011; Toftaker and Tjelmeland 2013) implement the concept of a moving search neighborhood to reduce the computational cost compared with Markov random fields, but the convergence is often unstable, especially in three-dimensional problems.

4.5 Multiple-Point Statistics

The information contained in two-point statistics methods, such as indicator variograms, spatial covariance functions, or first-order transition probabilities is limited and cannot capture complex geological shapes, such as channels or lobes. Multiple-point statistics (MPS) indicates a group of geostatistical algorithms that allow generating stochastic realizations of random variables with complex spatial correlations and structures, such as connectivity and stacking patterns of subsurface properties. In MPS, the spatial correlation of the model variables and the conditional probabilities for the simulation are inferred from a training image that represents a conceptual geological model, such as a channel system (Mariethoz and Caers 2014).

Many MPS algorithms are based on the concept of sequential simulations (Section 3.5). MPS algorithms sequentially simulate samples of the random variable according to a random or raster path of locations, and use the previously simulated samples as conditioning data. The new sample is generated according to the similarity of the previously simulated samples with the patterns observed in the training image. MPS algorithms include pixel-based methods, where random samples are sequentially generated at the model locations one at a time, and pattern-based methods, where groups of samples at adjacent locations (i.e. patterns) are sequentially simulated. Pixel-based algorithms can reproduce the direct measurements but might fail to reproduce high-order statistics of complex training images. Instead, pattern-based algorithms generally better reproduce the spatial correlation of the training image but might not be able to exactly reproduce the direct measurements.

Several algorithms have been proposed, including the extended normal equation simulation (ENESim; Guardiano and Srivastava 1993), single normal equation simulation (SNESim; Strebelle and Journel 2001), filter-based simulation (FilterSim; Zhang et al. 2006), simulation of patterns (SimPat; Arpat and Caers 2007), direct sampling (DS; Mariethoz et al. 2010), distance-based pattern simulation (DisPat; Honarkhah and Caers 2010), and cross-correlation simulation (CCS; Tahmasebi et al. 2012). The mathematical formulation and the implementation details of these algorithms are described in Mariethoz and Caers (2014).

Guardiano and Srivastava (1993) present a geostatistical approach, the ENESim, based on an extended concept of indicator kriging that allows the reproduction of high-order statistics based on a prior conceptual model, namely the training image, by sequentially sampling all the model locations according to a random path. At each model location, the algorithm searches for all the occurrences with the same data configuration in the training image, computes the conditional probabilities of all the possible outcomes, generates a sample according to the computed conditional probabilities, and assigns the value to the model location. The original algorithm is computationally expensive because it requires scanning the training image for data configuration occurrences, at each model location.

The SNESim algorithm (Strebelle and Journel 2001; Strebelle 2002) implements the same concept with the integration of a search tree that stores all the conditional probabilities, before sequentially simulating the random variable. The SNESim method is one of the most popular MPS algorithms and it is often applied for the simulation of discrete variables. The ENESim and SNESim algorithms for the simulation of geostatistical realizations of facies π, assuming $F = 2$ possible facies values (namely, facies 0 and facies 1), can be described in the following steps (Figure 4.8):

1) We randomly pick a location $\boldsymbol{u}_{h+1}$, where the value of π is unknown, according to a random path (e.g. location $\boldsymbol{u}_5$ in Figure 4.8).
2) We construct the data template using measurements and simulated values at previously visited locations within the moving search neighborhood: the data template represents the spatial configuration of the data, including the values of the data and their locations, and it is represented by the vector $\boldsymbol{d}$ (the length of the vector $\boldsymbol{d}$ changes at each location, based on the number of available data points in the moving search neighborhood).
3) We scan the training image (or the search tree) to find all the data configurations that replicate the configuration of the data template $\boldsymbol{d}$ (e.g. four occurrences in Figure 4.8).
4) We compute (or extract from the search tree) the conditional probabilities of the facies π at location $\boldsymbol{u}_{h+1}$ being equal to the value k, i.e. $P(\pi = k|\boldsymbol{d})$ for $k = 1, \dots, F$, by counting the number of occurrences where facies $\pi = k$ divided by the total number of occurrences (e.g. $P(\pi = 0|\boldsymbol{d}) = 1/4$ and $P(\pi = 1|\boldsymbol{d}) = 3/4$, in Figure 4.8).
5) We sample a random value $\pi_{h+1} \in \{1, \dots, F\}$ from the probability mass function $P(\pi|\boldsymbol{d})$.
6) We add the simulated value π_{h+1} to the set of conditioning data.
7) We repeat steps 1–6 for all the remaining locations on the random path.

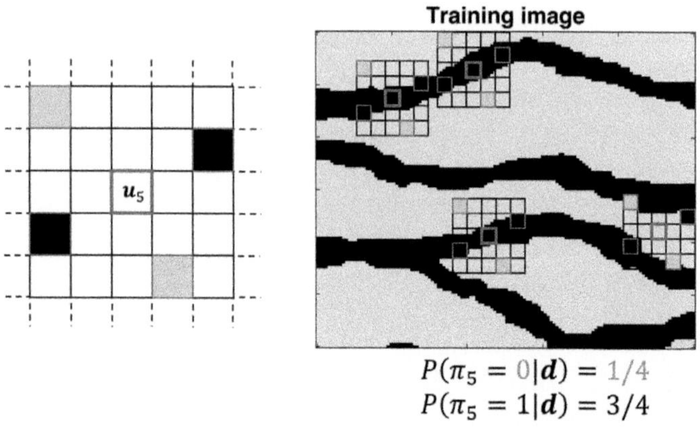

$$P(\pi_5 = 0|\boldsymbol{d}) = 1/4$$
$$P(\pi_5 = 1|\boldsymbol{d}) = 3/4$$

Figure 4.8 Schematic representation of ENESim and SNESim algorithms for a two-dimensional problem, with $F = 2$ facies (gray represents facies 0, black represents facies 1).

Practical implementations require the selection of a moving search neighborhood, the use of search trees to store pre-computed conditional probabilities, and the use of multi-grids to improve the computational efficiency for large datasets, as described in Strebelle and Journel (2001) and Strebelle (2002).

ENESim and SNESim are pixel-based algorithms where the model locations (i.e. the pixels) are simulated one at a time. For this reason, these methods might fail to reproduce the connectivity of the patterns observable in the training image. Arpat and Caers (2007) introduce the SimPat algorithm to reduce the computational time and improve the pattern reproduction, by replacing the conditional probabilities with a similarity measure of patterns. At a given location in the model, a pattern is randomly selected from the training image database according to the similarity distance between the training image patterns and the configuration in the data template centered at the model location; the selected pattern is then assigned to the model. FilterSim (Zhang et al. 2006) and DisPat (Honarkhah and Caers 2010) adopt the same approach, but reduce the computational cost by applying spatial filters and computing the similarity distance in a lower dimensional space. The CCS method (Tahmasebi et al. 2012) sequentially simulates patterns from the training image according to a cross-correlation function that quantifies the similarity between patterns. The model locations are sequentially visited according to a raster path and the new pattern is assigned by sampling among the patterns extracted from the training image, according to the cross-correlation of the new pattern and the previously simulated patterns at the adjacent locations, in the overlap regions (i.e. at the boundaries of the pattern). Tahmasebi et al. (2014) propose an extension of the CCS algorithm using multiple simulation grids to reduce the computational time for large model grids.

The DS method (Mariethoz et al. 2010) is a pixel-based algorithm that uses a similarity distance analogous to pattern-based algorithms. At each model location, the algorithm searches for patterns with the same data configuration in the training image; it selects the first pattern found, with similarity distance higher than a given threshold, and assigns the center of the selected pattern to the model location. According to this implementation, the training image is searched for each sampling at each location; however, only part of the training image is searched, until a matching pattern (i.e. a pattern with the same data configuration and similarity distance higher than the threshold) is found. If a suitable pattern is not found, then the pattern with the highest similarity is selected. This approach can be applied to discrete and continuous variables with the appropriate choice of the similarity metric as well as to co-simulations of discrete or continuous variables with non-linear relationships between the variables.

MPS is often used in reservoir modeling to generate realizations of discrete properties, such as facies or rock types, whereas for continuous properties, two-point statistics methods, such as SGS (Section 3.5.1) are generally preferred. MPS has also been successfully applied to fracture modeling (Guardiano and Srivastava 1993) and pore network reconstruction (Okabe and Blunt 2005).

Example 4.7 An example of application of MPS is shown in Figure 4.9. The training image represents a fluvio-deltaic reservoir with four channels in the east–west direction. Figure 4.9 shows three unconditional random realizations obtained using DS. Generally, MPS realizations better reproduce geologically realistic features than SIS (Section 4.2)

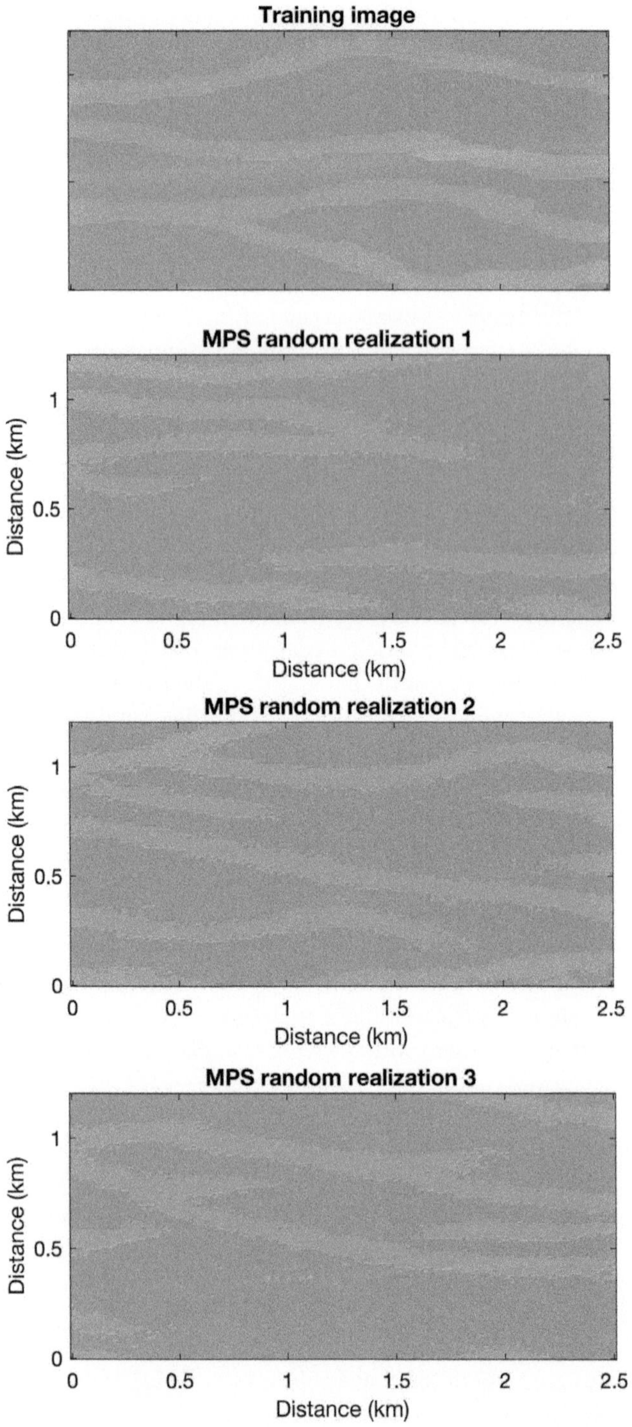

Figure 4.9 Training image and three multiple-point statistics random realizations obtained using the direct sampling method, in a fluvio-deltaic reservoir. Green represents the background shale and yellow represents the sand channels.

and TGS (Section 4.3). Recent developments in MPS algorithms further improve the continuity and connectivity of geobodies and extend the applications to more complex geological environments (Mariethoz and Caers 2014).

Similar to the continuous case, as an alternative approach, we can adopt iterative simulation algorithms, where an initial model generated from a prior distribution is perturbed until the values match the available data and the realization satisfies the desired correlation model. These methods generally use predefined perturbation functions and are based on stochastic optimization approaches such as Markov chain Monte Carlo methods (Section 5.6.1), genetic algorithms, simulated annealing, and other optimization-based approaches (Section 5.7). For example, Deutsch (1992) uses simulated annealing methods to reproduce MPS properties for integrating geological information and reservoir data.

Object-based simulation methods or Boolean methods are iterative optimization methods that place objects with predetermined shapes, associated with geological features, in a three-dimensional space with position constraints provided by direct measurements of the discrete property. The geometrical description might include the shape, size, direction, and sinuosity of the geobodies of interest. These methods have been mostly applied in fluvio-deltaic reservoirs and they produce geologically realistic images of geobodies with long-range connectivity and spatial relationships between geological features, such as channels and crevasse splays, that are difficult to achieve with pixel-based and pattern-based techniques. However, data conditioning is challenging owing to the increased complexity of the optimization problem. Unconditional object-based simulations are often used to generate training images for MPS algorithms. Process-based methods, mimicking the physical processes in the subsurface, can also be used to generate training images; however, these methods are computationally expensive and often require complex calibrations.

4.6 Application

In seismic reservoir characterization, a common modeling strategy is to sequentially simulate facies and petrophysical properties. First, facies realizations are generated using a geostatistical approach for discrete properties with direct measurements as conditioning data, then petrophysical property realizations are generated using a geostatistical approach for continuous properties using the facies realization as conditioning data (Pyrcz and Deutsch 2014). For example, we can generate facies realizations using SIS and then generate porosity realizations using SGS. For each simulation, realizations of petrophysical properties are often generated independently within each facies according to a facies-dependent prior distribution of the continuous property and a spatial correlation model, and then they are merged into the final realization based on the facies realization. For example, for each simulation, we generate as many porosity realizations as the number of facies, and at each location we assign the porosity value corresponding to the value of the porosity realization associated with the observed facies.

Alternatively, the sequential Gaussian mixture simulation (Section 3.5.2) approach can be applied for the simultaneous simulation of discrete and continuous properties; however, this approach might generate artifacts associated with isolated values of the discrete

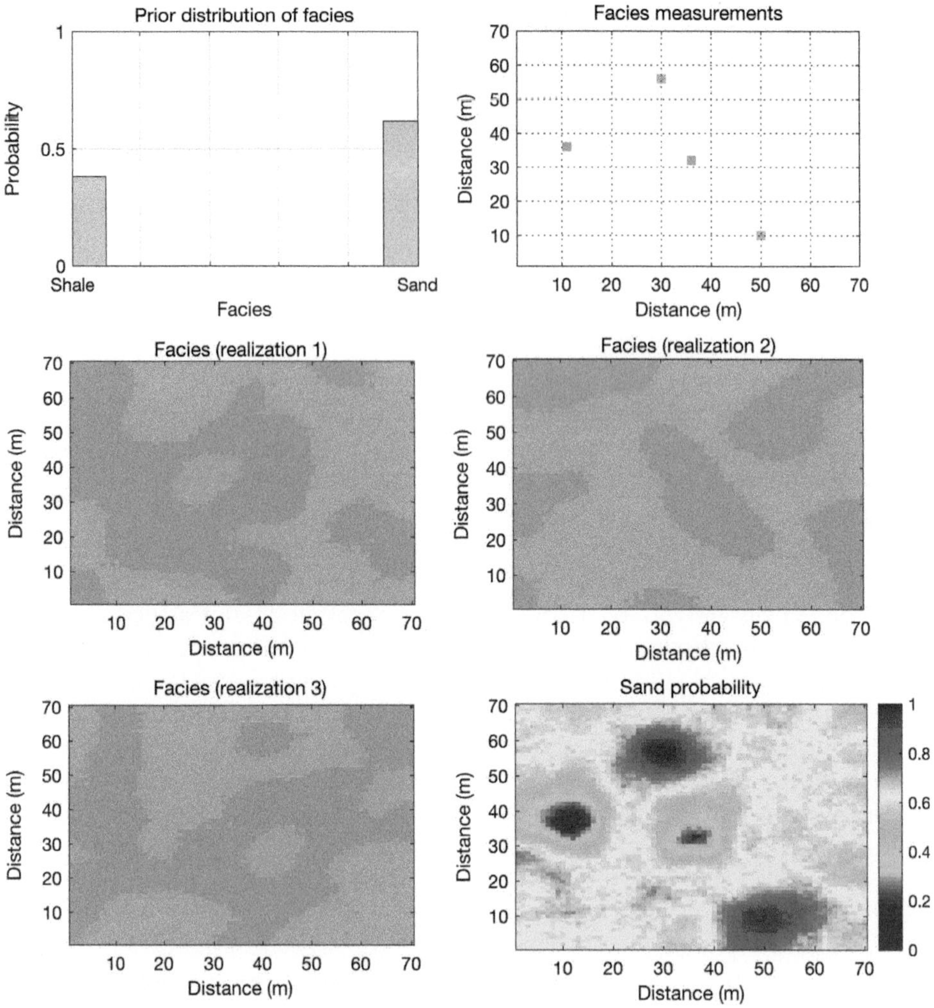

Figure 4.10 Application of sequential Gaussian mixture simulation to reservoir facies modeling: prior distribution of facies, spatial configuration of facies samples, three random realizations (shale in green and sand in yellow), and average map of 100 realizations corresponding to probability map of high-porosity sand.

property, especially for sparse datasets, and it might require an additional post-processing step, such as post-conditioning by kriging, to eliminate the singularities (Grana et al. 2012b).

We illustrate an application of geostatistical simulation of facies based on the sequential Gaussian mixture simulation algorithm using the example in Figure 3.19. We assume two facies, namely low-porosity shale (green facies) and high-porosity sand (yellow facies), with prior probability of 0.4 and 0.6, respectively (Figure 4.10). The locations of the four available measurements are illustrated in Figure 4.10: two samples belong to high-porosity sand and two samples belong to low-porosity shale. Figure 4.10 shows three random realizations of facies and the average map of a set of 100 facies realizations. We recall that for a binary property, the mean of the realizations coincides with the probability of one of the two facies, i.e. high-porosity sand in this application.

5

Seismic and Petrophysical Inversion

The term inversion is commonly used in reservoir geophysics and might refer to a large set of problems. In general, a geophysical inverse problem is the estimation of a set of properties given a set of geophysical measurements, assuming that the physical relation linking the model properties to the measurements is known. The most common inverse problem in reservoir geophysics is seismic inversion for the prediction of seismic attributes, such as velocities or impedances, from seismic data, including amplitudes and travel times (Tarantola 2005; Sen and Stoffa 2013). Similarly, the prediction of rock and fluid properties, such as porosity, mineral fractions, and fluid saturations, can be formulated as an inverse problem and it is generally referred to as petrophysical inversion or rock physics inversion (Doyen 2007; Bosch et al. 2010). Apart from seismic data, inversion methods also apply to various types of other geophysical measurements, including gravity, magnetic, and electromagnetic data. Our focus in this chapter is on seismic measurements.

The physics that links rock and fluid properties to seismic measurements is generally known and includes seismic wave propagation models (Russell 1988; Sheriff and Geldart 1995; Yilmaz 2001; Aki and Richards 2002) and rock physics models (Avseth et al. 2010; Dvorkin et al. 2014; Mavko et al. 2020). The solution of the inverse problem is generally not unique, owing to the uncertainty in the measurements and the approximations of the physical models. Several methods have been proposed for solving geophysical inverse problems, and include deterministic and probabilistic methods (Tarantola 2005; Sen and Stoffa 2013; Aster et al. 2018; Menke 2018). Most of these approaches have first been applied to the seismic inversion problem and then extended to the petrophysical inversion. A review of the available methods combining rock physics and seismic inversion has been proposed by Doyen (2007) and Bosch et al. (2010). Avseth et al. (2010) and Simm and Bacon (2014) provide insights on the geological interpretation and practical application of seismic reservoir modeling techniques.

A suitable inversion method should aim at estimating the most likely model as well as quantifying the uncertainty in the model prediction. Probabilistic approaches to inverse problems provide a natural framework for seismic and petrophysical inversion. In a probabilistic approach, the solution of the inverse problem can be expressed as a probability density function (PDF) of the model variables or a set of model realizations sampled from the PDF that captures the uncertainty in the model. The use of probabilistic inversion and stochastic optimization methods, especially for seismic inversion, was introduced in

Seismic Reservoir Modeling: Theory, Examples, and Algorithms, First Edition. Dario Grana, Tapan Mukerji, and Philippe Doyen.

geophysics approximately three decades ago (Duijndam 1988a, b; Sen and Stoffa 1991; Mallick 1995; Mosegaard and Tarantola 1995). Some of the precursor papers on the use of probabilistic methods for reservoir characterization problems were developed in conjunction with the advances of geostatistical algorithms (Doyen 1988; Bortoli et al. 1993; Haas and Dubrule 1994; Doyen and den Boer 1996; Mosegaard 1998) with the goal of generating multiple realizations of reservoir properties that match the seismic dataset, using conditional simulations and stochastic optimization techniques. The parallel progress in the fields of rock physics and seismic modeling led to inversion methodologies to estimate petrophysical properties from seismic data or attributes (Bosch 1999; Torres-Verdín et al. 1999; Mukerji et al. 2001; Mazzotti and Zamboni 2003; Bornard et al. 2005; Coléou et al. 2005). Such methods generally require a rock physics model to establish a relationship between petrophysical and elastic variables. For example, Mukerji et al. (2001) and Eidsvik et al. (2004a) introduce statistical rock physics models to estimate reservoir parameters from prestack seismic data and to evaluate the associated uncertainty. Rock physics models are also used in probabilistic approaches for the joint estimation of petrophysical properties conditioned on well logs and seismic data in Bachrach (2006), Gunning and Glinsky (2007), Connolly and Kemper (2007), Spikes et al. (2007), González et al. (2008), and Bosch et al. (2009). These works show that the probabilistic framework is the most suitable approach for quantifying uncertainty in reservoir models.

Bayesian inversion methods have been introduced in geophysics to assess the posterior distribution of the model variables given a set of measured geophysical data (Tarantola and Valette 1982; Scales and Tenorio 2001; Ulrych et al. 2001; Buland and Omre 2003; Gunning and Glinsky 2004; Tarantola 2005; Hansen et al. 2006). Buland and Omre (2003) propose an analytical solution of the seismic inversion problem based on the linearization of the seismic forward model and a Gaussian assumption for the prior distribution of the model variables and of the measurement errors. The Bayesian linearized approach has then been extended to seismic litho-fluid prediction and petrophysical properties estimation, which include more complex forward models, such as rock physics and petrophysics models, and more sophisticated statistical assumptions, such as Markov models and Gaussian mixtures, as proposed in Larsen et al. (2006), Buland et al. (2008), Ulvmoen and Omre (2010), Grana and Della Rossa (2010), Rimstad and Omre (2010), Rimstad et al. (2012), Kemper and Gunning (2014), Grana et al. (2017a), Fjeldstad and Grana (2018), and Grana (2018). Monte Carlo and Markov chain Monte Carlo (McMC) methods have also been applied in Sambridge and Mosegaard (2002), Eidsvik et al. (2004a), Connolly and Hughes (2016), Liu and Grana (2018), and de Figueiredo et al. (2019a, b). In this chapter, we first focus on Bayesian inversion methods based on analytical solutions, and then discuss numerical solutions of the Bayesian inversion problem, for both seismic and petrophysical applications.

The MATLAB codes for the examples included in this chapter are provided in the SeReM package and described in the Appendix, Section A.3.

5.1 Seismic Modeling

We first review the main principles of seismic modeling. Several approaches can be adopted to compute the seismic response of a sequence of geological layers with different elastic properties (velocities and densities) for multiple acquisition angles, including convolutional models based on Zoeppritz equations or their approximations, Kennett's invariant imbedding method, the Born weak scattering approximation, and the full waveform seismic model (Zoeppritz 1919; Kennett 1984; Russell 1988; Sheriff and Geldart 1995; Yilmaz 2001; Aki and Richards 2002). In this section, we focus on the convolutional model based on the linearized approximation of Zoeppritz equations (Shuey 1985; Aki and Richards 2002) for its mathematical tractability.

A seismogram can be approximated as a convolution of a source wavelet and a series of reflection coefficients, i.e. the reflectivity magnitudes of the elastic contrasts at the layer interfaces (Aki and Richards 2002). Seismic reflection coefficients depend on the elastic properties of the layers above and below the subsurface interfaces. An isotropic, elastic medium is completely described by three material parameters (Aki and Richards 2002), for example P-wave and S-wave velocity, and density $[V_P(t), V_S(t), \rho(t)]$, where t is the seismic two-way travel time.

The reflection seismic response $d(t, \theta)$ at the two-way travel time t for a given reflection angle θ of a sequence of geological layers can be written as:

$$d(t,\theta) = w(t,\theta) * r_{PP}(t,\theta) = \int w(u,\theta) r_{PP}(t-u,\theta) du, \tag{5.1}$$

where $*$ is the convolution operator, $w(t, \theta)$ is a wavelet, and $r_{PP}(t, \theta)$ is the PP-reflection coefficient series or reflectivity coefficients. The reflection coefficients $r_{PP}(t, \theta)$ can be exactly calculated using the Zoeppritz equations as a function of time t and the reflection angle θ or can be accurately approximated using linear approximations for small reflection angles and small changes of the elastic properties (i.e. weak elastic contrasts) across the interface (Aki and Richards 2002).

We adopt a linear approximation based on the amplitude variation-versus-offset (AVO) attributes, namely the intercept R, the gradient G, and the curvature F (Aki and Richards 2002):

$$\begin{cases} R = \dfrac{1}{2}\left(\dfrac{\Delta V_P}{\overline{V_P}} + \dfrac{\Delta \rho}{\overline{\rho}}\right) \\ G = \left[\dfrac{1}{2}\dfrac{\Delta V_P}{\overline{V_P}} - 2\dfrac{\overline{V_S}^2}{\overline{V_P}^2}\left(\dfrac{\Delta \rho}{\overline{\rho}} + 2\dfrac{\Delta V_S}{\overline{V_S}}\right)\right] \\ F = \dfrac{1}{2}\dfrac{\Delta V_P}{\overline{V_P}} \end{cases}, \tag{5.2}$$

where:

$$\begin{cases} \Delta V_P = V_P(t_{i+1}) - V_P(t_i); \quad \overline{V_P} = \dfrac{V_P(t_{i+1}) + V_P(t_i)}{2} \\ \Delta V_S = V_S(t_{i+1}) - V_S(t_i); \quad \overline{V_S} = \dfrac{V_S(t_{i+1}) + V_S(t_i)}{2} \\ \Delta \rho = \rho(t_{i+1}) - \rho(t_i); \quad \overline{\rho} = \dfrac{\rho(t_{i+1}) + \rho(t_i)}{2} \end{cases}, \tag{5.3}$$

with t_i and t_{i+1} indicating two consecutive times across the interface, respectively. The approximation of the reflection coefficients $r_{PP}(\theta)$ is then given by:

$$r_{PP}(\theta) = R + G\sin^2\theta + F\left(\tan^2\theta - \sin^2\theta\right)$$
$$= \frac{1}{2}\left(1+\tan^2\theta\right)\frac{\Delta V_P}{\overline{V_P}} - 4\frac{\overline{V_S}^2}{\overline{V_P}^2}\sin^2\theta\frac{\Delta V_S}{\overline{V_S}} + \frac{1}{2}\left(1-4\frac{\overline{V_S}^2}{\overline{V_P}^2}\sin^2\theta\right)\frac{\Delta\rho}{\overline{\rho}}, \tag{5.4}$$

where θ is the reflection angle, and where we have omitted the travel time variable t to simplify the notation.

The single-interface reflection coefficient in Eq. (5.4) can be extended to a time-continuous reflectivity function (Stolt and Weglein 1985) as:

$$c_{PP}(t,\theta) = c_P(\theta)\frac{\partial}{\partial t}\ln V_P(t) + c_S(\theta)\frac{\partial}{\partial t}\ln V_S(t) + c_\rho(\theta)\frac{\partial}{\partial t}\ln \rho(t), \tag{5.5}$$

where $c_P(\theta)$, $c_S(\theta)$, and $c_\rho(\theta)$ are coefficients that depend on the reflection angle θ and the average velocities:

$$\begin{cases} c_P(\theta) = \frac{1}{2}\left(1+\tan^2\theta\right) \\ c_S(\theta) = -4\frac{\overline{V_S}^2}{\overline{V_P}^2}\sin^2\theta \\ c_\rho(\theta) = \frac{1}{2}\left(1-4\frac{\overline{V_S}^2}{\overline{V_P}^2}\sin^2\theta\right). \end{cases} \tag{5.6}$$

Therefore, if the elastic properties of the layers are known, namely P-wave and S-wave velocity and density, then we can compute the seismic response as:

$$d(t,\theta) = w(t,\theta) * c_{PP}(t,\theta), \tag{5.7}$$

where the function $c_{PP}(t, \theta)$ is computed according to Eqs. (5.5) and (5.6). This formulation can be combined with any rock physics model (Chapter 2) to predict the seismic response of a sequence of layers based on their petrophysical properties, such as porosity, mineral volumes, and fluid saturations. It should be noted that the single-interface formulation ignores scattering of wave energy between multiple interfaces, and only accounts for the primary reflections.

Example 5.1 We illustrate the application of the convolutional forward model for the computation of synthetic seismic data using a set of well logs from the Norwegian Sea. The well log data are shown in Figure 5.1 and include P-wave and S-wave velocity and density. We assume three angle stacks, namely near, mid, and far angles, corresponding to 12°, 24°, and 36°, respectively. We also assume a Ricker wavelet with a dominant frequency of 45 Hz for all the angle stacks. However, in practical applications, the dominant frequency of the wavelet depends on the reflection angle and far angles generally have lower dominant frequencies. We first compute the corresponding travel times based on the depth and velocity measurements. We then compute the reflection coefficients using the discretized version of the linearized approximation in Eqs. (5.5) and (5.6), assuming an average $\overline{V_S}/\overline{V_P}$ of 0.54, and we calculate the seismograms using the angle-dependent convolutional model in Eq. (5.7). The resulting seismic dataset is shown in Figure 5.2, where the vertical

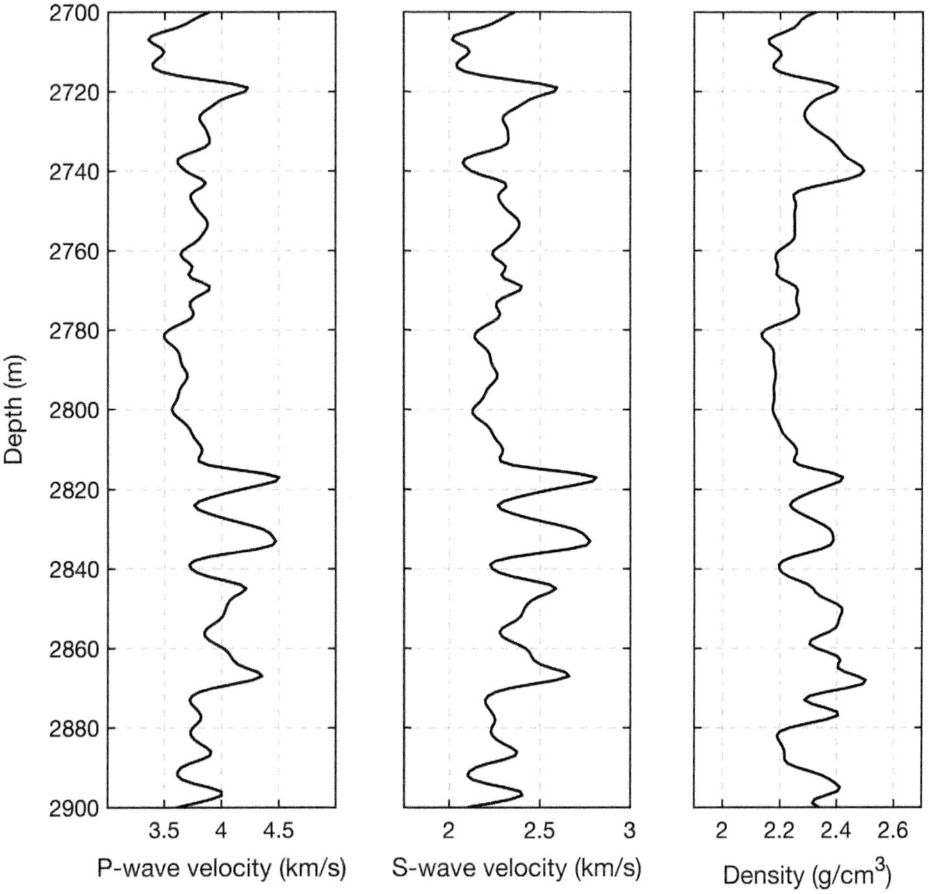

Figure 5.1 Well log data from a borehole in the Norwegian Sea: P-wave velocity, S-wave velocity, and density.

axis represents the two-way time. This example shows how the reflection amplitudes vary with the angle; for example, at approximately $t = 1.81$ s, the amplitude decreases as the reflection angle increases.

5.2 Bayesian Inversion

The main goal of seismic reservoir characterization is to predict a model of rock and fluid properties (elastic or petrophysical variables) from borehole measurements and seismic data. Seismic attributes depend on rock and fluid properties; for example, if porosity increases, velocity generally decreases, and if gas saturation increases relative to water saturation, then velocity generally decreases (Chapter 2). If $\boldsymbol{d}$ represents the measured

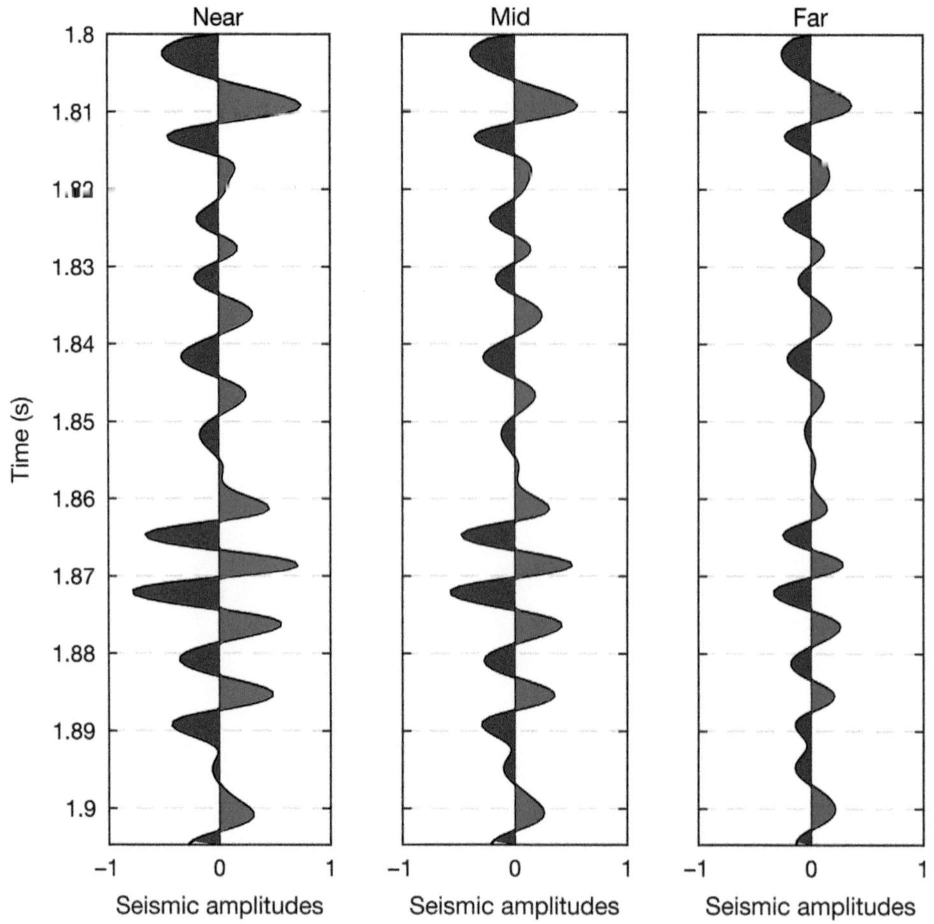

Figure 5.2 Synthetic seismic data for three partial angle stacks: near (12°), mid (24°), and far (36°), computed from the well log data in Figure 5.1.

seismic data and $\boldsymbol{m}$ represents the subsurface properties of interest, then we can link the model to the data through a set of geophysical equations $\boldsymbol{f}$:

$$\boldsymbol{d} = \boldsymbol{f}(\boldsymbol{m}) + \boldsymbol{\varepsilon}, \tag{5.8}$$

where $\boldsymbol{\varepsilon}$ represents the noise in the data. The geophysical operator $\boldsymbol{f}$ can include rock physics models such as granular media, inclusion, and empirical models to link petrophysical properties to elastic attributes (Chapter 2) and seismic models such as the convolution of Zoeppritz equations or AVO approximations of the reflectivity coefficients to link elastic attributes to their seismic response (Section 5.1). These models can be linear or non-linear. The model variables $\boldsymbol{m}$ can then be computed as the solution of an inverse problem (Tarantola 2005).

We operate in a Bayesian setting to assess the posterior distribution $P(\boldsymbol{m}|\boldsymbol{d})$ of the model variables $\boldsymbol{m}$ conditioned on the data $\boldsymbol{d}$:

$$P(\boldsymbol{m}|\boldsymbol{d}) = \frac{P(\boldsymbol{d}|\boldsymbol{m})P(\boldsymbol{m})}{P(\boldsymbol{d})}, \tag{5.9}$$

where $P(\boldsymbol{m})$ is the prior distribution of the model variables, $P(\boldsymbol{d}|\boldsymbol{m})$ is the likelihood function of the data, and $P(\boldsymbol{d})$ is the probability of the data and is a normalization constant.

The formulation in Eq. (5.9) can be applied to discrete and continuous random variables. In geophysical applications, examples of inverse problems with continuous random variables include elastic inversion of seismic data to predict a set of elastic properties, such as velocities or impedances (Section 5.3), and petrophysical inversion of seismic data to predict a set of petrophysical variables, such as porosity, mineral volumes, and fluid saturations (Sections 5.4 and 5.5). Examples of inverse problems with discrete random variables include facies or rock type classification of seismic data (Chapter 6). In the discrete case, the prior model is a probability mass function, whereas in the continuous case, the prior model is a PDF.

First, we describe the application of Bayesian inverse theory to seismic inversion problems using the AVO linearized method, then we extend the approach to rock physics models discussing Gaussian and non-parametric assumptions.

5.3 Bayesian Linearized AVO Inversion

The first part of the seismic characterization problem is the prediction of elastic properties, such as P-wave and S-wave velocity (or P- and S-impedance) and density from seismic data. We present a Bayesian linearized inversion approach, namely Bayesian linearized AVO inversion, for the prediction of elastic properties $\boldsymbol{m}$ from seismic data $\boldsymbol{d}$, developed by Buland and Omre (2003). In the formulation by Buland and Omre (2003), the model variables are the logarithm of P-wave and S-wave velocity and density $[\ln V_P, \ln V_S, \ln \rho]$ but the inverse problem could also be formulated in terms of the logarithm of P- and S-impedance and density. The forward operator $\boldsymbol{f}$ is a linear operator in the logarithm of the elastic properties and it is given by the convolution of the source wavelet and the reflectivity coefficients computed based on the elastic properties through a linearization of the Zoeppritz equations. The operator is linear because convolution is a linear operator and the reflectivity coefficients are computed through a linearized model. We recall that reflection seismic data are defined at the layer interfaces, whereas the elastic properties are defined within the layers.

5.3.1 Forward Model

We assume that the model variables are represented by the vector $\boldsymbol{m} = [\ln \boldsymbol{V}_P, \ln \boldsymbol{V}_S, \ln \boldsymbol{\rho}]^T$ of length $n_m = 3(n_t + 1)$, where each of the three vectors of elastic properties include $n_t + 1$ time samples (since $n_t + 1$ layers correspond to n_t interfaces, hence, n_t reflections and amplitude values). Then a discrete version of the continuous reflectivity series **c** of n_t time samples and for n_θ reflection angles can be derived from Eqs. (5.5) and (5.6) as:

$$\mathbf{c} = \mathbf{AD}\boldsymbol{m} = \mathbf{A}\boldsymbol{m}', \tag{5.10}$$

where **A** is a block matrix containing the discrete time samples of the coefficients $c_P(\theta)$, $c_S(\theta)$, and $c_\rho(\theta)$ in Eq. (5.6), **D** is a first-order differential matrix, and $\boldsymbol{m}'$ is the vector of derivatives of $\boldsymbol{m}$ with respect to time t. In Eq. (5.10), $\boldsymbol{m}'$ is a vector of length $3n_t$, and **A** is a block matrix of dimensions $(n_\theta n_t \times 3n_t)$. The blocks of the matrix **A** are diagonal matrices of dimensions $(n_t \times n_t)$, making **A** a sparse matrix. For example, if we assume 5 different reflection angles and 100 time samples (at 100 interfaces) in the vertical profile, then **A** is a matrix of dimensions (500×300). The matrix **A** is constructed as follows:

$$\mathbf{A} = \begin{bmatrix} A_P(\theta_1) & A_S(\theta_1) & A_\rho(\theta_1) \\ \vdots & \vdots & \vdots \\ A_P(\theta_{n_\theta}) & A_S(\theta_{n_\theta}) & A_\rho(\theta_{n_\theta}) \end{bmatrix}, \tag{5.11}$$

where $A_P(\theta_i)$, $A_S(\theta_i)$, and $A_\rho(\theta_i)$ are diagonal matrices of dimensions $(n_t \times n_t)$ with diagonal elements equal to $c_P(\theta_i)$, $c_S(\theta_i)$, and $c_\rho(\theta_i)$, for $i = 1, \ldots, n_\theta$. If the ratio $\overline{V_S}^2/\overline{V_P}^2$ is assumed to be constant along the profile, then the reflection coefficients only depend on the angle θ_i; therefore, for each angle θ_i, the reflection coefficients are constant for all time samples t_j for $j = 1, \ldots, n_t$. The first-order differential matrix **D** has dimensions $(3n_t \times n_m)$ and includes, on the diagonal, three blocks $\mathbf{D}_b$ of dimensions $(n_t \times (n_t + 1))$ of the following form:

$$\mathbf{D}_b = \begin{bmatrix} -1 & 1 & 0 & \cdots & 0 \\ 0 & -1 & 1 & \ddots & \vdots \\ \vdots & \ddots & \ddots & \ddots & 0 \\ 0 & \cdots & 0 & -1 & 1 \end{bmatrix}. \tag{5.12}$$

The seismic response $\boldsymbol{d}$ can then be obtained as a discrete convolution of the angle-dependent wavelets and the reflection coefficients **c**, and it can be written as a matrix–vector multiplication:

$$\boldsymbol{d} = \mathbf{Wc} + \boldsymbol{\varepsilon} = \mathbf{WA}\boldsymbol{m}' + \boldsymbol{\varepsilon}, \tag{5.13}$$

where **W** is a block-diagonal matrix containing the discretized wavelets (one wavelet for each reflection angle) and $\boldsymbol{\varepsilon}$ is the data error assumed to be distributed according to a Gaussian distribution $\boldsymbol{\varepsilon} \sim \mathcal{N}(\boldsymbol{\varepsilon}; \mathbf{0}, \boldsymbol{\Sigma}_\varepsilon)$ with **0** mean and covariance matrix $\boldsymbol{\Sigma}_\varepsilon$. In Eq. (5.13), the seismic data $\boldsymbol{d}$ and the error $\boldsymbol{\varepsilon}$ are vectors of length $n_\theta n_t$, and **W** is a band matrix of dimensions $(n_\theta n_t \times n_\theta n_t)$. The blocks $\mathbf{W}_i$ (for $i = 1, \ldots, n_\theta$) of the matrix **W** are band matrices of dimensions $(n_t \times n_t)$, making **W** a sparse matrix.

To clarify the notation, we rewrite Eq. (5.13) making the time and angle dependence explicit:

$$\begin{bmatrix} d(t_1,\theta_1) \\ \vdots \\ d(t_{n_t},\theta_1) \\ d(t_1,\theta_2) \\ \vdots \\ \vdots \\ d(t_{n_t},\theta_{n_\theta}) \end{bmatrix} = \begin{bmatrix} \mathbf{W}_1(\theta_1) & \cdots & \mathbf{0} \\ \vdots & \ddots & \vdots \\ \mathbf{0} & \cdots & \mathbf{W}_{n_\theta}(\theta_{n_\theta}) \end{bmatrix} \begin{bmatrix} c(t_1,\theta_1) \\ \vdots \\ c(t_{n_t},\theta_1) \\ c(t_1,\theta_2) \\ \vdots \\ \vdots \\ c(t_{n_t},\theta_{n_\theta}) \end{bmatrix} + \begin{bmatrix} \varepsilon(t_1,\theta_1) \\ \vdots \\ \varepsilon(t_{n_t},\theta_1) \\ \varepsilon(t_1,\theta_2) \\ \vdots \\ \vdots \\ \varepsilon(t_{n_t},\theta_{n_\theta}) \end{bmatrix}, \tag{5.14}$$

where $t_1, ..., t_{n_t}$ are the time samples and $\theta_1, ..., \theta_{n_\theta}$ are the reflection angles. The block $\mathbf{W}_i$ of dimensions $(n_t \times n_t)$ contains the wavelet for the angle θ_i and can be written as:

$$\mathbf{W}_i(\theta_i) = \begin{bmatrix} w_i(s_1) & \cdots & 0 & \cdots & 0 \\ \vdots & \ddots & \vdots & \vdots & \vdots \\ \vdots & \vdots & w_i(s_1) & \vdots & \vdots \\ \vdots & \vdots & \vdots & \ddots & \vdots \\ w_i(s_{n_w}) & \vdots & \vdots & \vdots & w_i(s_1) \\ \vdots & \ddots & \vdots & \vdots & \vdots \\ \vdots & \vdots & w_i(s_{n_w}) & \vdots & \vdots \\ \vdots & \vdots & \vdots & \ddots & \vdots \\ 0 & \cdots & 0 & \cdots & w_i(s_{n_w}) \end{bmatrix}, \tag{5.15}$$

where w_i is the source wavelet associated with the reflection angle θ_i, for $i = 1, ..., n_\theta$, and $s_1, ..., s_{n_w}$ are the time samples of the wavelet. The resulting seismic data $\boldsymbol{d}$ is a vector of length $n_\theta n_t$, where the seismograms are concatenated. For example, if we assume 5 reflection angles and 100 time samples in the vertical profile, the vector $\boldsymbol{d}$ includes 500 elements, corresponding to 5 concatenated seismograms of 100 samples.

5.3.2 Inverse Problem

We assume that the elastic properties V_P, V_S, and ρ are distributed according to a log-Gaussian distribution (Section 1.4.4) at each time sample t. Therefore, the vector of the logarithm of the elastic properties, $\boldsymbol{m}(t) = [\ln V_P(t), \ln V_S(t), \ln \rho(t)]^T$, is distributed according to a trivariate Gaussian distribution, at each time sample t. However, elastic properties are not spatially independent in the subsurface (Section 3.1). Because the seismic travel time in the vertical direction is a function of the distance between adjacent samples, then the elastic properties at time sample t are correlated to the elastic properties at time sample s. Therefore, we cannot model the profile of elastic properties as a sequence of independent log-Gaussian distributions, but we must account for their temporal correlation.

For this reason, we assume that the concatenated vector $\boldsymbol{m} = [\ln \boldsymbol{V}_P, \ln \boldsymbol{V}_S, \ln \boldsymbol{\rho}]^T$ of length $n_m = 3(n_t + 1)$ is distributed as a multivariate Gaussian distribution $\boldsymbol{m} \sim \mathcal{N}(\boldsymbol{m}; \boldsymbol{\mu}_m, \boldsymbol{\Sigma}_m)$ with prior mean $\boldsymbol{\mu}_m$ and prior covariance matrix $\boldsymbol{\Sigma}_m$. The prior mean $\boldsymbol{\mu}_m$ is the vector of length n_m of the prior means of the logarithm of elastic properties:

$$\boldsymbol{\mu}_m = \left[\boldsymbol{\mu}_P, \boldsymbol{\mu}_S, \boldsymbol{\mu}_\rho\right]^T, \tag{5.16}$$

where $\boldsymbol{\mu}_P$, $\boldsymbol{\mu}_S$, and $\boldsymbol{\mu}_\rho$ represent the vectors of the prior means of the logarithm of P-wave and S-wave velocity and density. The prior mean vector in Eq. (5.16) is assumed to include low-frequency models of the elastic properties, owing to the band-limited nature of seismic data. Such models can be computed by filtering and interpolating well logs measured at the boreholes or from prior geophysical information. The prior covariance matrix $\boldsymbol{\Sigma}_m$ includes the covariances of the logarithm of elastic properties and time-covariances at times t and s based on a time correlation model:

$$\boldsymbol{\Sigma}_m = \mathrm{cov}_{\boldsymbol{m}(t),\boldsymbol{m}(s)}. \tag{5.17}$$

A time correlation function $\nu(\tau)$ is the equivalent of a space correlation function (Section 3.2) in the temporal domain, where the distance $\tau = |t - s|$ is the time lag between two time samples t and s. For example, the prior covariance matrix $\boldsymbol{\Sigma}_m$ can be defined as the Kronecker product $\boldsymbol{\Sigma}_m = \boldsymbol{\Sigma}_0 \otimes \boldsymbol{\Sigma}_t$ of the (time-invariant) covariance matrix of the logarithm of elastic properties $\boldsymbol{\Sigma}_0$ and the time-dependent correlation matrix $\boldsymbol{\Sigma}_t$ defined by the time correlation function $\nu(|t - s|)$ as in Buland and Omre (2003). The matrix $\boldsymbol{\Sigma}_0$ has dimensions (3×3) and it can be written in the form:

$$\boldsymbol{\Sigma}_0 = \begin{bmatrix} \sigma_P^2 & \sigma_{P,S} & \sigma_{P,\rho} \\ \sigma_{P,S} & \sigma_S^2 & \sigma_{S,\rho} \\ \sigma_{P,\rho} & \sigma_{S,\rho} & \sigma_\rho^2 \end{bmatrix} = \begin{bmatrix} \sigma_P^2 & \sigma_P\sigma_S\varrho_{P,S} & \sigma_P\sigma_\rho\varrho_{P,\rho} \\ \sigma_P\sigma_S\varrho_{P,S} & \sigma_S^2 & \sigma_S\sigma_\rho\varrho_{S,\rho} \\ \sigma_P\sigma_\rho\varrho_{P,\rho} & \sigma_S\sigma_\rho\varrho_{S,\rho} & \sigma_\rho^2 \end{bmatrix}, \tag{5.18}$$

where σ, σ^2, and ϱ represent the standard deviation, variance, and correlation, and the subscripts P, S, and ρ represent the logarithm of P-wave and S-wave velocity and density, respectively. The time-dependent covariance matrix $\boldsymbol{\Sigma}_t$ is a symmetric positive-definite matrix of dimensions $((n_t + 1) \times (n_t + 1))$ built upon the time correlation function $\nu(\tau)$ with correlation length t_l:

$$\boldsymbol{\Sigma}_t = \begin{bmatrix} 1 & \cdots & \nu(t_l) & \cdots & \cdots & \cdots & 0 \\ \vdots & \ddots & \vdots & \ddots & \vdots & \vdots & \vdots \\ \nu(t_l) & \vdots & 1 & \vdots & \nu(t_l) & \vdots & \vdots \\ \vdots & \ddots & \vdots & \ddots & \vdots & \ddots & \vdots \\ \vdots & \vdots & \nu(t_l) & \vdots & 1 & \vdots & \nu(t_l) \\ \vdots & \vdots & \vdots & \ddots & \vdots & \ddots & \vdots \\ 0 & \cdots & \cdots & \cdots & \nu(t_l) & \cdots & 1 \end{bmatrix} \tag{5.19}$$

and it generally takes the form of a band matrix owing to the finite correlation length t_l of the correlation function, causing it to decay close to 0 beyond the correlation length. The resulting covariance matrix $\mathbf{\Sigma}_m = \mathbf{\Sigma}_0 \otimes \mathbf{\Sigma}_t$ is symmetric and positive definite, of dimensions $(n_m \times n_m)$.

Because we assume that $\boldsymbol{m}$ is prior distributed acording to a Gaussian distribution and based on the linearity of the differential operator, the time derivative $\boldsymbol{m}'$ is also distributed according to a Gaussian distribution $\boldsymbol{m}' \sim \mathcal{N}(\boldsymbol{m}'; \boldsymbol{\mu}'_m, \mathbf{\Sigma}''_m)$ with mean $\boldsymbol{\mu}'_m$ equal to the time derivative of $\boldsymbol{\mu}_m$ and covariance matrix $\mathbf{\Sigma}''_m$ equal to the second derivative of the covariance matrix $\mathbf{\Sigma}_m$ (Buland and Omre 2003). Because of the linear approximation of the seismic forward operator in Eqs. (5.10)–(5.15), the seismic data vector $\boldsymbol{d}$ is also distributed according to a Gaussian distribution $\boldsymbol{d} \sim \mathcal{N}(\boldsymbol{d}; \boldsymbol{\mu}_d, \mathbf{\Sigma}_d)$ with mean:

$$\boldsymbol{\mu}_d = \mathbf{W}\mathbf{A}\boldsymbol{\mu}'_m \tag{5.20}$$

and covariance matrix:

$$\mathbf{\Sigma}_d = \mathbf{W}\mathbf{A}\mathbf{\Sigma}''_m\mathbf{A}^T\mathbf{W}^T + \mathbf{\Sigma}_\varepsilon. \tag{5.21}$$

Therefore, the posterior distribution of the elastic model $\boldsymbol{m}$ conditioned on the seismic data $\boldsymbol{d}$ is distributed according to a Gaussian distribution $\boldsymbol{m}|\boldsymbol{d} \sim \mathcal{N}\left(\boldsymbol{m};\, \boldsymbol{\mu}_{m|d}, \mathbf{\Sigma}_{m|d}\right)$ with posterior mean $\boldsymbol{\mu}_{m|d}$ given by:

$$\boldsymbol{\mu}_{m|d} = \boldsymbol{\mu}_m + \left(\mathbf{W}\mathbf{A}\mathbf{\Sigma}'_m\right)^T\mathbf{\Sigma}_d^{-1}(\boldsymbol{d} - \boldsymbol{\mu}_d) \tag{5.22}$$

and posterior covariance matrix $\mathbf{\Sigma}_{m|d}$ given by:

$$\mathbf{\Sigma}_{m|d} = \mathbf{\Sigma}_m - \left(\mathbf{W}\mathbf{A}\mathbf{\Sigma}'_m\right)^T\mathbf{\Sigma}_d^{-1}\mathbf{W}\mathbf{A}\mathbf{\Sigma}'_m, \tag{5.23}$$

where $\mathbf{\Sigma}'_m$ is the first derivative of the covariance matrix $\mathbf{\Sigma}_m$. Equations (5.22) and (5.23) can be derived using the analytical expressions of the parameters of conditional Gaussian distributions (Section 1.7), as shown in Buland and Omre (2003). Since the posterior distribution is Gaussian, then the maximum a posteriori solution (i.e. the mode) is equal to the posterior mean (Eq. 5.22). However, because the model $\boldsymbol{m}$ represents the logarithm of elastic properties, the posterior distribution of P-wave and S-wave velocity and density is log-Gaussian, hence the maximum a posteriori and the mean are not equal.

Using log-Gaussian transformations (Section 1.4.4), the expression of the maximum a posteriori $\hat{\boldsymbol{m}}$ of the elastic parameters becomes:

$$\hat{\boldsymbol{m}} = \exp\left(\boldsymbol{\mu}_{m|d} - \boldsymbol{\sigma}^2_{m|d}\right) \tag{5.24}$$

and the expression of the posterior mean $\overline{\boldsymbol{m}}$ is:

$$\overline{\boldsymbol{m}} = \exp\left(\boldsymbol{\mu}_{m|d} + \frac{\boldsymbol{\sigma}^2_{m|d}}{2}\right) \tag{5.25}$$

The confidence intervals can also be analytically estimated. For example, the 0.95 confidence interval can be computed as:

$$\boldsymbol{I}_{0.95} = \left[\exp\left(\boldsymbol{\mu}_{m|d} - 1.96\boldsymbol{\sigma}_{m|d}\right), \exp\left(\boldsymbol{\mu}_{m|d} + 1.96\boldsymbol{\sigma}_{m|d}\right)\right]. \tag{5.26}$$

Typically, in seismic applications, the posterior variance is much smaller than the posterior mean, component-wise, and the posterior log-Gaussian distributions are close to being symmetric; hence, the maximum a posteriori and posterior mean are similar. However, this is not the case for highly skewed log-Gaussian distributions, for which the posterior mean in Eq. (5.25) might fall outside the confidence interval in Eq. (5.26).

Example 5.2 We apply the Bayesian linearized AVO inversion to the synthetic seismic dataset in Figure 5.2. We compute the low-frequency model by filtering the actual well log data (Figure 5.1) using a low-pass filter with cut-off frequency of 5 Hz. The logarithms of the filtered well logs represent the locally variable mean of the Gaussian prior distribution. The variance is computed based on the residuals of the well logs and the filtered logs. The prior distribution is shown in Figure 5.3 (blue lines). The time correlation function is an exponential function with a correlation length of $t_l = 10$ ms, corresponding to approximately 18 m. We apply the Bayesian AVO linearized inversion assuming a signal-to-noise ratio of 5.

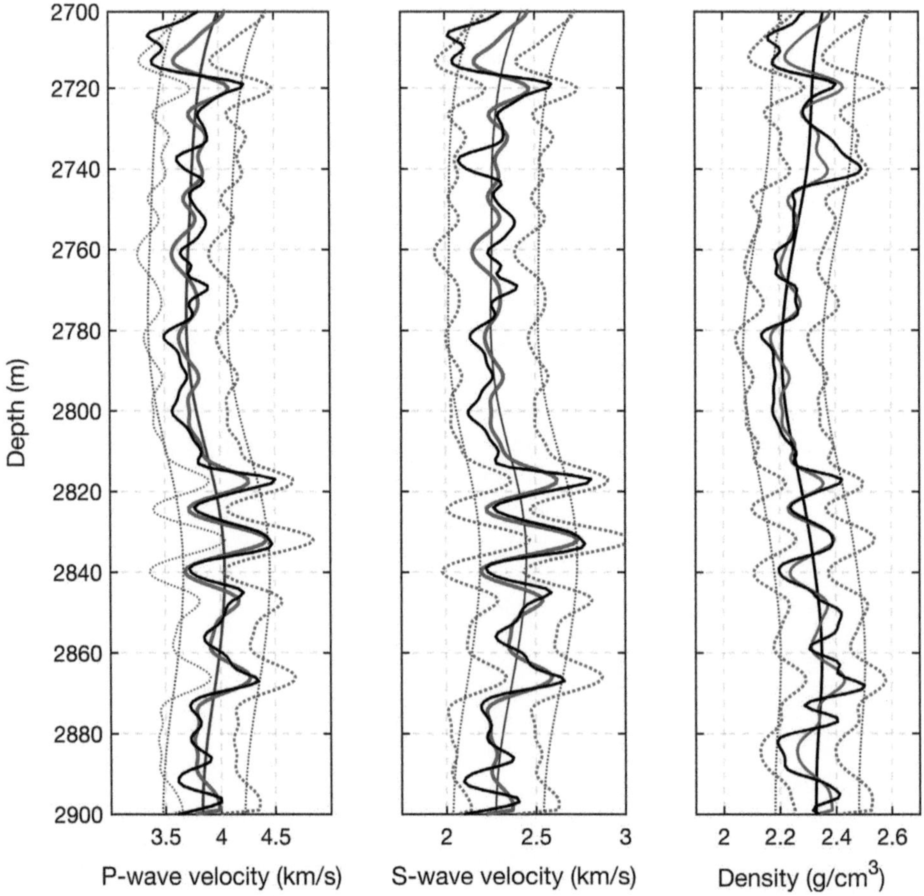

Figure 5.3 Bayesian linearized AVO inversion results: posterior distribution of P-wave velocity, S-wave velocity, and density. The solid blue curves represent the prior mean; the solid red curves represent the maximum a posteriori. The dotted curves represent the 0.95 confidence intervals of the prior and posterior distributions. The solid black curves represent the actual well logs.

The result of the inversion is the posterior distribution of the logarithm of the elastic properties. Figure 5.3 shows the maximum a posteriori and the 0.95 confidence interval of the elastic properties (red lines). The correlation between the maximum a posteriori and the actual well logs is approximately 0.87, whereas the coverage ratio of the confidence interval (i.e. the fraction of true samples falling inside the confidence interval) is approximately 0.96, showing a slight overestimation of the posterior uncertainty.

Synthetic and real-case applications with sensitivity analysis on the signal-to-noise ratio can be found in Buland and Omre (2003). The extension of the method to two- and three-dimensional problems is relatively straightforward when the method is applied trace by trace. In many practical applications, the lateral continuity of the elastic properties is inherited from the lateral continuity of the seismic data, but the resulting spatial continuity model might be biased by seismic processing, such as the migration operator, which is a spatial filter whose correlation length is associated with the Fresnel zone (Sheriff and Geldart 1995; Yilmaz 2001). Because the result is a probability distribution of elastic properties, at each time sample where seismic data are available, it is possible to sample multiple stochastic realizations of elastic properties from the distribution, using geostatistical methods (Chapter 3). Buland et al. (2003) provide an efficient method using fast Fourier transform (FFT) to include spatial coupling of the model parameters in the linearized AVO inversion and impose lateral continuity.

5.4 Bayesian Rock Physics Inversion

The second part of the seismic reservoir characterization problem is the prediction of rock properties such as porosity, mineral volumes, and fluid saturations from elastic properties. We first assume that a model of elastic properties estimated from seismic data is available and we predict the petrophysical properties of interest. The model of elastic properties can be obtained using seismic inversion methods, either deterministic or probabilistic. In the case of probabilistic inversion methods, such as Bayesian linearized AVO inversion (Section 5.3), we initially ignore the uncertainty in the elastic properties and only use the most likely model (e.g. the maximum a posteriori or the mean models in Eqs. (5.24) and (5.25)). We then extend the proposed approach to include the uncertainty propagation in Section 5.5. The prediction of petrophysical properties relies on a rock physics model (Chapter 2). Different rock physics relations are available and the choice of the model depends on the geological environment. For example, Raymer's equation and the stiff sand model can be applied in highly consolidated granular media, whereas the soft sand model is more suitable for unconsolidated rocks (Dvorkin et al. 2014).

In the following, $\boldsymbol{m}$ represents the set of elastic attributes, for example P-wave and S-wave velocity and density, and $\boldsymbol{r}$ represents the set of rock properties, such as porosity, lithology, and saturations. The set of rock properties varies depending on the specific applications. In a clastic reservoir, we generally aim to predict porosity, clay volume, and fluid saturations, whereas in a carbonate reservoir, we might be interested in the volume of calcite. The rock physics model is represented by the function $\mathbf{g}$, as:

$$\boldsymbol{m} = \mathbf{g}(\boldsymbol{r}) + \boldsymbol{e}, \tag{5.27}$$

where $\boldsymbol{e}$ is the data error assumed to be distributed according to a Gaussian distribution $\boldsymbol{e} \sim \mathcal{N}(\boldsymbol{e}; \mathbf{0}, \boldsymbol{\Sigma}_e)$ with $\mathbf{0}$ mean and covariance matrix $\boldsymbol{\Sigma}_e$.

The analytical solution of the Bayesian inverse problem relies on two assumptions: the linearity of the forward model and the Gaussian distribution of the prior model (Section 1.7). We analyze four different cases: (i) linear rock physics model and Gaussian distribution of petrophysical properties; (ii) linear rock physics model and Gaussian mixture distribution of petrophysical properties; (iii) non-linear rock physics model and Gaussian mixture distribution of petrophysical properties; and (iv) non-linear rock physics model and non-parametric distribution of petrophysical properties.

5.4.1 Linear – Gaussian Case

We first focus on linear or linearized rock physics models. If $\mathbf{G}$ is the matrix associated with the linear rock physics model $\mathbf{g}$, then Eq. (5.27) can be rewritten as a matrix–vector multiplication:

$$\boldsymbol{m} = \mathbf{G}\boldsymbol{r} + \boldsymbol{e}. \tag{5.28}$$

For example, if the model variable $\boldsymbol{r}$ includes porosity, clay volume, and water saturation $\boldsymbol{r} = [\phi, v_c, s_w]^T$, and the measurement $\boldsymbol{m}$ includes P-wave and S-wave velocity and density $\boldsymbol{m} = [V_P, V_S, \rho]^T$, with associated errors $\boldsymbol{e} = [e_P, e_S, e_\rho]^T$, then we can express the rock physics model as a multilinear regression in the form:

$$\begin{cases} V_P = \alpha_P\phi + \beta_P v_c + \gamma_P s_w + \delta_P + e_P \\ V_S = \alpha_S\phi + \beta_S v_c + \gamma_S s_w + \delta_S + e_S \\ \rho = \alpha_\rho\phi + \beta_\rho v_c + \gamma_\rho s_w + \delta_\rho + e_\rho \end{cases}, \tag{5.29}$$

where α, β, γ, and δ are the coefficients of the multilinear regression, and the subscripts P, S, and ρ represent P-wave and S-wave velocity and density (Grana 2016). Equivalently, the rock physics model can be written in the matrix–vector form as:

$$\begin{bmatrix} V_P \\ V_S \\ \rho \end{bmatrix} = \begin{bmatrix} \alpha_P & \beta_P & \gamma_P \\ \alpha_S & \beta_S & \gamma_S \\ \alpha_\rho & \beta_\rho & \gamma_\rho \end{bmatrix} \begin{bmatrix} \phi \\ v_c \\ s_w \end{bmatrix} + \begin{bmatrix} \delta_P \\ \delta_S \\ \delta_\rho \end{bmatrix} + \begin{bmatrix} e_P \\ e_S \\ e_\rho \end{bmatrix}. \tag{5.30}$$

The additive constant $\boldsymbol{\delta} = [\delta_P, \delta_S, \delta_\rho]^T$ in Eq. (5.30) can be subtracted from the measurements $\boldsymbol{m}$, and the linear inverse problem can be written in the form of Eq. (5.28), where the measurement vector $\boldsymbol{m}$ is replaced by $\tilde{\boldsymbol{m}} = \boldsymbol{m} - \boldsymbol{\delta}$.

Grana (2016) proposes a mathematical approach for the linearization of slightly non-linear rock physics models, using first-order Taylor series approximations. Taylor series aim to approximate an arbitrary function at a given value of the independent variable using a polynomial in which the coefficients depend on the derivatives of the function. For simple functions, a good approximation can be achieved using a limited number of terms. Taylor series have been used in geophysics to approximate complex non-linear functions, especially in seismic imaging (Ursin and Stovas 2006). In rock physics inversion, we are interested in linear approximations of rock physics models; therefore, the Taylor series expansion of the rock physics equation is truncated after the first term. The first-order Taylor series approximation of the rock physics model $\mathbf{g}$ at the value $\boldsymbol{r}_0$ is then:

$$\boldsymbol{m} \cong \mathbf{g}(\boldsymbol{r}_0) + \mathbf{J}_{r_0}(\boldsymbol{r} - \boldsymbol{r}_0) + \boldsymbol{e}, \tag{5.31}$$

where $\mathbf{J}_{r_0}$ is the Jacobian of the function $\mathbf{g}$ evaluated at the value $\boldsymbol{r}_0$. The value $\boldsymbol{r}_0$ can be, for example, the mean of the prior distribution. The linearization of the forward problem in Eq. (5.28) based on the approximation in Eq. (5.31) can then be rewritten as:

$$\boldsymbol{m} \cong \mathbf{J}_{r_0}\boldsymbol{r} + (\mathbf{g}(\boldsymbol{r}_0) - \mathbf{J}_{r_0}\boldsymbol{r}_0) + \boldsymbol{e} = \mathbf{G}\boldsymbol{r} + \boldsymbol{c} + \boldsymbol{e}, \tag{5.32}$$

where $\boldsymbol{c} = \mathbf{g}(\boldsymbol{r}_0) - \mathbf{J}_{r_0}\boldsymbol{r}_0$ is a constant. Grana (2016) illustrates the derivation of the linearization of the rock physics model in terms of porosity, clay volume, and water saturation for the following rock physics models: Raymer, stiff sand, and spherical inclusion models.

If we assume that the prior distribution of the petrophysical variable $\boldsymbol{r}$ is Gaussian distributed, $\boldsymbol{r} \sim \mathcal{N}(\boldsymbol{r}; \boldsymbol{\mu}_r, \boldsymbol{\Sigma}_r)$, with prior mean $\boldsymbol{\mu}_r$ and prior covariance matrix $\boldsymbol{\Sigma}_r$; the forward operator $\mathbf{g}$ is linear; and the data error $\boldsymbol{e}$ is Gaussian $\boldsymbol{e} \sim \mathcal{N}(\boldsymbol{e}; \mathbf{0}, \boldsymbol{\Sigma}_e)$, with $\mathbf{0}$ mean and covariance matrix $\boldsymbol{\Sigma}_e$, and independent of $\boldsymbol{r}$; then the posterior distribution $\boldsymbol{r}|\boldsymbol{m}$ is also Gaussian $\boldsymbol{r}|\boldsymbol{m} \sim \mathcal{N}(\boldsymbol{r}; \boldsymbol{\mu}_{r|m}, \boldsymbol{\Sigma}_{r|m})$ with conditional mean $\boldsymbol{\mu}_{r|m}$ given by:

$$\boldsymbol{\mu}_{r|m} = \boldsymbol{\mu}_r + \boldsymbol{\Sigma}_r\mathbf{G}^T\left(\mathbf{G}\boldsymbol{\Sigma}_r\mathbf{G}^T + \boldsymbol{\Sigma}_e\right)^{-1}(\boldsymbol{m} - \mathbf{G}\boldsymbol{\mu}_r) \tag{5.33}$$

and conditional covariance matrix $\boldsymbol{\Sigma}_{r|m}$ given by:

$$\boldsymbol{\Sigma}_{r|m} = \boldsymbol{\Sigma}_r - \boldsymbol{\Sigma}_r\mathbf{G}^T\left(\mathbf{G}\boldsymbol{\Sigma}_r\mathbf{G}^T + \boldsymbol{\Sigma}_e\right)^{-1}\mathbf{G}\boldsymbol{\Sigma}_r, \tag{5.34}$$

where $\mathbf{G}$ is the matrix associated with the linear rock physics model. As for the Bayesian linearized AVO inversion (Section 5.3), it is possible to include a spatial correlation function in the covariance model. Generally, the linearization of the rock physics model is accurate for porosity and mineral volumes but it might fail for water saturation, especially for homogeneous saturations of water and gas. Lang and Grana (2018) propose a Bayesian linearized AVO inversion for petrophysical properties based on a linearization of the rock physics model in terms of porosity and elastic moduli and density of solid and fluid phases, to improve the accuracy of the linear approximation.

The main limitation of this approach is the Gaussian assumption of the model properties. In general, in rock physics inversion, the Gaussian assumption is not valid for two reasons: (i) petrophysical properties generally represent volumetric fractions that are bounded variables defined in the domain [0, 1], and (ii) in many applications, petrophysical properties are multimodal. It is a common practice to overcome the first assumption by truncating the Gaussian distributions, i.e. bounding the random variables from below and above, and neglecting the probability of the tails outside the physical limits. However, sequences of multiple lithologies often show multimodal distributions of porosity, volumetric fractions, and fluid saturations. For example, the distribution of effective porosity in a sequence of sand and shale is generally multimodal with two main modes: high porosity in sand and low porosity in shale.

5.4.2 Linear – Gaussian Mixture Case

Gaussian mixture models (Section 1.4.5), with adequate truncations or transformations for bounded properties, are particularly suitable to describe the multimodal behavior of rock and fluid properties, for their analytical tractability. One of the advantages of Gaussian mixture models in geophysical applications is the possibility to identify the components of the mixtures with geological classifications, such as facies, geobodies, or flow units (Grana and Della Rossa 2010).

Multiple examples of Gaussian mixture models for geophysical applications are shown in Figure 5.4, where the parameters of the mixtures are estimated from a set of well logs. First, we show the surface of a bivariate joint Gaussian mixture model with two components in the domain of porosity and P-wave velocity. The joint distribution shows two modes that can be identified with the facies, namely sand and shale. Then, we show the projections of a quadrivariate joint Gaussian mixture model of porosity, clay volume, and P-wave and S-wave velocity. For illustration purposes, we only display the bivariate projections in the petrophysical domain of porosity and volume of clay, and in the petro-elastic domain of porosity and P-wave velocity; however, the distribution of all the possible combinations of properties can be computed. The projections represent the bivariate marginal distributions

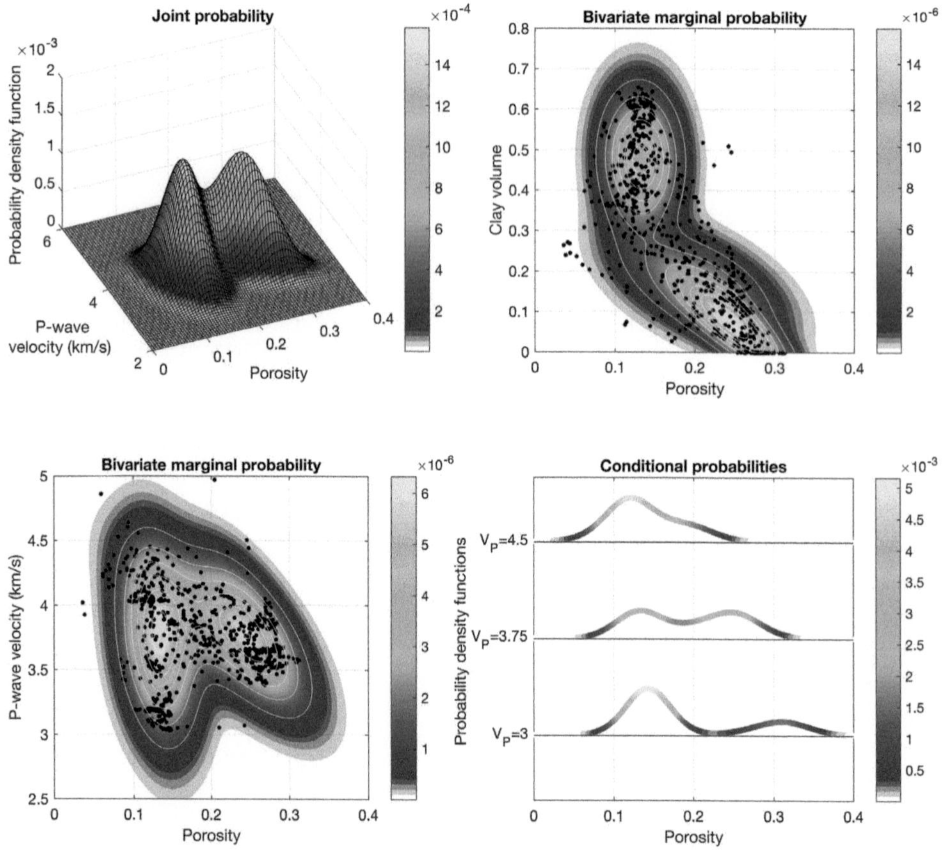

Figure 5.4 Examples of Gaussian mixture models: a bivariate joint distribution of porosity and P-wave velocity with two components; a bivariate marginal distribution of porosity and volume of clay; a bivariate marginal distribution of porosity and P-wave velocity; and three conditional distributions of porosity conditioned on P-wave velocity.

of the quadrivariate joint Gaussian mixture model. The joint distribution has three components associated with three litho-fluid facies, namely hydrocarbon sand, water sand, and shale; however, in the marginal distributions, two of the three components largely overlap owing to the limited fluid effect. We then show the conditional distribution of porosity for three values of P-wave velocity, computed from the bivariate marginal distribution. The conditional distribution of a Gaussian mixture is still a Gaussian mixture but the conditional weights of some components might be negligible, hence, the number of components is less than or equal to the number of components of the joint distribution.

Bayesian approaches for geophysical inverse problems with Gaussian mixture models have been presented in Grana and Della Rossa (2010), Grana (2016), and Grana et al. (2017a). We assume that the prior distribution of the model variable $\boldsymbol{r}$ is distributed according to a Gaussian mixture distribution of F components:

$$\boldsymbol{r} \sim \sum_{k=1}^{F} \lambda_k \mathcal{N}\left(\boldsymbol{r}; \boldsymbol{\mu}_{r|k}, \boldsymbol{\Sigma}_{r|k}\right), \tag{5.35}$$

where the distributions $\mathcal{N}(\boldsymbol{r}; \boldsymbol{\mu}_{r|k}, \boldsymbol{\Sigma}_{r|k})$ represent the kth Gaussian component with means $\boldsymbol{\mu}_{r|k}$ and covariance matrices $\boldsymbol{\Sigma}_{r|k}$, and the coefficients λ_k represent the weights of the linear combination, for $k = 1, \ldots, F$, with the constraint that the sum of the weights is equal to 1 (i.e. $\sum_{k=1}^{F} \lambda_k = 1$). For linear rock physics models, if we assume that the distribution of the data error is Gaussian $\boldsymbol{e} \sim \mathcal{N}(\boldsymbol{e}; \boldsymbol{0}, \boldsymbol{\Sigma}_e)$, with $\boldsymbol{0}$ mean and covariance matrix $\boldsymbol{\Sigma}_e$, and independent of $\boldsymbol{r}$, then the posterior distribution $\boldsymbol{r}|\boldsymbol{m}$ is also a Gaussian mixture:

$$\boldsymbol{r}|\boldsymbol{m} \sim \sum_{k=1}^{F} \lambda_{k|m} \mathcal{N}\left(\boldsymbol{r}; \boldsymbol{\mu}_{r|m,k}, \boldsymbol{\Sigma}_{r|m,k}\right) \tag{5.36}$$

with conditional means $\boldsymbol{\mu}_{r|m,k}$ given by:

$$\boldsymbol{\mu}_{r|m,k} = \boldsymbol{\mu}_{r|k} + \boldsymbol{\Sigma}_{r|k}\mathbf{G}^T\left(\mathbf{G}\boldsymbol{\Sigma}_{r|k}\mathbf{G}^T + \boldsymbol{\Sigma}_e\right)^{-1}\left(\boldsymbol{m} - \mathbf{G}\boldsymbol{\mu}_{r|k}\right), \tag{5.37}$$

conditional covariance matrices $\boldsymbol{\Sigma}_{r|m,k}$ given by:

$$\boldsymbol{\Sigma}_{r|m,k} = \boldsymbol{\Sigma}_{r|k} - \boldsymbol{\Sigma}_{r|k}\mathbf{G}^T\left(\mathbf{G}\boldsymbol{\Sigma}_{r|k}\mathbf{G}^T + \boldsymbol{\Sigma}_e\right)^{-1}\mathbf{G}\boldsymbol{\Sigma}_{r|k}, \tag{5.38}$$

and conditional weights $\lambda_{k|m}$ given by:

$$\lambda_{k|m} = \frac{\lambda_k \mathcal{N}\left(\boldsymbol{m}; \mathbf{G}\boldsymbol{\mu}_{r|k}, \mathbf{G}\boldsymbol{\Sigma}_{r|k}\mathbf{G}^T + \boldsymbol{\Sigma}_e\right)}{\sum_{h=1}^{F} \lambda_h \mathcal{N}\left(\boldsymbol{m}; \mathbf{G}\boldsymbol{\mu}_{r|h}, \mathbf{G}\boldsymbol{\Sigma}_{r|h}\mathbf{G}^T + \boldsymbol{\Sigma}_e\right)}, \tag{5.39}$$

for $k = 1, \ldots, F$.

The maximum a posterior estimate $\hat{\boldsymbol{r}}$ corresponds to the conditional mean $\boldsymbol{\mu}_{r|m,\hat{k}}$ of the component $\hat{k}$ with the highest probability:

$$\hat{\boldsymbol{r}} = \boldsymbol{\mu}_{r|m,\hat{k}}, \tag{5.40}$$

where $\hat{k} = \text{argmax}_{k=1,\ldots,F}\, \lambda_{k|m}$; whereas the posterior mean $\overline{\boldsymbol{r}}$ is the linear combination of the conditional means weighted by the conditional weights:

$$\overline{\boldsymbol{r}} = \sum_{k=1}^{F} \lambda_{k|m}\, \boldsymbol{\mu}_{r|m,k}. \tag{5.41}$$

Equation (5.39) assumes that the weights are spatially independent and their spatial continuity structure relies entirely on the spatial correlation of the measurements $\boldsymbol{m}$. A spatial correlation model can be included in the prior model for the mixture components (Grana et al. 2017a), but under this assumption, the conditional weights $\lambda_{k|m}$ must be numerically evaluated (Section 6.2).

Example 5.3 We demonstrate the Bayesian linearized rock physics inversion for the prediction of petrophysical properties from elastic measurements using the well log data of P-wave and S-wave velocity and density shown in Figure 5.1. The corresponding petrophysical curves computed from neutron porosity, gamma ray, and resistivity are shown in Figure 5.5, and include porosity, clay volume, and water saturation. The interval under examination includes a sequence of clean sand and shaley sand with an average porosity of 0.2. In the upper part of the interval, there are two oil sand layers.

The rock physics model is a linear approximation of the Raymer–Dvorkin's model (Section 2.1.1). The prior distribution of the petrophysical properties is a Gaussian mixture model with three components, corresponding to three litho-fluid facies: oil sand, water sand, and shaley sand. The parameters of the prior distribution, namely weights, means, and covariance matrices, have been estimated from the actual measurements (Figure 5.5). The prior parameters are assumed to be constant in the entire interval. The results of the inversion are shown in Figure 5.6, where the posterior distribution conditioned on the elastic properties is a Gaussian mixture and it is represented using the background color. The maximum a posteriori estimates of the model variables in Figure 5.6 (red lines) show that porosity and clay volume predictions are accurate and precise. In particular, the uncertainty in the porosity predictions is relatively small. The uncertainty in the water saturation estimates is large and the predictions are less accurate than the other properties owing to the lower sensitivity of velocities to fluid changes. Overall, the posterior distribution captures the multimodal behavior of the petrophysical properties; however, the predictions overestimate the correlation between the model properties, owing to the linearization of the rock physics operator, and fail to capture non-linear relations in the lower part of the interval.

5.4.3 Non-linear – Gaussian Mixture Case

For non-linear rock physics models, the posterior distribution must be numerically evaluated. Grana and Della Rossa (2010) propose a solution based on Monte Carlo simulations. According to this approach, we randomly sample n_s values from the prior Gaussian mixture distribution (Eq. 5.35) to obtain a set of samples of petrophysical properties $\{\boldsymbol{r}_i\}_{i=1,\ldots,n_s}$, and we apply the rock physics model (Eq. 5.27) to obtain the corresponding set of samples of

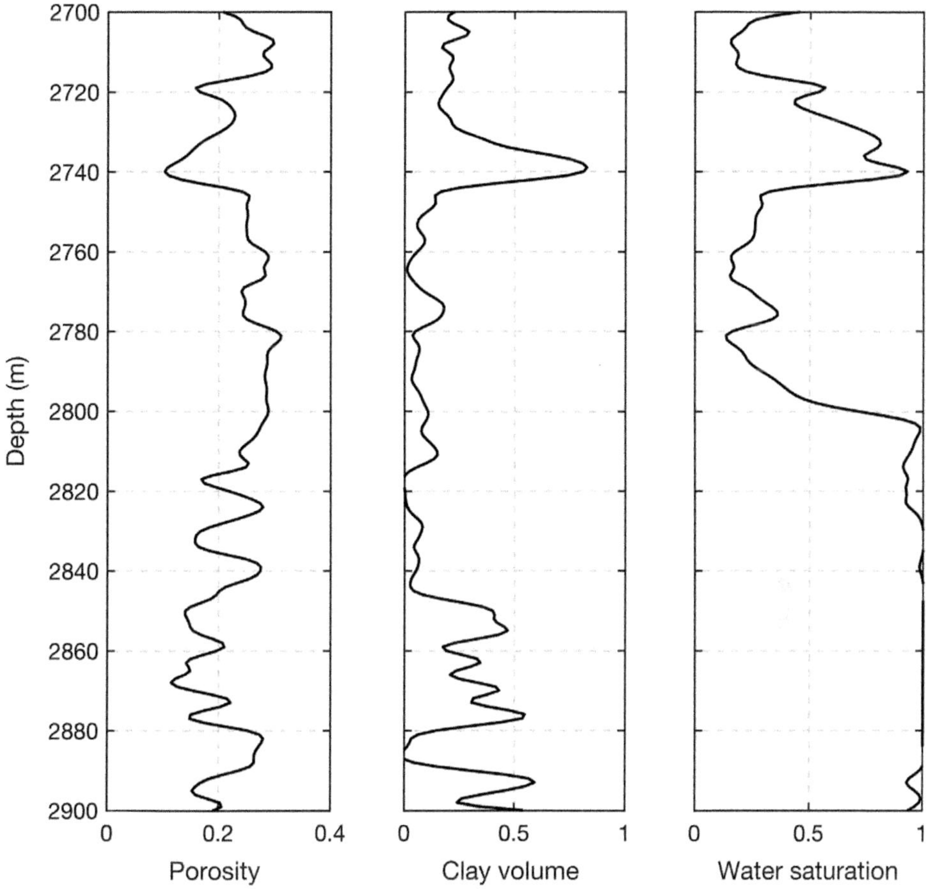

Figure 5.5 Petrophysical curves computed from well log data in Figure 5.1, measured in a borehole in the Norwegian Sea: porosity, clay volume, and water saturation.

elastic properties $\{\boldsymbol{m}_i\}_{i=1,\dots,n_s}$, assuming independent data errors randomly generated according to a Gaussian distribution $\boldsymbol{e} \sim \mathcal{N}(\boldsymbol{e}; \boldsymbol{0}, \boldsymbol{\Sigma}_e)$ with $\boldsymbol{0}$ mean and covariance matrix $\boldsymbol{\Sigma}_e$. Each component of the mixture is identified with one of the F classes of a geological classification, for example lithological or litho-fluid facies. The Monte Carlo samples and their facies classification are used as a training dataset to estimate the joint distribution of petrophysical and elastic properties $(\boldsymbol{r}, \boldsymbol{m})$. We assume that the joint distribution is distributed according to a Gaussian mixture model with F components:

$$(\boldsymbol{r}, \boldsymbol{m}) \sim \sum_{k=1}^{F} \lambda_k \mathcal{N}\left(\boldsymbol{r}; \boldsymbol{\mu}_{(r,m)|k}, \boldsymbol{\Sigma}_{(r,m)|k}\right), \tag{5.42}$$

where the joint mean $\boldsymbol{\mu}_{(r,m)|k}$ is the vector of the means of $\boldsymbol{r}$ and $\boldsymbol{m}$ in each facies:

$$\boldsymbol{\mu}_{(r,m)|k} = \begin{bmatrix} \boldsymbol{\mu}_{r|k} \\ \boldsymbol{\mu}_{m|k} \end{bmatrix}, \tag{5.43}$$

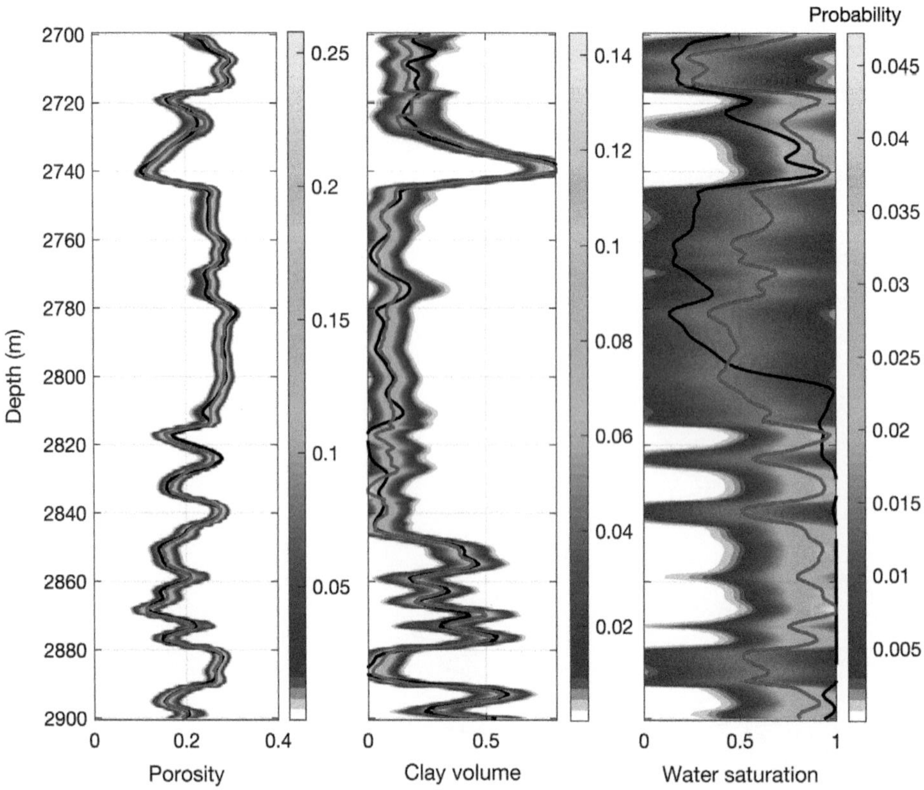

Figure 5.6 Bayesian linearized rock physics inversion results: posterior distribution of porosity, clay volume, and water saturation. The background color represents the posterior Gaussian mixture distribution and the solid red curves represent the maximum a posteriori predictions. The solid black curves represent the actual well logs.

and the joint covariance matrix $\mathbf{\Sigma}_{(r,m)|k}$ is a block matrix of the form:

$$\mathbf{\Sigma}_{(r,m)|k} = \begin{bmatrix} \mathbf{K}_{r|k} & \mathbf{K}_{r,m|k} \\ \mathbf{K}_{m,r|k} & \mathbf{K}_{m|k} \end{bmatrix}, \tag{5.44}$$

where $\mathbf{K}_{r|k}$ is the conditional covariance matrix of $\boldsymbol{r}$, $\mathbf{K}_{m|k}$ is the conditional covariance matrix of $\boldsymbol{m}$, and $\mathbf{K}_{r,m|k} = (\mathbf{K}_{m,r|k})^T$ is the cross-covariance of $\boldsymbol{r}$ and $\boldsymbol{m}$, in each facies. The weights λ_k of the mixture represent the proportions of the facies $k = 1, \ldots, F$. The parameters of the joint distribution can be inferred from the training dataset using maximum likelihood estimation methods such as the expectation–maximization algorithm (Hastie et al. 2009; Grana and Della Rossa 2010). Then the conditional distribution $\boldsymbol{r}|\boldsymbol{m}$ is also a Gaussian mixture:

$$\boldsymbol{r}|\boldsymbol{m} \sim \sum_{k=1}^{F} \lambda_{k|m} \mathcal{N}\left(\boldsymbol{r}; \boldsymbol{\mu}_{r|m,k}, \mathbf{\Sigma}_{r|m,k}\right), \tag{5.45}$$

with conditional means $\boldsymbol{\mu}_{r|m,k}$ given by:

$$\boldsymbol{\mu}_{r|m,k} = \boldsymbol{\mu}_{r|k} + \mathbf{K}_{r,m|k}\left(\mathbf{K}_{m|k}\right)^{-1}\left(\boldsymbol{m} - \boldsymbol{\mu}_{m|k}\right), \tag{5.46}$$

conditional covariance matrices $\boldsymbol{\Sigma}_{r|m,k}$ given by:

$$\boldsymbol{\Sigma}_{r|m,k} = \boldsymbol{\Sigma}_{r|k} - \mathbf{K}_{r,m|k}\left(\mathbf{K}_{m|k}\right)^{-1}\mathbf{K}_{m,r|k}, \tag{5.47}$$

and conditional weights $\lambda_{k|m}$ given by:

$$\lambda_{k|m} = \frac{\lambda_k \mathcal{N}\left(\boldsymbol{m}; \boldsymbol{\mu}_{m|k}, \mathbf{K}_{m|k}\right)}{\sum_{h=1}^{F} \lambda_h \mathcal{N}\left(\boldsymbol{m}; \boldsymbol{\mu}_{m|h}, \mathbf{K}_{m|h}\right)}, \tag{5.48}$$

for $k = 1, \ldots, F$. This approach generally gives accurate results when the rock physics model is not highly non-linear, as shown in Grana and Della Rossa (2010). A case study based on this approach is shown in Section 8.1.1.

5.4.4 Non-linear - Non-parametric Case

For highly non-linear models, we can adopt a non-parametric approach based on kernel density estimation (Silverman 1986; Bowman and Azzalini 1997) proposed in Mukerji et al. (2001), Doyen (2007), and Grana (2018), where the joint distribution is described by a non-parametric PDF. Kernel density estimation is a non-parametric technique that estimates the PDF of interest by summing the contribution of a kernel function centered at each data point. A kernel function K is a non-negative symmetric function with integral 1. Examples of valid functions include the Gaussian kernel and the Epanechnikov kernel (Silverman 1986; Bowman and Azzalini 1997). In the univariate case, given a random variable X and a set of n measurements $\{x_i\}_{i=1,\ldots,n}$, the PDF $f(X)$ can be estimated at each value x as:

$$f(x) = \frac{1}{nh}\sum_{i=1}^{n} K\left(\frac{x - x_i}{h}\right), \tag{5.49}$$

where h is the kernel bandwidth, i.e. a scaling factor that controls the smoothness of the estimated function. In the proposed approach, K is the Epanechnikov kernel:

$$K(x) = \begin{cases} 0 & x < -1 \\ \frac{3}{4}\left(1 - x^2\right) & -1 \leq x \leq 1 \\ 0 & x > 1 \end{cases}. \tag{5.50}$$

Equation (5.49) can be extended to the multivariate case of m random variables $X_1, \ldots, X_m$. Given a set of measurements $\{x_{1_i}, \ldots, x_{m_i}\}_{i=1,\ldots,n}$, the joint PDF $f(X_1, \ldots, X_m)$ for each value $(x_1, \ldots, x_m)$ is given by:

$$f(x_1, \ldots, x_m) = \frac{1}{n(h_1 \ldots h_m)}\sum_{i=1}^{n} K\left(\frac{x_1 - x_{1_i}}{h_1}\right) \ldots K\left(\frac{x - x_{m_i}}{h_m}\right), \tag{5.51}$$

with kernel bandwidths $h_1, \ldots, h_m$. Kernel density estimation can be seen as a histogram-smoothing method, with a normalization constant associated with the bin size to guarantee that the resulting function is a valid PDF. The smoothness of the resulting PDF depends on the choice of the kernel and the kernel bandwidths. If the kernel bandwidth is small, the resulting distribution is noisy, whereas, if the kernel bandwidth is large, the resulting distribution is smooth.

Similar to the method in Section 5.4.3, we build a training dataset using Monte Carlo simulations. We sample n_s values from a general prior distribution and apply the rock physics model (Eq. 5.27) to obtain a set of samples of petrophysical and elastic properties $\{\boldsymbol{r}_i, \boldsymbol{m}_i\}_{i=1,\ldots,n_s}$. The prior distribution can be any parametric or non-parametric distribution. For example, we could use a Beta distribution (Section 1.4.6) for water saturation to model the high-likelihood values at the bounds of the saturation domain. The set of samples is then used to compute the joint distribution of $(\boldsymbol{r}, \boldsymbol{m})$ assuming a non-parametric PDF estimated using kernel density estimation.

If $\boldsymbol{r} = (r_1, \ldots, r_R)$ represents a multivariate random variable with R variables and $\boldsymbol{m} = (m_1, \ldots, m_M)$ represents a multivariate random variable with M variables, and $\{r_{1_i}, \ldots, r_{R_i}, m_{1_i}, \ldots, m_{M_i}\}_{i=1,\ldots,n_s}$ represent the set of n_s Monte Carlo samples, then the joint distribution $P(\boldsymbol{r}, \boldsymbol{m})$ can be estimated as:

$$P(\boldsymbol{r}, \boldsymbol{m}) = \frac{1}{n_s \prod_{u=1}^{R} h_{r_u} \prod_{v=1}^{M} h_{m_v}} \sum_{i=1}^{n_s} \prod_{u=1}^{R} K\left(\frac{r_u - r_{u_i}}{h_{r_u}}\right) \prod_{v=1}^{M} K\left(\frac{m_v - m_{v_i}}{h_{m_v}}\right), \tag{5.52}$$

where K is the kernel function and $\boldsymbol{h}_r = [h_{r_1}, \ldots, h_{r_R}]$ and $\boldsymbol{h}_m = [h_{m_1}, \ldots, h_{m_M}]$ are the vectors of kernel bandwidths of each variable. Then, the conditional distribution $P(\boldsymbol{r}|\boldsymbol{m})$ can be numerically evaluated by definition:

$$P(\boldsymbol{r}|\boldsymbol{m}) = \frac{P(\boldsymbol{r}, \boldsymbol{m})}{\int P(\boldsymbol{r}, \boldsymbol{m}) d\boldsymbol{r}}. \tag{5.53}$$

Equation (5.53) corresponds to a normalization of the joint distribution $P(\boldsymbol{r}, \boldsymbol{m})$ in Eq. (5.52), for each given value of $\boldsymbol{m}$. Because the joint and conditional distributions are numerically evaluated in a discretized domain for all the possible combinations of values of the multivariate variables $\boldsymbol{r}$ and $\boldsymbol{m}$, one of the limitations of the proposed approach is the memory requirement for a large number of variables and for fine discretizations of the variable domains.

Example 5.4 We apply the Bayesian non-linear and non-parametric inversion to the dataset in Figure 5.1 for the prediction of petrophysical properties. We first build a training dataset of 1000 samples using a Monte Carlo approach. We sample from a prior non-parametric distribution estimated from the petrophysical curves (porosity, clay volume, and water saturation) in Figure 5.5 and we apply the Raymer–Dvorkin's model (Section 2.1.1) to compute the corresponding elastic properties, including P-wave and S-wave velocity and density. We add an uncorrelated random error to the rock physics model predictions. We then estimate the six-variate joint distribution of petrophysical and elastic properties by applying the kernel density estimation approach to the Monte Carlo samples. The joint PDF is numerically computed on a discretized six-dimensional grid. The kernel bandwidths are assumed to be equal to 1/10 of the length of the domain of each property. For each sample of the elastic measurements, we then numerical evaluate the conditional distribution of petrophysical variables conditioned on elastic properties, by normalizing the joint PDF evaluated at the grid point that minimizes the L2 norm of the difference with the conditioning elastic measurement. The results are shown in Figure 5.7 and are consistent with the Bayesian linearized Gaussian mixture inversion for porosity and clay volume (Example 5.3, Figure 5.6). The posterior distribution of water saturation better captures the bimodality of the fluid spatial distribution; however, the maximum a posteriori estimate is noisier than the linearized Gaussian mixture case (Example 5.3, Figure 5.6).

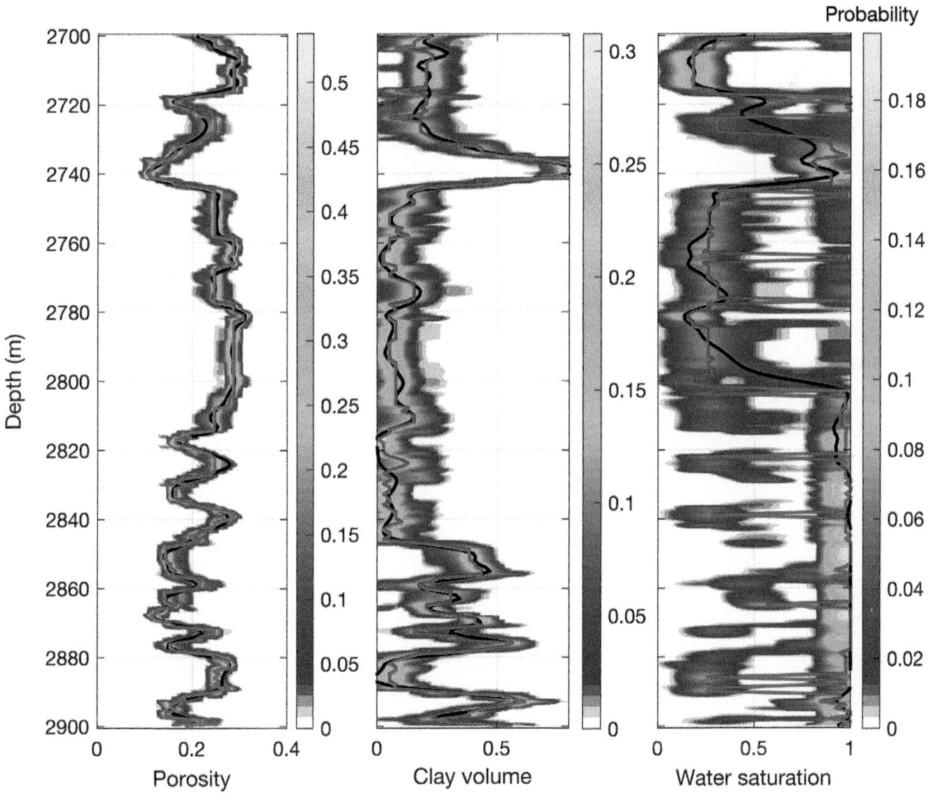

Figure 5.7 Bayesian non-linear and non-parametric rock physics inversion results: posterior distribution of porosity, clay volume, and water saturation. The background color represents the posterior non-parametric distribution and the solid red curves represent the maximum a posteriori predictions. The solid black curves represent the actual well logs.

5.5 Uncertainty Propagation

The Bayesian rock physics inversion methods presented in Section 5.4 do not account for the uncertainty in the estimation of elastic properties. The result of the Bayesian AVO linearized inversion in Section 5.3 is the posterior probability distribution of elastic properties; however, in the Bayesian rock physics inversions in Section 5.4, we only use the posterior estimation of elastic properties, i.e. the mean or the maximum a posteriori. We then propose to propagate the uncertainty from the measured seismic data to the petrophysical property estimation by combining the probability distribution obtained in the Bayesian linearized AVO inversion (Section 5.3) with the probability distribution obtained in the Bayesian rock physics inversion (Section 5.4).

From the Bayesian linearized AVO inversion, we obtain the posterior probability distribution $P(\boldsymbol{m}|\boldsymbol{d})$ of elastic properties $\boldsymbol{m}$ given seismic data $\boldsymbol{d}$. Similarly, from the Bayesian rock physics inversion, we obtain the posterior probability distribution $P(\boldsymbol{r}|\boldsymbol{m})$ of rock properties $\boldsymbol{r}$ given elastic attributes $\boldsymbol{m}$, according to Gaussian, Gaussian mixture, or non-parametric PDFs. Then, the probability distribution $P(\boldsymbol{r}|\boldsymbol{d})$ of rock properties $\boldsymbol{r}$ given seismic data $\boldsymbol{d}$ can be obtained using the Chapman Kolmogorov equation (Papoulis and Pillai 2002):

$$P(\boldsymbol{r}|\boldsymbol{d}) = \int P(\boldsymbol{r}|\boldsymbol{m})P(\boldsymbol{m}|\boldsymbol{d})d\boldsymbol{m}, \tag{5.54}$$

where we assume that $P(\boldsymbol{r}|\boldsymbol{m}) = P(\boldsymbol{r}|\boldsymbol{m}, \boldsymbol{d})$. This assumption is generally verified in seismic reservoir characterization problems, because rock properties $\boldsymbol{r}$ depend on seismic data $\boldsymbol{d}$ only through elastic attributes $\boldsymbol{m}$.

In the general case, the integral cannot be solved analytically, but must be computed numerically. The conditional probability $P(\boldsymbol{r}|\boldsymbol{m})$ must be evaluated for all the possible combinations of values of elastic properties. The numerical evaluations of the two probabilities in Eq. (5.54) can be stored in matrices: the conditional probability $P(\boldsymbol{r}|\boldsymbol{m})$ can be stored in a matrix where the rows represent all the possible combinations of values of petrophysical properties and the columns represent all the possible combinations of values of elastic properties, whereas the conditional probability $P(\boldsymbol{m}|\boldsymbol{d})$ can be stored in a matrix where the rows represent all the possible combinations of values of elastic properties and the columns represent the sequence of conditioning data along the seismic trace. The integral in Eq. (5.54) can then be computed as a matrix multiplication to obtain a numerical evaluation of the posterior PDF $P(\boldsymbol{r}|\boldsymbol{d})$. The resulting posterior PDF is not necessarily Gaussian and the statistical estimators, such as mean, mode, variance, and confidence interval, must be numerically computed. An example of application can be found in Grana and Della Rossa (2010) and in Section 8.1.1.

Example 5.5 We apply the proposed approach based on the Chapman Kolmogorov equation to the joint seismic–petrophysical inversion problem using the seismic dataset in Figure 5.2. We first solve the Bayesian linearized AVO inversion to obtain the probability distribution of the elastic properties $P(\boldsymbol{m}|\boldsymbol{d})$, as shown in Example 5.2. We then compute the rock physics likelihood function $P(\boldsymbol{r}|\boldsymbol{m})$, as shown in Examples 5.3 and 5.4; however, in this case, the likelihood function must be evaluated for all the possible combinations of values of elastic properties, whereas in Examples 5.3 and 5.4, it was computed only for the conditioning data observed in the well logs. The two probabilities are evaluated in a discretized domain; hence, the integral can be computed as a matrix multiplication. Figures 5.8 and 5.9 show the predicted posterior probability distributions $P(\boldsymbol{r}|\boldsymbol{d})$ of petrophysical properties conditioned on seismic data, assuming Gaussian mixture (Figure 5.8) and non-parametric (Figure 5.9) distributions for the rock physics likelihood function. Compared with the Bayesian linearized rock physics inversion of elastic data in Figures 5.6 and 5.7 (Examples 5.3 and 5.4), the uncertainty is generally larger and the predicted models are smoother. The water saturation prediction in the non-parametric case preserves the bimodal behavior observed in the actual data.

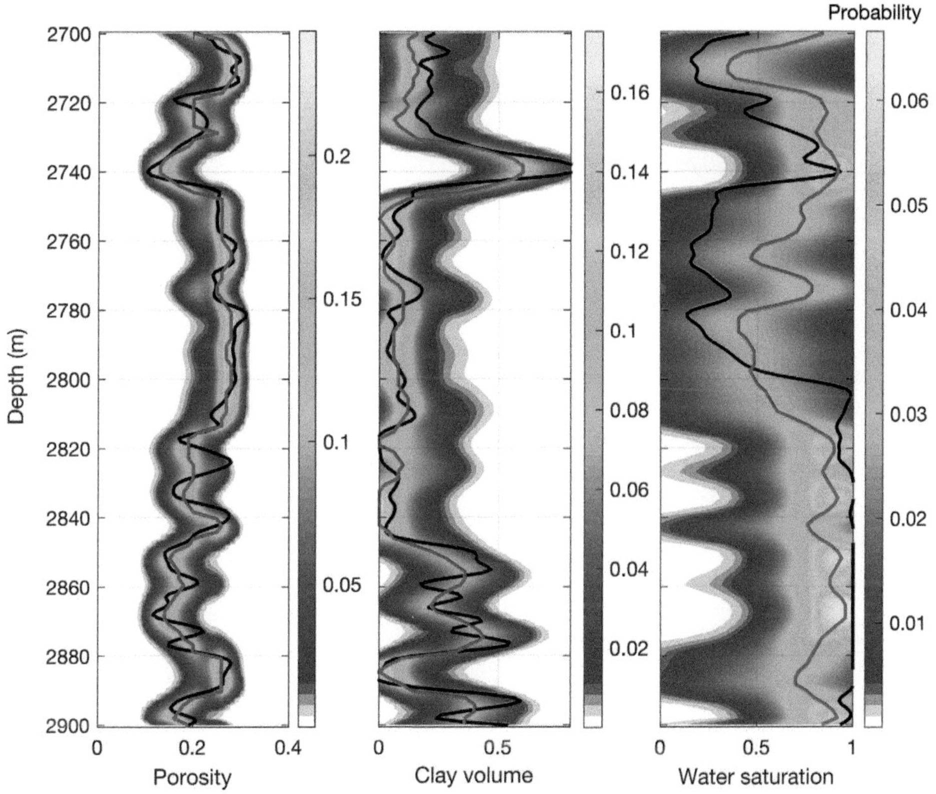

Figure 5.8 Uncertainty propagation (Gaussian mixture case): posterior distribution of porosity, clay volume, and water saturation. The background color represents the posterior non-parametric distribution and the solid red curves represent the maximum a posteriori predictions. The solid black curves represent the actual well logs.

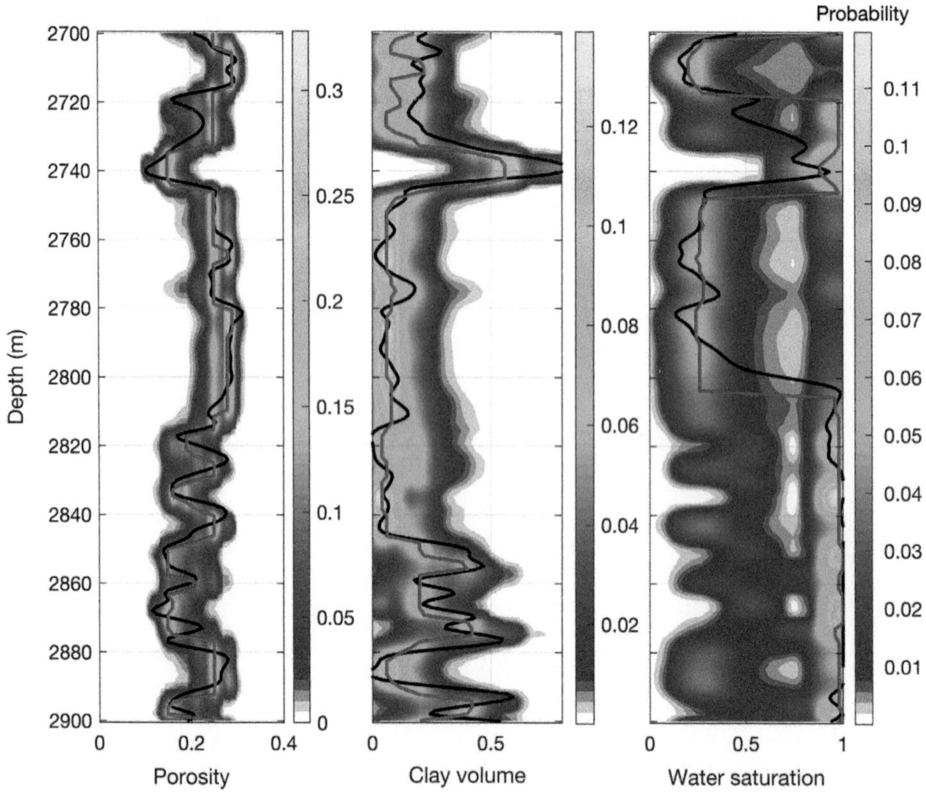

Figure 5.9 Uncertainty propagation (non-parametric case): posterior distribution of porosity, clay volume, and water saturation. The background color represents the posterior non-parametric distribution and the solid red curves represent the maximum a posteriori predictions. The solid black curves represent the actual well logs.

5.6 Geostatistical Inversion

The Bayesian methods discussed in Sections 5.3–5.5 are based on the evaluation of the posterior distribution of the model variables. However, geophysical inverse problems generally require the introduction of complex a priori information, which limits the derivation of analytical solutions. Geostatistical methods allow the integration of complex a priori information in the form of covariance functions and training images (Chapter 3). By combining geostatistical methods and inverse theory, it is possible to generate realizations from the posterior distribution of the inverse problem, including prior spatial models describing the spatial continuity of the model variables (Hansen et al. 2006; Doyen 2007).

Stochastic sampling and optimization methods are generally based on the iterative application of a forward operator to a proposed petrophysical model and the posterior distribution is numerically assessed based on a set of models that minimize the objective function of the inverse problem or maximize the likelihood function of the data. These methods are often referred to as geostatistical inversion methods because they generally require geostatistical methods to sample the initial model and stochastic methods to perturb the geostatistical realizations.

Early foundations of geostatistical inversion were laid by Doyen (1988), Bortoli et al. (1993), and Haas and Dubrule (1994). These geostatistical inversions are based on a trace-by-trace approach, using sequential Gaussian simulation (Section 3.5.1), with variogram-based two-point statistics models for the spatial continuity of the model variables. A random path is defined over all spatial locations of the seismic traces. At each location, vertical traces (often called pseudo-logs) of model properties (petrophysical and/or elastic properties) are simulated using sequential Gaussian simulation, honoring the spatial variogram model and available well log data. The simulated traces are used to compute synthetic seismograms, generally using the convolutional model (Section 5.1). An acceptance/rejection criterion based on cross-correlation between the simulated and observed seismic data is applied. If accepted, the trace of model properties is retained and is used as conditioning data in the geostatistical simulation of the model properties at the next trace location in the random path. If rejected, then the trace location is visited in the next iteration, after other locations have been simulated. The final realization of model properties honors the spatial continuity model by construction and is consistent with the seismic data within the acceptance criterion. This idea is extended to geostatistical inversions using multiple-point statistics (Section 4.5) and rock physics models in González et al. (2008), where sequential Gaussian simulation is replaced by a sequential multiple-point simulation algorithm.

In general, the seismic and petrophysical inversions in Eqs. (5.8) and (5.27) can be formulated in a joint problem as:

$$\boldsymbol{d} = \boldsymbol{f}(\boldsymbol{m}) + \boldsymbol{e} = \boldsymbol{f}(\mathbf{g}(\boldsymbol{r})) + \boldsymbol{e}, \tag{5.55}$$

where $\boldsymbol{d}$ represents the seismic data, $\boldsymbol{m}$ represents the elastic attributes, $\boldsymbol{f}$ represents the seismic forward model, $\boldsymbol{r}$ represents the petrophysical properties, $\mathbf{g}$ represents the rock physics model, and $\boldsymbol{e}$ is the observation error term. The functions $\boldsymbol{f}$ and $\mathbf{g}$ can be linear or non-linear.

Several stochastic sampling and optimization methods have been proposed to solve the inverse problem associated with Eq. (5.55) (Doyen 2007; Bosch et al. 2010). Generally, spatial models of petrophysical properties are generated by sampling from a prior distribution using geostatistical algorithms (Chapter 3). Then, the forward geophysical operator, including linear or non-linear rock physics and seismic models, is applied to compute the predicted response of the model in the seismic data domain. The solution is iteratively obtained by perturbing the initial model or its probability distribution until the predictions match the measured data. This approach can be implemented in a Monte Carlo simulation, where, at each iteration, a proposed model of petrophysical properties is accepted or rejected according to the likelihood function of the predicted data. This method is generally called Monte Carlo acceptance/rejection sampling and it is relatively easy to implement, since it only requires a geostatistical sampling algorithm and the forward operator (i.e. the geophysical model). Connolly and Hughes (2016) describe a Monte Carlo-based inversion method in which they predict facies and reservoir properties from seismic data by independently sampling the model variables from a prior distribution and accepting or rejecting the realization according to the likelihood of the predictions.

5.6.1 Markov Chain Monte Carlo Methods

Monte Carlo methods based on independent sampling are generally slow to converge for large inverse problems with spatially correlated variables. In Markov chain Monte Carlo (McMC) methods, the proposed model depends on the outcome of the previous iteration. McMC methods design a random walk in the model space using a Markov chain, i.e. a random process in which the probability of moving from the previous state to the current state is defined by a transition probability that depends only on the previous state.

A McMC method is an iterative process in which we sample a proposal solution $\boldsymbol{m}$ from a prior distribution $P(\boldsymbol{m})$, then a new sample is generated from a proposal distribution conditioned on the solution at the previous iteration and is accepted or rejected according to a likelihood criterion. If accepted the proposal solution is the new sample, otherwise the previous solution is retained. In applications in geophysical inverse problems, this process is used to draw samples from the posterior distribution of the model variables conditioned on the measured data. The sequence of samples, i.e. the chain, asymptotically reaches a stationary distribution that is the posterior distribution of the Bayesian inverse problem. There are different McMC algorithms, including Metropolis–Hastings, Metropolis, and Gibbs sampling algorithms.

The Metropolis–Hastings algorithm is the most common McMC algorithm in seismic inversion problems to compute the posterior probability $P(\boldsymbol{m}|\boldsymbol{d})$ of the model variables conditioned on the seismic data. We first sample a starting model $\boldsymbol{m}_0$ from a prior distribution $P(\boldsymbol{m})$ that is the initial state of the chain. Then, at each iteration, the model $\boldsymbol{m}_{i+1}$ is generated conditioned on the model $\boldsymbol{m}_i$ at the previous iteration i. In particular, the sampling is performed as follows: first, we draw a candidate model $\boldsymbol{m}'$ from a proposal distribution $g(\boldsymbol{m}'|\boldsymbol{m}_i)$, then we accept the candidate $\boldsymbol{m}'$ as the new model with probability P_a given by:

$$P_a = \min\left\{\frac{P(\boldsymbol{d}|\boldsymbol{m}')P(\boldsymbol{m}')g(\boldsymbol{m}_i|\boldsymbol{m}')}{P(\boldsymbol{d}|\boldsymbol{m}_i)P(\boldsymbol{m}_i)g(\boldsymbol{m}'|\boldsymbol{m}_i)}, 1\right\}. \tag{5.56}$$

If accepted, the updated model is the new proposed solution $\boldsymbol{m}_{i+1} = \boldsymbol{m}'$; otherwise, we retain the solution of the previous iteration $\boldsymbol{m}_{i+1} = \boldsymbol{m}_i$. The proposal distribution $g(\boldsymbol{m})$ is a generic PDF that is often assumed to be Gaussian. When the proposal distribution is the prior probability, the acceptance ratio reduces to the ratio of the likelihood probabilities $P(\boldsymbol{d}|\boldsymbol{m}')/P(\boldsymbol{d}|\boldsymbol{m}_i)$.

In the Metropolis algorithm, the PDF $g(\boldsymbol{m})$ is symmetric (i.e. $g(\boldsymbol{m}_i|\boldsymbol{m}') = g(\boldsymbol{m}'|\boldsymbol{m}_i)$), and the acceptance probability simply becomes:

$$P_a = \min\left\{\frac{P(\boldsymbol{d}|\boldsymbol{m}')P(\boldsymbol{m}')}{P(\boldsymbol{d}|\boldsymbol{m}_i)P(\boldsymbol{m}_i)}, 1\right\} = \min\left\{\frac{P(\boldsymbol{m}'|\boldsymbol{d})}{P(\boldsymbol{m}_i|\boldsymbol{d})}, 1\right\}; \tag{5.57}$$

therefore, if $P(\boldsymbol{m}'|\boldsymbol{d}) > P(\boldsymbol{m}_i|\boldsymbol{d})$, $\boldsymbol{m}'$ is always accepted; otherwise, $\boldsymbol{m}'$ is accepted with probability $P(\boldsymbol{m}'|\boldsymbol{d})/P(\boldsymbol{m}_i|\boldsymbol{d})$.

The Gibbs sampling is a special case of the Metropolis–Hastings algorithm and it involves a proposal from the full conditional distribution, which always has an acceptance ratio of 1, i.e. the proposal is always accepted. In practical applications, there is generally a "burn-in" time before convergence, in which early samples are rejected until the chain loses dependence on the starting model $\boldsymbol{m}_0$. It is a common practice to adopt multiple chains to increase the robustness of the posterior inference.

McMC methods have been applied to seismic and petrophysical inverse problems in Mosegaard and Tarantola (1995), Mosegaard (1998), Sambridge and Mosegaard (2002), Eidsvik et al. (2004a), and Jeong et al. (2017). De Figueiredo et al. (2019a, b) extend the application of McMC methods to joint seismic–petrophysical inversion problems for Gaussian mixture and non-parametric distributions (Section 6.3).

5.6.2 Ensemble Smoother Method

Ensemble-based methods are stochastic approaches to predict the solution of an inverse problem by updating an ensemble of prior realizations of the model variables based on the sequential or simultaneous assimilation of the measured data. Ensemble-based methods include the ensemble smoother (ES; van Leeuwen and Evensen 1996), the ensemble Kalman filter (EnKF; Evensen 2009), the ensemble randomized maximum likelihood method (EnRML; Chen and Oliver 2012), and the ensemble smoother with multiple data assimilation (ES-MDA; Emerick and Reynolds 2013). Ensemble-based methods are commonly used in reservoir engineering and production optimization problems, where they have been successfully applied to dynamic systems for fluid flow modeling. Applications of ensemble-based methods to geophysical inverse problems can be found in Liu and Grana (2018), Thurin et al. (2019), and Gineste et al. (2020). In ensemble-based methods, the prior distribution of the model parameters and the distribution of the observation errors are assumed to be Gaussian. The data assimilation corresponds to a Bayesian updating step in which the conditional distribution of the model parameters is approximated by estimating the sample conditional mean and sample covariance matrix from the ensemble of models.

We illustrate the application of the ES-MDA to the joint seismic–petrophysical inversion problem (Eq. 5.55) using a non-linear forward operator including a convolutional model of a wavelet and the full Zoeppritz equations (Zoeppritz 1919) combined with a rock physics model, as proposed in Liu and Grana (2018). The model variables are assumed to be Gaussian, otherwise a logit transformation can be applied to perform the inversion in the transformed domain and back-transform the results in the original model space. Data are assimilated multiple times with an inflated measurement error covariance matrix (Emerick and Reynolds 2013). The ES-MDA method can be summarized in the following steps:

1) First, we define the number n_e of models in the ensemble and a sequence of inflation coefficients $\{\alpha_i\}_{i=1,\dots,n_a}$ with $\sum_{i=1}^{n_a} \frac{1}{\alpha_i} = 1$, where n_a is the number of data assimilations (i.e. the number of iterations).
2) Then, we generate an ensemble of n_e prior realizations $\{\boldsymbol{m}_j^i\}_{j=1,\dots,n_e}$ of the model properties using geostatistical algorithms for continuous variables (Chapter 3), where the superscript i indicates the iteration number and it is initially set equal to 1.
3) At each iteration i (for $i = 1, \dots, n_a$), we apply a perturbation to the measured data $\boldsymbol{d}$ as follows:

$$\mathbf{d}_{p_j}^i = \mathbf{d} + \sqrt{\alpha_i}\boldsymbol{\Sigma}_e^{1/2}\mathbf{z}_{p_j}^i, \tag{5.58}$$

where $\boldsymbol{\Sigma}_e$ is the $(n_d \times n_d)$ covariance matrix of the measurement errors and the random vector $\mathbf{z}_{p_j}^i$ is distributed according to a Gaussian distribution $\mathbf{z}_{p_j}^i \sim \mathcal{N}(\mathbf{z}; \mathbf{0}, \mathbf{I}_{n_d})$, for $j = 1, ..., n_e$, with n_d being the number of data measurements and $\mathbf{I}_{n_d}$ being the $(n_d \times n_d)$ identity matrix.

4) We then apply the forward geophysical operators $\mathbf{g}$ and $\boldsymbol{f}$ to the ensemble models $\{\boldsymbol{m}_j^i\}_{j=1,...,n_e}$ to compute the predicted data $\{\mathbf{d}_j^i\}_{j=1,...,n_e}$.
5) The ensemble models $\boldsymbol{m}_j^i$ are then updated to obtain the ensemble models $\boldsymbol{m}_j^{i+1}$ according to the following expression:

$$\boldsymbol{m}_j^{i+1} = \boldsymbol{m}_j^i + \boldsymbol{\Sigma}_{m,d}^i\left(\boldsymbol{\Sigma}_{d,d}^i + \alpha_i\boldsymbol{\Sigma}_e\right)^{-1}\left(\mathbf{d}_{p_j}^i - \mathbf{d}_j^i\right), \tag{5.59}$$

for $j = 1, ..., N_e$, where $\boldsymbol{\Sigma}_{m,d}^i$ is the $(n_m \times n_d)$ cross-covariance matrix of models $\boldsymbol{m}^i$ and predicted data $\mathbf{d}^i$ (with n_m being the number of model variables), $\boldsymbol{\Sigma}_{d,d}^i$ is the $(n_d \times n_d)$ covariance matrix of the predicted data $\mathbf{d}^i$, and $\boldsymbol{\Sigma}_e$ is the $(n_d \times n_d)$ covariance matrix of the measurement errors. Equation (5.59) can be interpreted as a Bayesian updating step under the assumption that the model vector $\boldsymbol{m}$ is Gaussian; however, because the covariance matrices $\boldsymbol{\Sigma}_{m,d}^i$ and $\boldsymbol{\Sigma}_{d,d}^i$ cannot be analytically computed owing to the non-linearity of the forward operator, we approximate these covariance matrices using the sample covariance matrices estimated from the ensemble as:

$$\boldsymbol{\Sigma}_{m,d}^i = \frac{1}{n_e - 1}\sum_{j=1}^{n_e}\left(\boldsymbol{m}_j^i - \overline{\boldsymbol{m}}^i\right)\left(\mathbf{d}_j^i - \overline{\mathbf{d}}^i\right) \tag{5.60}$$

$$\boldsymbol{\Sigma}_{d,d}^i = \frac{1}{n_e - 1}\sum_{j=1}^{n_e}\left(\mathbf{d}_j^i - \overline{\mathbf{d}}^i\right)\left(\mathbf{d}_j^i - \overline{\mathbf{d}}^i\right), \tag{5.61}$$

where $\overline{\boldsymbol{m}}^i$ and $\overline{\mathbf{d}}^i$ are the ensemble means of the models and data predictions at the ith iteration. The $(n_m \times n_d)$ matrix $\mathbf{K}^i = \boldsymbol{\Sigma}_{m,d}^i\left(\boldsymbol{\Sigma}_{d,d}^i + \alpha_i\boldsymbol{\Sigma}_e\right)^{-1}$ is generally called the Kalman gain.

6) The ensemble models are iteratively updated according to steps 3–5, until the fixed number of data assimilations n_a is reached.

The algorithm updates each model of the ensemble to minimize the mismatch between the measured and predicted data; therefore, we can obtain multiple model realizations that match the data and quantify the associated uncertainty through the empirical covariance matrix of the updated ensemble members. A sensitivity analysis on the number of models is generally necessary, because small ensemble sizes might lead to severe underestimations of the uncertainty when the ensemble collapses to a single model.

Since the ES-MDA assumes a Gaussian distribution of the model parameters, for applications to petrophysical variables, we apply a logit transformation to each variable $\boldsymbol{m}$ to map the bounded values to the domain of real numbers:

$$\boldsymbol{w} = \ln\left(\frac{\boldsymbol{m} - m_{\min}}{m_{\max} - \boldsymbol{m}}\right), \tag{5.62}$$

where $m_{\min}$ is the lower bound, equal to 0 for porosity and clay volume and equal to the irreducible water saturation value for water saturation, and $m_{\max}$ is the upper bound, equal to 1 for clay volume and water saturation and equal to the critical porosity for porosity. We then apply the ES-MDA in the transformed domain and transform the results back in the petrophysical (bounded) domain using the inverse transformation:

$$\boldsymbol{m} = \frac{m_{\max}\exp(\boldsymbol{w}) + m_{\min}}{\exp(\boldsymbol{w}) + 1}, \tag{5.63}$$

as shown in Liu and Grana (2018).

Because of the dimension of geophysical data in seismic reservoir characterization studies, the required number of ensemble models for the uncertainty quantification is extremely large and the application to real datasets is unfeasible. To solve this problem, Liu and Grana (2020) propose to apply ES-MDA in a lower-dimensional data space, obtained by reparameterizing the data using the deep convolutional auto-encoder (Goodfellow et al. 2016) or other dimensionality reduction algorithms.

Example 5.6 We apply the ensemble-based inversion to the seismic dataset in Figure 5.2 to solve the joint seismic–petrophysical inverse problem. The model properties of interest are porosity, clay volume, and water saturation. The rock physics model includes the Voigt–Reuss–Hill average for the mineral properties, the stiff sand model to relate elastic moduli to porosity and clay content, and Gassmann's equations (Chapter 2) to relate elastic moduli to pore fluid saturations. The seismic model is a convolution of the wavelet and the full Zoeppritz equations for reflected P-wave amplitudes (Aki and Richards 2002).

We first generate an initial ensemble of 1000 vertical profiles of porosity, clay volume, and water saturation by sampling from the prior distribution according to an exponential vertical correlation function with parameters estimated from the well log data. The ensemble of prior models is shown in Figure 5.10.

We run the ES-MDA with four data assimilations with equal inflation factors. Figures 5.11 and 5.12 show the posterior ensemble of petrophysical properties and the corresponding elastic properties, respectively. The results are consistent with the Bayesian inversion results in Figures 5.8 and 5.9 (Example 5.5). The predicted models are slightly more accurate in terms of linear correlation between the model predictions and the actual data, whereas the uncertainty is slightly underestimated in terms of coverage ratios compared to the results of the Bayesian inversions in Example 5.5.

5.6.3 Gradual Deformation Method

The gradual deformation (GD) method is a stochastic optimization method where an initial realization of the model is gradually perturbed (Hu 2000). The GD method is based on the idea that if we generate two independent realizations, $\boldsymbol{y}_1$ and $\boldsymbol{y}_2$, of a spatially correlated Gaussian random field with $\boldsymbol{0}$ mean and the same spatial covariance model, then we can

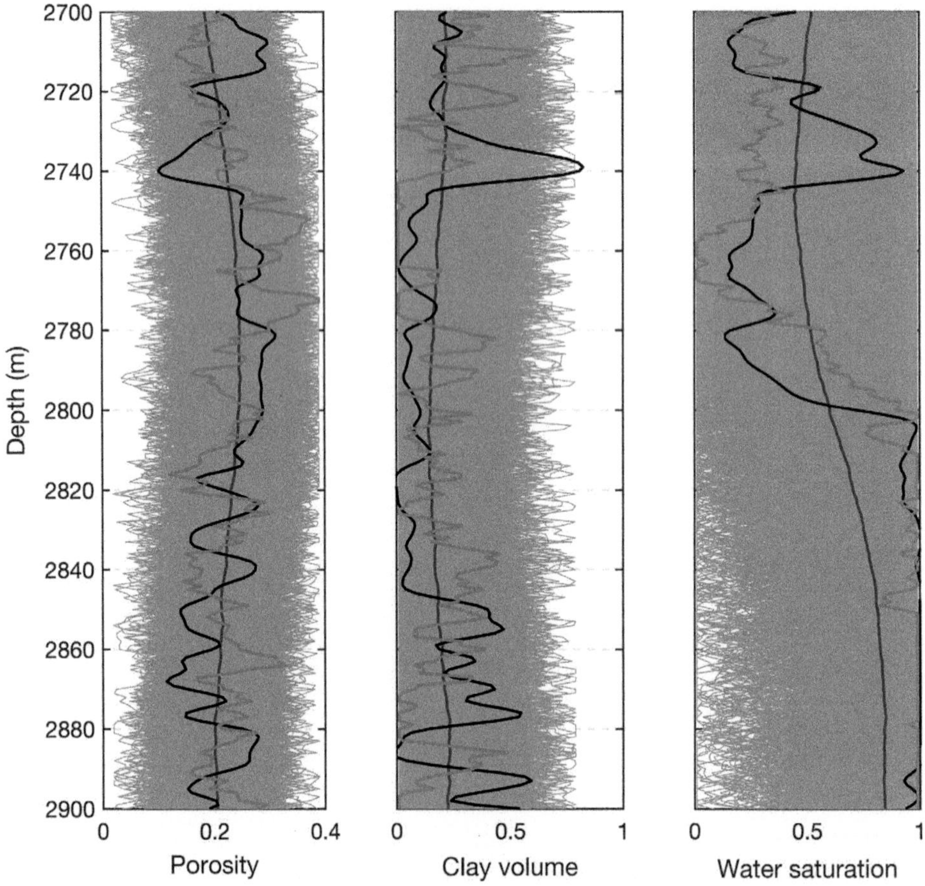

Figure 5.10 Prior ensemble of petrophysical properties: porosity, clay volume, and water saturation. The gray curves represent 1000 prior realizations, the green curves represent one random realization, and the blue curves represent the prior mean. The black curves represent the actual well logs.

generate a new Gaussian random field with mean $\boldsymbol{\mu}_y$ and the same spatial covariance model as:

$$\mathbf{y}(\theta) = \boldsymbol{\mu}_y + \mathbf{y}_1 \cos(\theta) + \mathbf{y}_2 \sin(\theta), \tag{5.64}$$

where the parameter θ belongs to the interval $[0, \pi/2]$. We can then sample multiple values of the deformation parameter θ to generate multiple realizations with mean $\boldsymbol{\mu}_y$ and the same covariance model as the two initial realizations. When $\theta = 0$, the new realization $\mathbf{y}$ coincides with $\mathbf{y}_1 + \boldsymbol{\mu}_y$, whereas, when $\theta = \pi/2$, the new realization $\mathbf{y}$ coincides with $\mathbf{y}_2 + \boldsymbol{\mu}_y$. The spatially correlated realizations $\mathbf{y}_1$ and $\mathbf{y}_2$ can be generated using the fast Fourier transform – moving average method (Section 3.6) or any other geostatistical method. One of the advantages of the GD method is that it can be integrated in an iterative stochastic optimization approach for geophysical inverse problems, by generating, at each iteration, multiple realizations as a function of the parameter θ and choosing the realization that best matches the measured data, according to an objective function. For example, the GD approach can be applied in geophysical inverse problems to stochastically perturb the initial

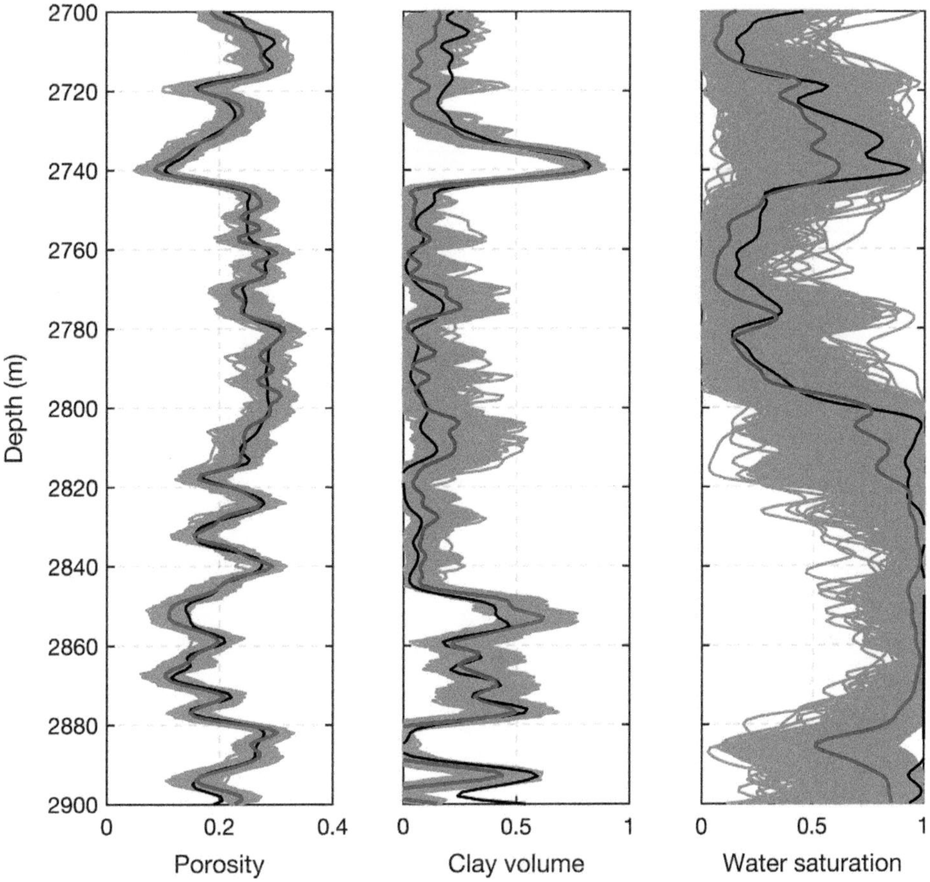

Figure 5.11 Posterior ensemble of petrophysical properties: porosity, clay volume, and water saturation. The gray curves represent 1000 posterior realizations, the red curves represent the posterior mean, and the black curves represent the actual well logs.

realization to maximize the likelihood of the data predictions or minimize the misfit between measured and predicted data (Le Ravalec 2005). The GD approach for inverse problems in the form of Eq. (5.55) can be summarized as follows:

1) We first generate two unconditional realizations $\boldsymbol{u}$ and $\boldsymbol{v}$ of the model variables, with **0** mean and the same spatial covariance function.
2) We simulate multiple realizations of the Gaussian random field $\boldsymbol{m}^i(\theta)$ with prior mean $\boldsymbol{\mu}_m$ generated using the two unconditional realizations $\boldsymbol{u}$ and $\boldsymbol{v}$ by applying the GD approach in Eq. (5.64), where the superscript i indicates the iteration number and it is initially set equal to 1.
3) We compute the data predictions $\mathbf{d}^i(\theta)$ using the geophysical forward operator in Eq. (5.55).
4) We solve a one-dimensional optimization on the deformation parameter $\theta \in [0, \pi/2]$ to select the model $\hat{\boldsymbol{m}}^i = \boldsymbol{m}^i(\hat{\theta})$ that best matches the data according to a predefined objective function, such as the L2-norm of the data residuals, $J^i(\theta) = \|\mathbf{d}^i(\theta) - \mathbf{d}\|^2$.

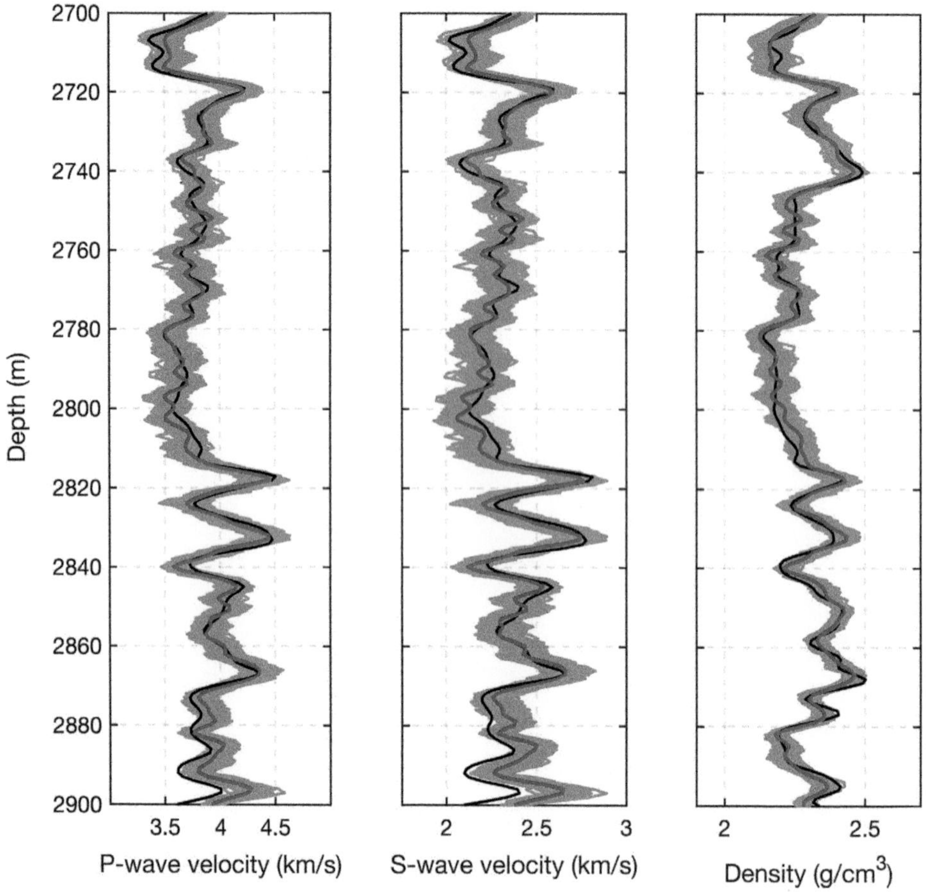

Figure 5.12 Posterior ensemble of elastic properties: P-wave velocity, S-wave velocity, and density. The gray curves represent 1000 posterior realizations, the red curves represent the posterior mean, and the black curves represent the actual well logs.

5) We generate a new unconditional realization $\boldsymbol{w}$ of the model variables, with the same spatial covariance function as the two unconditional realizations $\boldsymbol{u}$ and $\boldsymbol{v}$.
6) We simulate multiple realizations of the Gaussian random field $\boldsymbol{m}^{i+1}(\theta)$ generated as a deformation of the realizations $\hat{\boldsymbol{m}}^{i}$ and $\boldsymbol{w}$.
7) We iteratively repeat steps 3–6 until convergence.

By adopting the GD approach, we reduce the M-dimensional optimization problem, where M is the dimension of the model variables, to a sequence of one-dimensional optimizations, whose solution is a realization of the prior random field. The GD method can be combined with the fast Fourier transform – moving average simulation method (Section 3.6) to locally perturb the realization. The GD principle can be extended to more than two independent realizations.

The probability perturbation method (PPM) is another stochastic optimization approach based on the stochastic perturbation of an initial model (Caers and Hoffman 2006). This method is based on the perturbation of the probability distribution rather than the perturbation of the realizations as in the GD approach. At each step, the PPM proposes a new probability as a convex combination of the conditional probability obtained in the previous iteration and the prior probability. Because the perturbation is done in the probability domain, the PPM is more commonly applied to discrete properties (Section 6.4) where the model space has a lower dimension than the continuous case, but it can be extended to continuous variables. Similar to the GD, the PPM can then be combined with geostatistical algorithms, including two-point and multiple-point statistics methods, to sample from the proposed probability distribution (Caers and Hoffman 2006). The approach is iteratively applied until convergence.

5.7 Other Stochastic Methods

Several other stochastic optimization methods have been applied to geophysical inverse problems, including simulated annealing, particle swarm optimization, and genetic algorithms (Sen and Stoffa 2013).

Simulated annealing is one of the popular methods in optimization, owing to its efficient and simple implementation. The simulated annealing optimization is based on a Monte Carlo approach. The term annealing refers to the field of thermodynamics, specifically to the process of cooling and annealing of metals. Simulated annealing uses the objective function of an optimization problem instead of the energy of the material of the thermodynamic system. We first sample an initial model $\boldsymbol{m}_0$ from a prior distribution $P(\boldsymbol{m})$; then, at each iteration, a perturbation is applied to generate a new model $\boldsymbol{m}_{i+1}$ that is accepted or rejected according to the value of the optimization function and the acceptance probability. At each iteration, we compute the predicted data $\boldsymbol{d}_{i+1}$ of the proposed solution $\boldsymbol{m}_{i+1}$ using the forward operator of the inverse problem, and we evaluate the objective function $J_{i+1}(\boldsymbol{m}_{i+1}) = \|\boldsymbol{d}^{i+1} - \boldsymbol{d}\|^2$. If the new solution $\boldsymbol{m}_{i+1}$ improves the objective function compared with the solution $\boldsymbol{m}_i$ with objective function $J_i(\boldsymbol{m}_i)$, then the new solution is accepted; otherwise it is accepted according to a Metropolis step (Eq. 5.57) with probability P_i:

$$P_i = \min\left\{\exp\left(-\frac{\Delta J_i}{T_i}\right), 1\right\}, \tag{5.65}$$

where $\Delta J = J_{i+1}(\boldsymbol{m}_{i+1}) - J_i(\boldsymbol{m}_i)$ and T_i is a non-negative parameter. In the analogy with the thermodynamic annealing system, the objective function J represents the energy of the system, and the parameter T is the temperature. The parameter T is a non-increasing function of the iteration number, at it gradually decreases according to a predefined numerical scheme. If $J_{i+1}(\boldsymbol{m}_{i+1}) \leq J_i(\boldsymbol{m}_i)$, then $P_i = 1$, and the new model $\boldsymbol{m}_{i+1}$ is accepted. Otherwise, we randomly sample a value u from a uniform distribution $U([0, 1])$, then if $u > P_i$, $\boldsymbol{m}_{i+1}$ is accepted, otherwise it is rejected and the previous solution is kept, i.e. $\boldsymbol{m}_{i+1} = \boldsymbol{m}_i$. For large values of T, in the initial iterations, the algorithm allows an exhaustive search of the model space to avoid local minima of the objective function, whereas, when the parameter T tends to 0, in the subsequent iterations, the algorithm favors solutions that decrease the value of the objective function. Several efficient implementations have been proposed and adapted for geophysical inverse problems (Sen and Stoffa 2013).

Particle swarm optimization and genetic algorithms are stochastic optimization methods that update a population of candidate solutions according to the value of the objective function. Particle swarm optimization is a stochastic evolutionary computation technique inspired by the behavior of a population of individuals (i.e. the particles) in nature. Genetic algorithms are inspired by the natural selection process observable in biological process. Both approaches iteratively modify the population of solutions based on subsequent evaluations of the objective function associated with the inverse problem. Similar to simulated annealing, these algorithms do not require the calculation of the Jacobian of the objective function. Derivative-free stochastic optimization methods have been successfully applied in reservoir modeling and exploration geophysics (Deutsch 1992; Mallick 1995, 1999; Fernández Martínez et al. 2012; Sen and Stoffa 2013).

6

Seismic Facies Inversion

In Chapter 4, we discussed geostatistical methods to interpolate and simulate multiple realizations of discrete properties, such as facies and rock types, conditioned on the available measurements. The integration of seismic data in geostatistical methods is challenging, because of the convolutional nature of the data. Classification methods are generally applied to group subsurface rocks in facies based on measured data. These methods are clustering and pattern recognition algorithms (Hastie et al. 2009), which aim to divide input data into the desired number of facies such that samples in the same group have similar values of the measured data (Mukerji et al. 2001; Doyen 2007; Avseth et al. 2010). Reservoir facies are often estimated using Bayesian classification methods based on inverted elastic properties obtained from elastic inversion of seismic data. In the first part of this chapter, we focus on Bayesian methods for facies classification. In the second part, we discuss the application of Bayesian inverse theory to joint facies and petrophysical inversion.

6.1 Bayesian Classification

Statistical methods are often applied to facies classification problems, because they provide the most likely facies classification as well as the posterior probability that can be used to quantify the uncertainty in the classification. Bayesian classification is a non-iterative method that estimates the facies posterior probability by combining the prior information about the facies model with the likelihood function linking the available data to the facies definition.

We assume that, at a given location, a vector $\mathbf{d}$ of measurements of continuous variables, such as geophysical properties, is available, and we aim to estimate the conditional probability $P(\pi|\mathbf{d})$ and the corresponding most likely facies $\hat{\pi}$. The assessment of $P(\pi|\mathbf{d})$ can be done using Bayes' theorem (Eq. 1.8) as:

$$P(\pi = k|\mathbf{d}) = \frac{P(\mathbf{d}|\pi = k)P(\pi = k)}{P(\mathbf{d})} = \frac{P(\mathbf{d}|\pi = k)P(\pi = k)}{\sum_{h=1}^{F} P(\mathbf{d}|\pi = h)P(\pi = h)} \tag{6.1}$$

for $k = 1, \ldots, F$, where F is the number of facies. In Eq. (6.1), $P(\mathbf{d}|\pi)$ is the likelihood function that links the measured data to the facies, $P(\pi)$ is the prior model representing the prior knowledge about the facies distribution (i.e. the facies proportions), and $P(\mathbf{d})$ is a normalizing constant that ensures that $P(\pi|\mathbf{d})$ is a valid probability mass function. In statistical

Seismic Reservoir Modeling: Theory, Examples, and Algorithms, First Edition. Dario Grana, Tapan Mukerji, and Philippe Doyen.

classification literature, $P(\mathbf{d}|\pi)$ is also called the class-conditioned distribution of the features $\mathbf{d}$ used for the classification. At each location, the most likely facies $\hat{\pi}$ is then obtained by computing the maximum of the probability $P(\pi|\mathbf{d})$:

$$\hat{\pi} = \text{argmax}_{k=1,\ldots,F} P(\pi = k|\mathbf{d}) \tag{6.2}$$

over all the possible values of the facies π.

If we are only interested in predicting the most likely facies $\hat{\pi}$ conditioned on the measured data, then it is not necessary to compute the normalizing constant $P(\mathbf{d})$, since the numerator of Eq. (6.1) is proportional to $P(\pi|\mathbf{d})$ and the argument of the maximum is invariant for normalizing constant. However, if we aim to estimate the probability of occurrence of the facies, then the normalizing constant $P(\mathbf{d})$ must be computed and requires the evaluation of the marginal distribution of the data, in each facies, calculated at the measurement value.

It is common to assume that the marginal distribution of the data, in each facies, is a multivariate Gaussian distribution $P(\mathbf{d}|\pi) = \mathcal{N}\left(\mathbf{d}; \boldsymbol{\mu}_{d|\pi}, \boldsymbol{\Sigma}_{d|\pi}\right)$ where the means $\boldsymbol{\mu}_{d|\pi}$ and the covariance matrices $\boldsymbol{\Sigma}_{d|\pi}$ are facies-dependent and are often estimated from a training dataset including core measurements or borehole data. Under this assumption, the conditional probability becomes:

$$P(\pi = k|\mathbf{d}) = \frac{\mathcal{N}\left(\mathbf{d};\, \boldsymbol{\mu}_{d|k}, \boldsymbol{\Sigma}_{d|k}\right) P(\pi = k)}{\sum_{h=1}^{F} \mathcal{N}\left(\mathbf{d};\, \boldsymbol{\mu}_{d|h}, \boldsymbol{\Sigma}_{d|h}\right) P(\pi = h)} \tag{6.3}$$

for $k = 1, \ldots, F$.

We illustrate the application of Bayesian classification through a series of simple examples. The MATLAB codes for these examples are provided in the SeReM package and described in the Appendix, Section A.4.

Example 6.1 In the first example, we aim to predict the facies value at a given location where a measurement of P-wave velocity, for example $V_P = 3.8$ km/s, is available. According to the prior geological information, only two facies are present in the area, high-porosity sand and low-porosity shale, with the same prior probability of occurrence equal to 0.5. Therefore, in this example, facies is represented by a binary variable π with values π = sand or π = shale, whereas the data are represented by a continuous variable $d = V_P$. The prior probability is then $P(\pi = \text{sand}) = P(\pi = \text{shale}) = 0.5$. From a set of available measurements of core samples, we estimate that the likelihood function of P-wave velocity in sand is a Gaussian distribution $\mathcal{N}\left(V_P; \mu_1, \sigma_1^2\right)$ with mean $\mu_1 = 3.5$ km/s and standard deviation $\sigma_1 = 0.5$ km/s, whereas the likelihood function in shale is a Gaussian distribution $\mathcal{N}\left(V_P; \mu_2, \sigma_2^2\right)$ with mean $\mu_2 = 4.5$ km/s and standard deviation $\sigma_2 = 0.5$ km/s. The likelihood functions are shown in Figure 6.1. By applying Eq. (6.3), we can compute the posterior probability of sand and shale conditioned on the available measurement, $V_P = 3.8$ km/s. The posterior probability $P(\pi = \text{sand}|V_P)$ of sand is:

$$P(\pi = \text{sand}|V_P) = \frac{\mathcal{N}(3.8; 3.5, 0.5^2) \times 0.5}{\mathcal{N}(3.8; 3.5, 0.5^2) \times 0.5 + \mathcal{N}(3.8; 4.5, 0.5^2) \times 0.5}$$

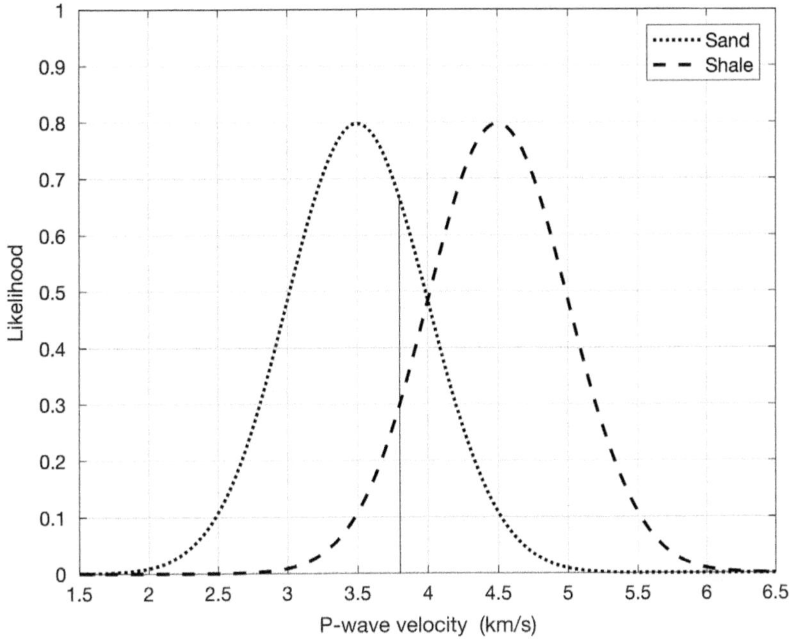

Figure 6.1 Likelihood function of P-wave velocity in sand and shale in Example 6.1: the dotted line represents the probability distribution in sand and the dashed line represents the probability distribution in shale. The vertical line shows the data measurement.

$$= \frac{0.6664 \times 0.5}{0.6664 \times 0.5 + 0.2995 \times 0.5} = 0.69$$

and the posterior probability $P(\pi = \text{shale}|V_P)$ of shale is:

$$P(\pi = \text{shale}|V_P) = \frac{0.2995 \times 0.5}{0.6664 \times 0.5 + 0.2995 \times 0.5} = 0.31 = 1 - P(\pi = \text{sand}|V_P).$$

In this case, the interpretation of the result is intuitive, because the prior proportions are equal; therefore, the posterior probability of facies is directly proportional to the likelihood function. Since, the measurement is closer to the mean of the likelihood function in sand than the mean in shale, we expect a higher posterior probability of sand than shale.

If additional geological information is available, such as a depositional model or the facies proportions in a borehole, and indicates that one of the two facies is more probable than the other, we can choose a more informative prior distribution and the posterior probabilities change accordingly. For example, if the prior probabilities are: $P(\pi = \text{sand}) = 0.8$ and $P(\pi = \text{shale}) = 0.2$, then the posterior probability $P(\pi = \text{sand}|V_P)$ of sand is:

$$P(\pi = \text{sand}|V_P) = \frac{0.6664 \times 0.8}{0.6664 \times 0.8 + 0.2995 \times 0.2} = 0.899$$

and the posterior probability $P(\pi = \text{shale}|V_P)$ of shale is:

$$P(\pi = \text{shale}|V_P) = \frac{0.2995 \times 0.2}{0.6664 \times 0.8 + 0.2995 \times 0.2} = 0.101.$$

If the prior probabilities are: $P(\pi = \text{sand}) = 0.2$ and $P(\pi = \text{shale}) = 0.8$, then the posterior probability of sand is $P(\pi = \text{sand}|V_P) = 0.3575$ and the posterior probability of shale is $P(\pi = \text{shale}|V_P) = 0.6425$.

Example 6.2 In this example, in addition to the data in Example 6.1, we introduce another rock property, namely density, and we aim to predict the facies value given two available measurements of P-wave velocity and density: $V_P = 3.8$ km/s and $\rho = 2.32$ g/cm^3. The likelihood function, for each facies, is a bivariate Gaussian distribution with means $\mu_1 = [3.5, 2.3]^T$ and $\mu_2 = [4.5, 2.5]^T$ in sand and shale, respectively. The variance of P-wave velocity is $\sigma^2_{V_P} = 0.25$ and the variance of density is $\sigma^2_\rho = 0.01$, in both facies. The correlation between P-wave velocity and density is 0.7 in sand and 0.8 in shale. We assume a lower correlation in sand to account for porosity changes and fluid effects. The likelihood functions are shown in Figure 6.2. The model parameters are estimated from a set of borehole measurements. We assume that the facies prior probabilities are $P(\pi = \text{sand}) = P(\pi = \text{shale}) = 0.5$. By applying Eq. (6.3) with the bivariate Gaussian likelihood functions in Figure 6.2, we compute the posterior probability of sand and shale conditioned on the measurements of P-wave velocity and density. The posterior probability $P(\pi = \text{sand}|V_P, \rho)$ of sand is:

$$P(\pi = \text{sand}|V_P, \rho) = \frac{3.5505 \times 0.5}{3.5505 \times 0.5 + 1.0476 \times 0.5} = 0.77$$

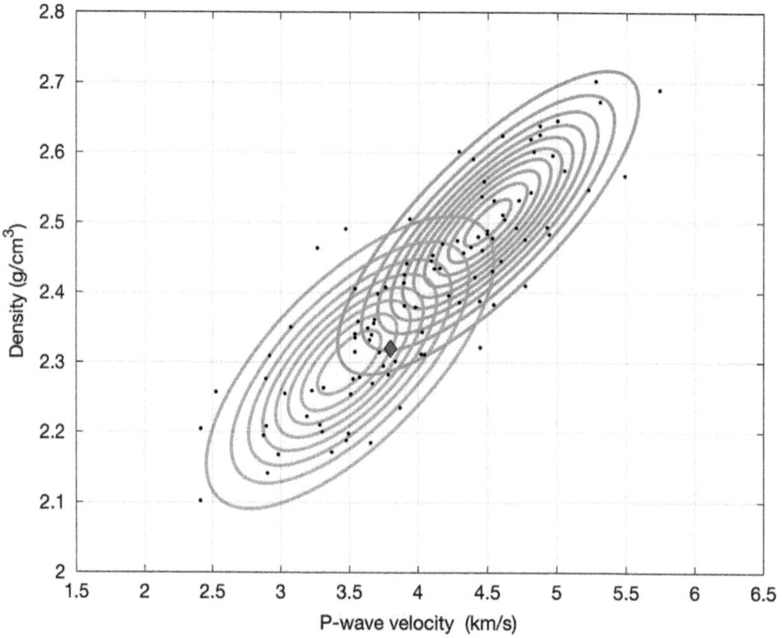

Figure 6.2 Contour plots of multivariate Gaussian likelihood functions of P-wave velocity and density, in each facies: sand in yellow and shale in green. The black dots represent the training dataset used to estimate the parameters of the Gaussian distributions. The red diamond represents the conditioning data measurement.

and the posterior probability $P(\pi = \text{shale}|V_P, \rho)$ of shale is:

$$P(\pi = \text{shale}|V_P, \rho) = \frac{1.0476 \times 0.5}{3.5505 \times 0.5 + 1.0476 \times 0.5} = 0.23.$$

If the prior probabilities are: $P(\pi = \text{sand}) = 0.25$ and $P(\pi = \text{shale}) = 0.75$, then the posterior probability of sand is $P(\pi = \text{sand}|V_P, \rho) = 0.53$ and the posterior probability of shale is $P(\pi = \text{shale}|V_P, \rho) = 0.47$ owing to the high likelihood of the measurement belonging to the sand facies.

Example 6.3 This example is similar to Example 6.2, but in this case, we assume that the available measurements do not follow a multivariate Gaussian distribution; therefore, we adopt a non-parametric approach for the estimation of the conditional probability $P(\mathbf{d}|\pi)$. We propose to estimate the distribution $P(\mathbf{d}|\pi = k)$ for $k = 1, \ldots, F$ using kernel density estimation (Silverman 1986; Bowman and Azzalini 1997) in a multidimensional domain. We consider two random variables $d_1 = V_P$ and $d_2 = \rho$, and adopt the formulation in Section 5.4.4 in each facies. The training dataset is the same set of borehole measurements as in Example 6.2. The estimated bivariate distributions are shown in Figure 6.3, for sand and shale. We assume that the prior probabilities are $P(\pi = \text{sand}) = P(\pi = \text{shale}) = 0.5$. Then, the posterior probability of sand is $P(\pi = \text{sand}|V_P, \rho) = 0.75$ and the posterior probability of shale is $P(\pi = \text{shale}|V_P, \rho) = 0.25$. The posterior probabilities are consistent with

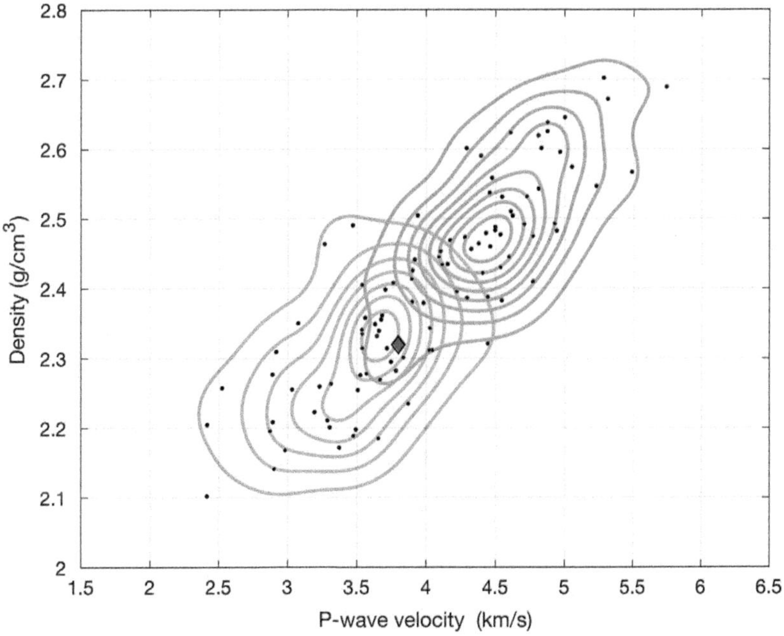

Figure 6.3 Contour plots of multivariate non-parametric likelihood functions of P-wave velocity and density in each facies: sand in yellow and shale in green. The black dots represent the training dataset used to estimate the non-parametric distributions using the kernel density estimation method. The red diamond represents the conditioning data measurement.

the results in Example 6.2, since the training dataset approximately follows Gaussian distributions in each facies (Figure 6.2).

Example 6.4 In this example, we extend the classification method in Example 6.2 to a set of multiple measurements at different depth locations and apply the Bayesian facies classification based on multivariate Gaussian distributions to a set of borehole measurements sampled every 1 m in an interval of 100 m. The available well logs at the borehole location include measurements of P-wave velocity and density as well as a reference facies classification (Figure 6.4). At each location, we assume uniform prior probabilities and the likelihood functions in Figure 6.2. The Bayesian classification (Eq. 6.3) is applied sample by sample to compute the posterior probability of sand and shale at each location in the interval. Then, the most likely value, i.e. the facies with the maximum probability, is predicted at each location. The posterior probabilities and the most likely facies profile are shown in Figure 6.4. Sand layers correspond to the subintervals with low P-wave velocity and low

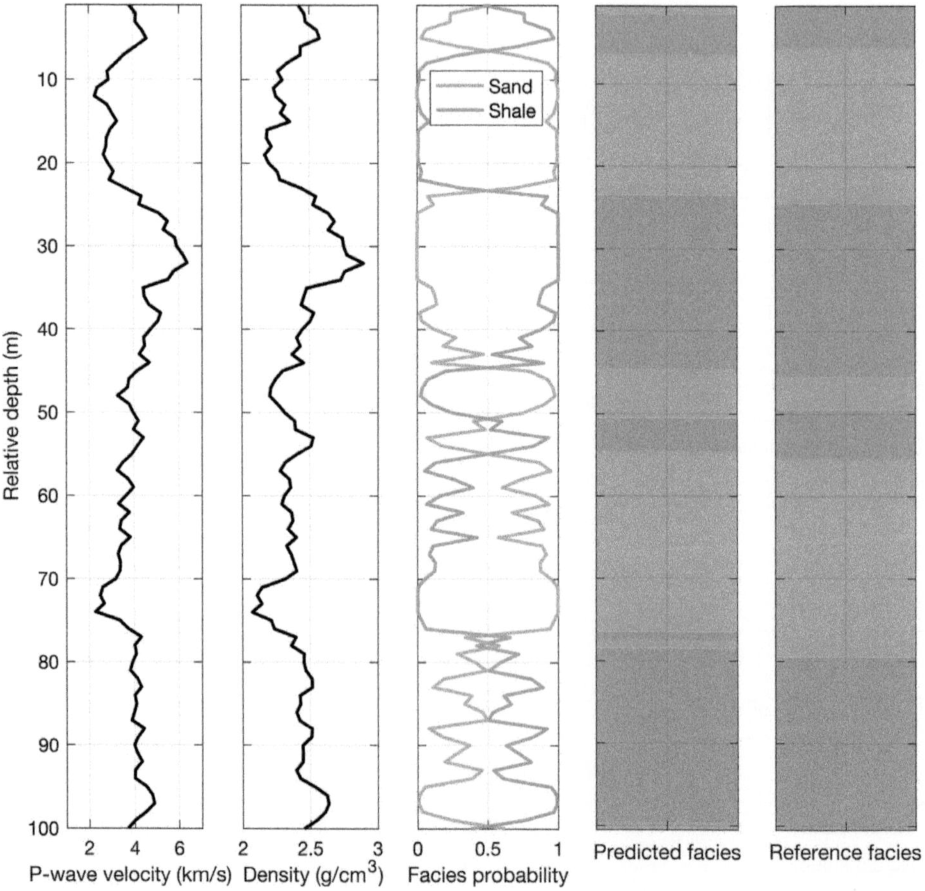

Figure 6.4 Bayesian facies classification along a borehole dataset, from left to right: P-wave velocity, density, facies posterior probability, predicted facies (sand in yellow and shale in green), and reference facies profile.

density, whereas shale layers correspond to the subintervals with high P-wave velocity and high density. The posterior probabilities of the two facies show a bimodal behavior within the thick layers, whereas they tend to be similar near the layer interfaces, owing to the gradual transitions of geophysical properties from one facies to another and to the data resolution.

The accuracy and precision of the facies classification can be quantified using contingency analysis, based on the absolute frequencies, reconstruction rate, recognition rate, and estimation index. In the absolute frequency table (also known as classification confusion matrix), we count the number of actual samples classified in the predicted facies (Table 6.1). In Example 6.4, 91 samples are classified in the correct facies, 43 in sand, and 48 in shale, whereas 9 samples are misclassified. The overall facies proportions are correctly reproduced, as we predict a posterior probability of sand equal to 0.48 and the actual sand fraction is 0.47. The reconstruction rate is obtained by normalizing the frequency table per row and it represents the fraction of samples of an actual facies that are classified in each predicted facies (Table 6.2). Sand samples in the actual classification are correctly classified in the predicted sand with probability 0.9149, whereas the probability of classifying actual sand samples in shale is 0.0851. The recognition rate is obtained by normalizing per columns and it represents the fraction of samples of a predicted facies that belong to each actual facies (Table 6.3). Predicted sand samples belong to the correct facies with probability

Table 6.1 Absolute frequencies of the facies classification.

	Predicted sand	Predicted shale	
Actual sand	43	4	47
Actual shale	5	48	53
	48	52	100

Table 6.2 Reconstruction rate of the facies classification.

	Predicted sand	Predicted shale	
Actual sand	0.9149	0.0851	1
Actual shale	0.0943	0.9057	1

Table 6.3 Recognition rate of the facies classification.

	Predicted sand	Predicted shale
Actual sand	0.8958	0.0769
Actual shale	0.1042	0.9231
	1	1

Table 6.4 Estimation index of the facies classification.

	Predicted sand	Predicted shale
Actual sand	0.0191	0.0082
Actual shale	−0.0098	−0.0174

0.8958, whereas the probability that predicted sand samples actually belong to shale is 0.1042. The estimation index is defined as the difference between the reconstruction rate and recognition rate (Table 6.4). On the main diagonal, a negative estimation index indicates underestimation, and a positive estimation index indicates overestimation, whereas the off-diagonal terms describe the misclassifications. A perfect classification corresponds to reconstruction and recognition matrices equal to the identity matrix and the estimation index matrix equal to the null matrix.

In many practical applications, the resolution of the available data might limit the ability to discriminate between different facies. In these cases, it is common to group different facies in a smaller number of broadly defined groups. For example, the number of facies that can be discriminated at the seismic scale is often smaller than the initial number of facies defined based on core samples.

The proposed Bayesian classification methods are often applied sample by sample, i.e. independently at adjacent measurements, as in Example 6.4. However, as discussed in Chapter 4, sampling should not be performed independently at each location, owing to the spatial correlation imposed by geological continuity. Spatially correlated sampling of discrete and continuous properties conditioned on seismic data is challenging owing to the convolutional nature of seismic amplitudes. Because of the complexity of the spatial correlation model of discrete and continuous properties, fully analytical solutions of the general Bayesian inverse problem are not available and the solution, i.e. the posterior distribution, must be numerically assessed. We then propose two approaches for mixed discrete and continuous problems, combining inverse theory and geostatistics.

6.2 Bayesian Markov Chain Gaussian Mixture Inversion

We formulate the joint inversion of seismic data for the prediction of facies and reservoir properties in a Bayesian setting. In this section, reservoir properties can refer to elastic or petrophysical properties. In the case of elastic properties, the forward operator is the seismic model, whereas, in the case of petrophysical properties, the forward operator includes both the seismic and the rock physics model. Therefore, the methodologies discussed in this chapter can be applied to both seismic and petrophysical inversion.

The distribution of the continuous model properties $\boldsymbol{m}$ depends on the underlying facies distribution $\boldsymbol{\pi}$ and it can be predicted from the measured seismic data $\boldsymbol{d}$. The forward problem can be formulated as:

$$\boldsymbol{d} = \mathbf{g}(\boldsymbol{m}(\boldsymbol{\pi})) + \boldsymbol{\varepsilon}, \tag{6.4}$$

where $\boldsymbol{\varepsilon}$ is the measurement error and $\mathbf{g}$ is the forward geophysical operator, and where for simplicity, we omitted the spatial coordinates.

The goal in the joint inversion of facies and reservoir properties is to assess the posterior probabilities $P(\boldsymbol{m}|\boldsymbol{d})$ and $P(\boldsymbol{\pi}|\boldsymbol{d})$ in a Bayesian setting. We first present the solution proposed in Grana et al. (2017a) where the prior model is a Gaussian mixture distribution for the continuous properties (Section 1.4.5) and a stationary first-order Markov chain model for the discrete property (Section 4.4). We assume that the forward operator is linear or that a linearization is available. The approach in Grana et al. (2017a) provides a semi-analytical solution, since the posterior distribution $P(\boldsymbol{m}|\boldsymbol{d})$ can be expressed in a closed form with analytical expressions for the means and the covariance matrices, whereas the posterior distribution $P(\boldsymbol{\pi}|\boldsymbol{d})$ must be numerically assessed using a Markov chain Monte Carlo (McMC) algorithm.

The formulation of the prior and posterior distributions is similar to the Bayesian rock physics inversion in Section 5.4.2, where the facies prior probability $P(\boldsymbol{\pi})$ and the facies posterior probability $P(\boldsymbol{\pi}|\boldsymbol{d})$ represent the weights of the prior and posterior probabilities of the reservoir properties $P(\boldsymbol{m})$ and $P(\boldsymbol{m}|\boldsymbol{d})$, respectively. However, in the approach proposed in Grana et al. (2017a), the weights are not calculated independently at each location as in Grana and Della Rossa (2010), but depend on the prior spatial correlation model.

We assume that the prior distribution $P(\boldsymbol{m})$ of the continuous reservoir properties is a Gaussian mixture model:

$$P(\boldsymbol{m}) = \sum_{\boldsymbol{\pi}\in\Omega_\pi^M} P(\boldsymbol{\pi})P(\boldsymbol{m}|\boldsymbol{\pi}) = \sum_{\boldsymbol{\pi}\in\Omega_\pi^M} P(\boldsymbol{\pi})\mathcal{N}\left(\boldsymbol{m};\boldsymbol{\mu}_{m|\pi},\boldsymbol{\Sigma}_{m|\pi}\right), \tag{6.5}$$

where $\boldsymbol{\pi}$ is a vector of M samples representing the facies vertical profile whose samples take values in $\Omega_\pi = \{1, ..., F\}$, Ω_π^M is the model domain of the facies vector $\boldsymbol{\pi}$, $\boldsymbol{\mu}_{m|\pi}$ is a vector of M samples of the prior conditional means of the reservoir properties, and $\boldsymbol{\Sigma}_{m|\pi}$ is the prior conditional covariance matrix. Differently from the approach proposed in Grana and Della Rossa (2010), in the prior model in Eq. (6.5), $\boldsymbol{\pi}$ represents the entire vertical facies profile. Grana et al. (2017a) assume that the reservoir properties are spatially correlated; therefore, the prior covariance matrix $\boldsymbol{\Sigma}_{m|\pi}$ is obtained by combining the spatially independent covariance matrix of the reservoir properties in each facies with a spatial correlation matrix obtained from a spatial correlation function that represents a model with full correlation between the residuals in all the Gaussian components of the mixture. For example, the prior covariance matrix $\boldsymbol{\Sigma}_{m|\pi}$ can be defined as the Kronecker product $\boldsymbol{\Sigma}_{m|\pi} = \boldsymbol{\Sigma}_{0|\pi} \otimes \boldsymbol{\Sigma}_{z|\pi}$ of the (space-invariant) covariance matrix of the model properties in each facies $\boldsymbol{\Sigma}_{0|\pi}$ and the space-dependent correlation matrix $\boldsymbol{\Sigma}_{z|\pi}$ defined by the correlation function $\nu(z)$. The correlation function $\nu(z)$ can be the same in each facies or facies-dependent.

The prior model of the facies is defined by a stationary first-order Markov chain:

$$P(\boldsymbol{\pi}) = P(\pi_1)\prod_{k=2}^{M} P(\pi_k|\pi_{k-1}), \tag{6.6}$$

with transition matrix $\mathbf{T}$ of dimensions $(F \times F)$ (Section 4.4), where $P(\pi_1)$ is the stationary distribution associated with $\mathbf{T}$ and represents the prior proportions of the facies.

If the linear forward operator $\mathbf{g}$ is associated with the matrix $\mathbf{G}$, and if the measurement errors $\boldsymbol{\varepsilon}$ are Gaussian distributed $\boldsymbol{\varepsilon} \sim \mathcal{N}(\boldsymbol{\varepsilon};\mathbf{0},\boldsymbol{\Sigma}_\varepsilon)$ with $\mathbf{0}$ mean and covariance matrix $\boldsymbol{\Sigma}_\varepsilon$, and independent of the model $\boldsymbol{m}$, then the posterior distribution of the reservoir properties $P(\boldsymbol{m}|\boldsymbol{d})$ of the Bayesian inverse problem in Eq. (6.4) is a Gaussian mixture model of the form:

$$P(\boldsymbol{m}|\boldsymbol{d}) = \sum_{\boldsymbol{\pi}\in\Omega_{\pi}^{M}} P(\boldsymbol{\pi}|\boldsymbol{d})P(\boldsymbol{m}|\boldsymbol{d},\boldsymbol{\pi}) = \sum_{\boldsymbol{\pi}\in\Omega_{\pi}^{M}} P(\boldsymbol{\pi}|\boldsymbol{d})\mathcal{N}\left(\boldsymbol{m};\boldsymbol{\mu}_{m|d,\pi},\boldsymbol{\Sigma}_{m|d,\pi}\right). \tag{6.7}$$

The result in Eq. (6.7) can be extended to any mixture model, including non-Gaussian mixtures. Indeed, Grana et al. (2017a) show that if the prior is a mixture distribution of some parametric PDFs, then, the posterior distribution is a mixture distribution of the posterior PDFs of the same type as the prior PDFs. In the Gaussian mixture case, the conditional means and covariance matrices of the Gaussian components can be analytically computed as:

$$\boldsymbol{\mu}_{m|d,\pi} = \boldsymbol{\mu}_{m|\pi} + \boldsymbol{\Sigma}_{m|\pi}\mathbf{G}^{T}\left(\mathbf{G}\boldsymbol{\Sigma}_{m|\pi}\mathbf{G}^{T} + \boldsymbol{\Sigma}_{\varepsilon}\right)^{-1}\left(\boldsymbol{d} - \mathbf{G}\boldsymbol{\mu}_{m|\pi}\right), \tag{6.8}$$

$$\boldsymbol{\Sigma}_{m|d,\pi} = \boldsymbol{\Sigma}_{m|\pi} - \boldsymbol{\Sigma}_{m|\pi}\mathbf{G}^{T}\left(\mathbf{G}\boldsymbol{\Sigma}_{m|\pi}\mathbf{G}^{T} + \boldsymbol{\Sigma}_{\varepsilon}\right)^{-1}\mathbf{G}\boldsymbol{\Sigma}_{m|\pi}, \tag{6.9}$$

as shown Section 5.4.2, whereas for other mixtures PDFs, the posterior parameters have to be numerically computed.

The posterior probability of the facies profile $P(\boldsymbol{\pi}|\boldsymbol{d})$ is given by:

$$P(\boldsymbol{\pi}|\boldsymbol{d}) = \frac{P(\boldsymbol{d}|\boldsymbol{\pi})P(\boldsymbol{\pi})}{P(\boldsymbol{d})}; \tag{6.10}$$

however, the probability $P(\boldsymbol{\pi}|\boldsymbol{d})$ in Eq. (6.10) cannot be exactly computed because the normalizing constant requires the evaluation of the integral of the numerator on the domain Ω_{π}^{M}, whose dimension is generally very large. For this reason, Grana et al. (2017a) propose a reliable approximation $P_k^*(\boldsymbol{\pi}|\boldsymbol{d})$ of the posterior distribution parameterized by an integer k, as:

$$P_k^*(\boldsymbol{\pi}|\boldsymbol{d}) \propto P_k^*(\boldsymbol{d}|\boldsymbol{\pi})P_k(\boldsymbol{\pi}). \tag{6.11}$$

The approximated probability $P_k^*(\boldsymbol{d}|\boldsymbol{\pi})$ describes the likelihood of the seismic data $\boldsymbol{d}$ conditioned on a facies sequence of length k, whereas the term $P_k(\boldsymbol{\pi})$ reformulates the prior model as a k-order Markov chain. The resulting approximation is a non-stationary k-order Markov chain and it can be exactly and efficiently calculated by the recursive forward–backward algorithm (Fjeldstad and Grana 2018). Accurate approximations can be obtained with small values of the integer k. In the approach proposed by Grana et al. (2017a), the authors use the approximation in Eq. (6.11) as a proposal distribution in a McMC approach, based on the Metropolis–Hastings algorithm (Section 5.6.1), to efficiently generate realizations of the posterior distribution $P(\boldsymbol{\pi}|\boldsymbol{d})$ in Eq. (6.10). Hence, realizations of the posterior distributions $P(\boldsymbol{m}|\boldsymbol{d})$ can also be generated using Eq. (6.7).

Example 6.5 We apply the Bayesian Markov chain Gaussian mixture inversion methodology to a borehole dataset in a clastic oil reservoir in the Norwegian Sea (Figure 6.5). The main reservoir layer, located in the upper part of the interval under study, shows an average porosity of 0.26 and it is partially saturated with oil with average irreducible water saturation of approximately 0.10. The reservoir is approximately 20 m thick and it is embedded

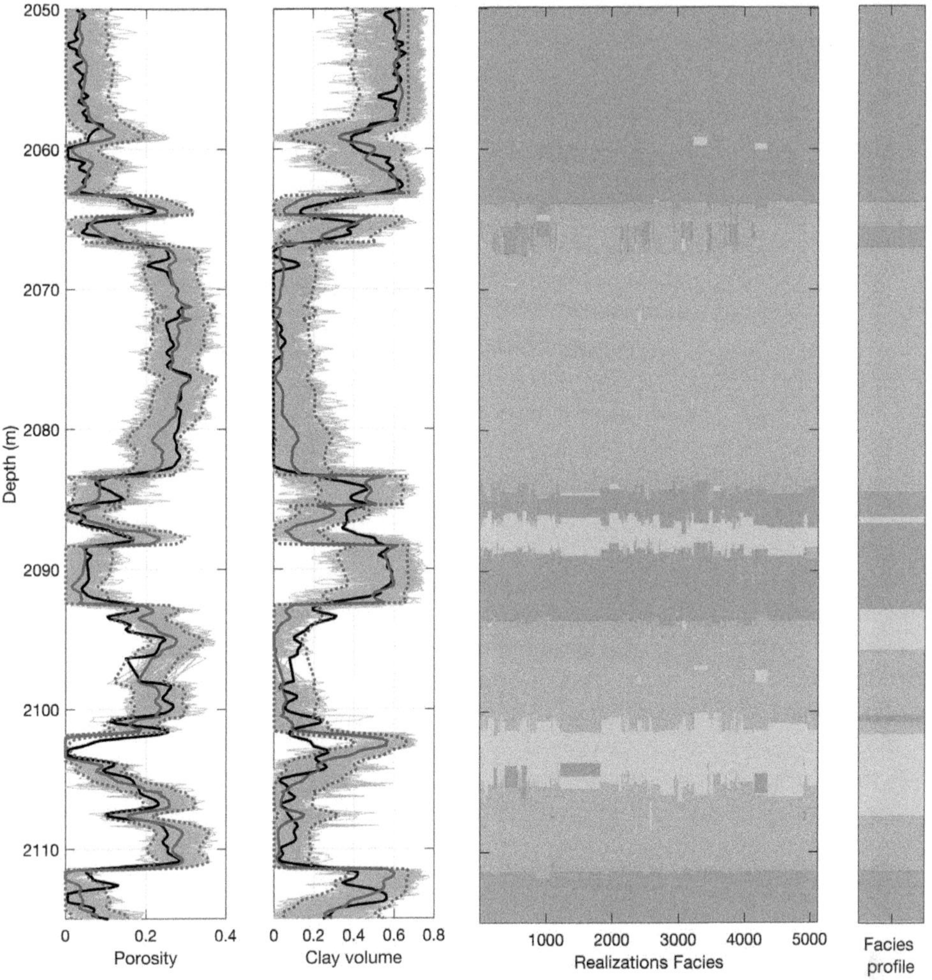

Figure 6.5 Bayesian Markov chain Gaussian mixture seismic inversion for facies, porosity, and volume of clay: posterior distribution of porosity and clay volume (the black lines represent the well logs, the solid red lines represent the maximum a posteriori, the dotted red lines represent the 0.80 confidence intervals, and the gray lines represent the posterior realizations); and 5000 facies realizations compared with the actual facies profile (shale in green, silt in gray, sand in yellow).

into two impermeable shaley layers. The lower part of the interval includes a sequence of alternating layers of sand, silt, and shale. Two synthetic seismograms corresponding to the near and far angles are available at the borehole location. We then predict the posterior distribution of facies and petrophysical properties, namely porosity and clay volume, conditioned on the seismic data.

We assume a stationary first-order Markov chain for the facies and a bivariate Gaussian mixture with three components (corresponding to sand, silt, and shale) for porosity and clay volume. The prior transition probabilities of the facies profile and the prior parameters of the Gaussian distributions are estimated from the actual borehole data, using a frequentist approach based on the actual facies profile. The prior linear correlation of porosity and clay volume is assumed to be $\rho = -0.8$, in all three facies. We assume an exponential correlation function with the same correlation length of 10 m for porosity and clay volume. The rock physics model is a multilinear regression that maps porosity and clay volume into P- and S-impedance with coefficients estimated from the well log data, and the seismic model is a convolutional model of a wavelet and a two-term linearized approximation of Zoeppritz equations. A modified logit transformation is applied to transform porosity and clay volume to two unbounded variables defined in the domain $(-\infty, +\infty)$.

Figure 6.5 shows the posterior distributions of facies and petrophysical properties and a subset of 5000 posterior realizations. The total number of proposed realizations is 10^6 with an acceptance ratio of approximately 0.10, assuming a truncation parameter $k = 3$ for the approximation in Eq. (6.11). The maximum a posteriori estimates of porosity and clay volume provide accurate predictions of the petrophysical properties. The posterior realizations follow the vertical correlation function and show multimodal marginal distributions. The facies realizations match the actual facies profile with a success rate of 0.85, despite some misclassifications of the silt layers, possibly due to misalignments in the well to seismic tie in sub-seismic resolution layers. The correlation between the maximum a posteriori and the actual measurements is 0.87 for porosity and 0.84 for clay volume. The coverage ratio of the 0.80 confidence interval is 0.88 for porosity and 0.86 for clay volume, showing a slight overestimation of the posterior uncertainty. Grana et al. (2017a) propose a comparison between the Gaussian mixture and the Gaussian models and show an improvement of the accuracy and precision of the results in the Gaussian mixture case.

In Example 6.5, the same linear rock physics model is assumed for all facies. However, the proposed method can be applied to facies-dependent rock physics models as well as non-linear relations. In the non-linear case, the computational cost of the likelihood evaluation increases compared with the linearized case. Fjeldstad and Grana (2018) propose an extension of the methodology including a lateral correlation model for the discrete property.

6.3 Multimodal Markov Chain Monte Carlo Inversion

The Bayesian Gaussian mixture inversion method for spatially correlated facies and reservoir properties requires the implementation of a McMC approach for the evaluation of the posterior probability of the facies. McMC methods have been applied in several geophysical applications (Mosegaard and Tarantola 1995; Mosegaard 1998; Sambridge and Mosegaard 2002; Eidsvik et al. 2004a; Zunino et al. 2014; Jeong et al. 2017). The use of McMC algorithms for Gaussian mixture random fields is challenging owing to the spatial correlation of the reservoir properties, which causes the number of components of the mixture distribution to exponentially increase with the number of samples in the data.

De Figueiredo et al. (2019a, b) propose a McMC method based on mixture distributions to sample from a high-dimensional multimodal posterior distribution, where each mode is associated with a specific facies configuration along the entire seismic trace. The sampling algorithm includes two steps, namely the "jump move" and "local move," for the update of the facies and continuous property realizations. The jump or local moves are chosen according to a predefined probability ϵ. In the jump move, we propose a new realization of facies and reservoir properties, whereas in the local move the facies realization of the previous iteration is preserved and only the continuous properties are updated conditioned on the facies. We first describe the Gaussian mixture case and then we extend the approach to the general case of any mixture of PDFs.

In the Gaussian mixture case (de Figueiredo et al. 2019a), the prior and posterior distributions are the same as in Eqs. (6.5) and (6.7). At each iteration i, given the previous realizations of facies $\boldsymbol{\pi}^{i-1}$ and continuous properties $\boldsymbol{m}^{i-1}$, the facies configuration $\boldsymbol{\pi}^i$ is updated according to either the jump move or the local move. In the jump move, we propose a new facies realization $\boldsymbol{\pi}^{new}$ using the Metropolis algorithm where $\boldsymbol{\pi}^{new}$ is accepted with probability r_{jump}:

$$r_{jump} = \min\left\{\frac{P(\boldsymbol{d}|\boldsymbol{\pi}^{new})P(\boldsymbol{\pi}^{new})}{P(\boldsymbol{d}|\boldsymbol{\pi}^{i-1})P(\boldsymbol{\pi}^{i-1})}, 1\right\}, \tag{6.12}$$

where $\boldsymbol{\pi}^{i-1}$ is the facies realization of the previous iteration $i-1$. In the local move, we preserve the facies realization $\boldsymbol{\pi}^i = \boldsymbol{\pi}^{i-1}$ obtained in the previous iteration. In both cases, the continuous property realization $\boldsymbol{m}^i$ is updated using a Gibbs sampling algorithm, where a new realization of the model properties is sampled from the Gaussian distribution $\mathcal{N}\left(\boldsymbol{m}; \boldsymbol{\mu}_{m|d,\pi^i}, \boldsymbol{\Sigma}_{m|d,\pi^i}\right)$ and it is always accepted.

In the general case (de Figueiredo et al. 2019b), the prior and posterior distributions are mixtures of general distributions, either parametric or non-parametric PDFs. In the jump move, given the previous realizations of facies $\boldsymbol{\pi}^{i-1}$ and continuous properties $\boldsymbol{m}^{i-1}$, we sample new realizations $\boldsymbol{\pi}^{new}$ and $\boldsymbol{m}^{new}$ from a proposal distribution $q(\boldsymbol{\pi}, \boldsymbol{m})$ and accept the new configuration according to the Metropolis–Hastings algorithm with probability r_{jump}:

$$r_{jump} = \min\left\{\frac{P(\boldsymbol{d}|\boldsymbol{m}^{new})P(\boldsymbol{m}^{new})q(\boldsymbol{\pi}^{i-1}, \boldsymbol{m}^{i-1})}{P(\boldsymbol{d}|\boldsymbol{m}^{i-1})P(\boldsymbol{m}^{i-1})q(\boldsymbol{\pi}^{new}, \boldsymbol{m}^{new})}, 1\right\}. \tag{6.13}$$

If the proposal distribution $q(\boldsymbol{\pi}, \boldsymbol{m})$ is the prior distribution, then the acceptance probability r_{jump} becomes:

$$r_{jump} = \min\left\{\frac{P(\boldsymbol{d}|\boldsymbol{m}^{new})}{P(\boldsymbol{d}|\boldsymbol{m}^{i-1})}, 1\right\}. \tag{6.14}$$

Alternatively, the proposal distribution $q(\boldsymbol{\pi}, \boldsymbol{m})$ can be an approximate posterior distribution as in Grana et al. (2017a). In the local move, we preserve the facies realization $\boldsymbol{\pi}^i = \boldsymbol{\pi}^{i-1}$ obtained in the previous iteration and sample a new realization $\boldsymbol{m}^{new}$ from a symmetric proposal distribution $q(\boldsymbol{m}|\boldsymbol{m}^{i-1}) = q(\boldsymbol{m}^{i-1}|\boldsymbol{m})$ with probability r_{local}:

$$r_{local} = \min\left\{\frac{P(\boldsymbol{d}|\boldsymbol{m}^{new})P(\boldsymbol{m}^{new}|\boldsymbol{\pi}^i)}{P(\boldsymbol{d}|\boldsymbol{m}^{i-1})P(\boldsymbol{m}^{i-1}|\boldsymbol{\pi}^i)}, 1\right\}. \tag{6.15}$$

The general algorithm for the prediction of the posterior distributions $P(\boldsymbol{\pi}|\boldsymbol{d})$ and $P(\boldsymbol{m}|\boldsymbol{d})$ of facies and reservoir properties conditioned on seismic data can be summarized in the following steps (de Figueiredo et al. 2019b):

1) We define the jump probability ϵ and we generate the initial facies realization $\boldsymbol{\pi}^0$ and the initial realization of reservoir properties $\boldsymbol{m}^0$.

2) We randomly sample a value u from a uniform distribution $U([0, 1])$. If $u < \epsilon$, we choose the jump move (A), otherwise we choose the local move (B).
   A) Jump move.
      A1. We sample a new facies realization $\boldsymbol{\pi}^{new}$ from the prior distribution $P(\boldsymbol{\pi})$ using a geostatistical algorithm for discrete properties.
      A2. We sample a new realization of the reservoir properties $\mathbf{m}^{new}$ from the proposal distribution $q(\mathbf{m}|\boldsymbol{\pi}^{new})$ using a geostatistical algorithm for continuous properties.
      A3. We accept the proposed realizations $\boldsymbol{\pi}^i = \boldsymbol{\pi}^{new}$ and $\mathbf{m}^i = \mathbf{m}^{new}$ with probability r_{jump} in Eq. (6.13), or we keep the previous realizations $\boldsymbol{\pi}^i = \boldsymbol{\pi}^{i-1}$ and $\mathbf{m}^i = \mathbf{m}^{i-1}$ with probability $1 - r_{jump}$.
   B) Local move.
      B1. We keep the facies realization $\boldsymbol{\pi}^i = \boldsymbol{\pi}^{i-1}$ obtained in the previous iteration $i - 1$.
      B2. We sample a new realization of the reservoir properties $\mathbf{m}^{new}$ from the proposal distribution $q(\mathbf{m}|\mathbf{m}^{i-1})$ using any geostatistical algorithm for continuous properties.
      B3. We accept the proposed realization $\mathbf{m}^i = \mathbf{m}^{new}$ with probability r_{local} in Eq. (6.15), or we keep the previous realization $\mathbf{m}^i = \mathbf{m}^{i-1}$ with probability $1 - r_{local}$.
3) After convergence, we estimate the posterior distributions $P(\boldsymbol{\pi}|\mathbf{d})$ and $P(\mathbf{m}|\mathbf{d})$ from the chain of realizations, excluding the realizations in the "burn-in" time.

Any geostatistical algorithm can be used for the simulation of facies and reservoir property realizations, including the algorithms presented in Chapters 3 and 4. The probability ϵ is generally less than 0.5 because the sampling space of the continuous properties is larger than the sampling space of the underlying discrete property (de Figueiredo et al. 2019a,b).

Example 6.6 We illustrate the multimodal McMC method, in the Gaussian mixture case, for a seismic inversion problem to predict the posterior distribution of facies and elastic properties. We apply the algorithm to a synthetic dataset including a reference facies profile of sand and shale layers and well logs of elastic properties, namely P- and S-impedance and density (Figure 6.6). The synthetic seismic data include partially stacked seismic traces corresponding to the reflection angles 8°, 18°, and 28°, and have a signal-to-noise ratio of 10. We use different wavelets for each angle to mimic different frequency bandwidths of each seismogram, assuming that the frequency content decreases as the angle increases. The seismic forward operator is the convolutional model of the wavelets and the linearized approximation of the Zoeppritz equations as a function of P- and S-impedance and density. The prior distribution of the facies is represented by a stationary first-order Markov chain with transition probabilities estimated from the actual facies profile. The prior distribution of the elastic properties is a trivariate Gaussian mixture model with two components (for sand and shale) with spatially invariant means and covariance matrices estimated from the well logs in each facies. The means of elastic properties are shown in Figure 6.6 (sand component in yellow and shale component in green).

Figure 6.6 shows the posterior distributions of facies and reservoir properties. For the facies, we show the posterior probability at each location and the predicted most likely model, whereas, for the elastic properties, we show the posterior distribution truncated within the 0.95 confidence interval and its maximum a posteriori. The posterior distribution captures the bimodal behavior of the elastic properties due to the sequence of alternating

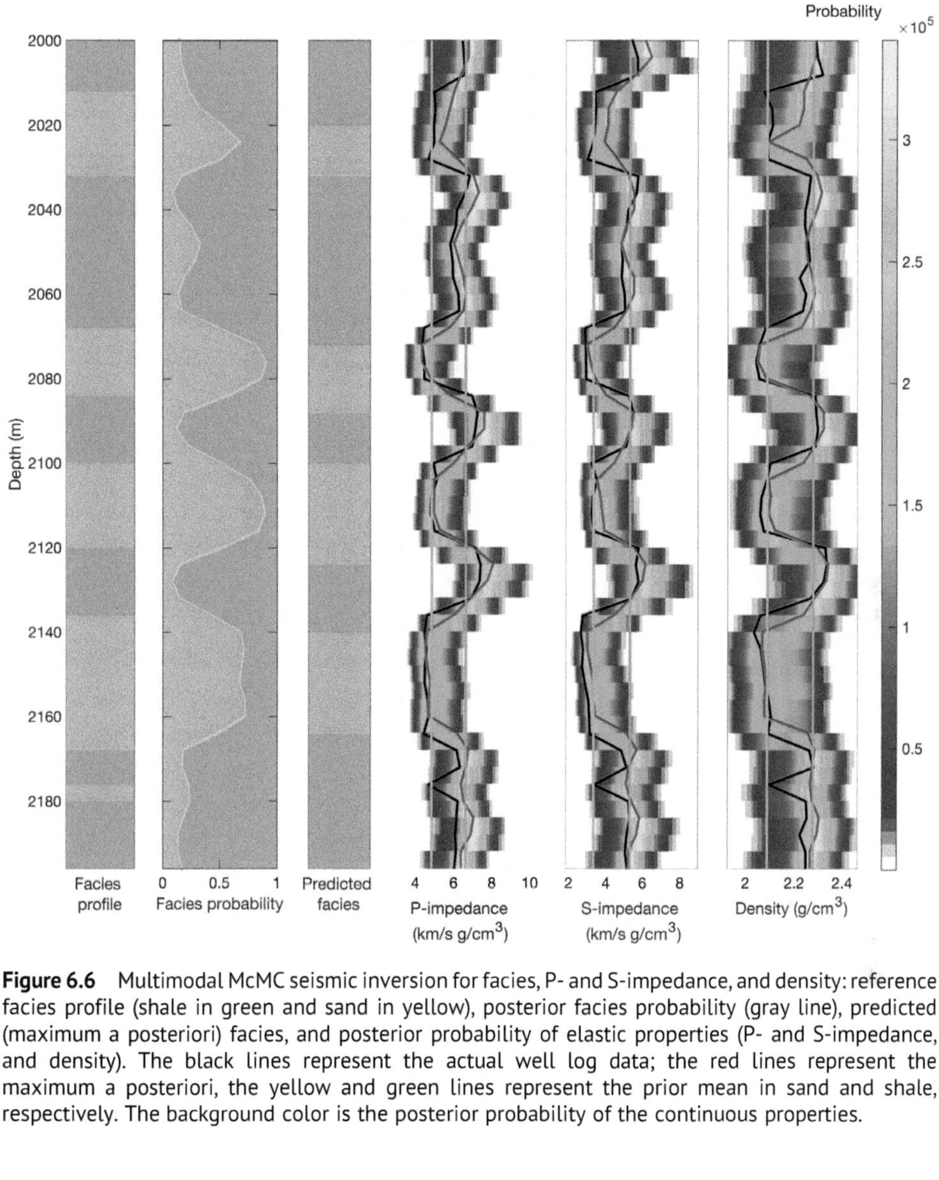

Figure 6.6 Multimodal McMC seismic inversion for facies, P- and S-impedance, and density: reference facies profile (shale in green and sand in yellow), posterior facies probability (gray line), predicted (maximum a posteriori) facies, and posterior probability of elastic properties (P- and S-impedance, and density). The black lines represent the actual well log data; the red lines represent the maximum a posteriori, the yellow and green lines represent the prior mean in sand and shale, respectively. The background color is the posterior probability of the continuous properties.

sand and shale layers. The posterior probabilities are estimated from the realizations obtained in 90000 iterations of the stationary regime, after a burn-in period of 10000 iterations. The acceptance ratio of the jump move is approximately 0.17.

6.4 Probability Perturbation Method

Another stochastic sampling and optimization approach commonly used in reservoir modeling is the probability perturbation method (PPM; Caers and Hoffman 2006). The PPM is a stochastic algorithm where the probability distribution of the model is gradually perturbed.

Conceptually, the PPM is similar to the gradual deformation method (Section 5.6.3), but the perturbation is applied to the probability distribution rather than the realizations. The PPM is generally applied to discrete variables but can be extended to continuous properties (Hu 2008). In the discrete case, the PPM is based on the idea that we can generate a new probability distribution $P^{new}(\boldsymbol{\pi})$ of a discrete random variable $\boldsymbol{\pi}$ as a convex combination of the indicator variable $I(\boldsymbol{\pi}_0)$ associated with a spatially correlated realization $\boldsymbol{\pi}_0$ with prior probability $P(\boldsymbol{\pi})$, and the prior distribution $P(\boldsymbol{\pi})$ as:

$$P^{new}(\boldsymbol{\pi} = k) = (1 - r)I(\boldsymbol{\pi}_0 = k) + rP(\boldsymbol{\pi} = k), \tag{6.16}$$

for $k = 1, \ldots, F$, where the parameter r belongs to the interval $[0, 1]$. We can then sample multiple values of the perturbation parameter r to generate multiple realizations of the discrete property $\boldsymbol{\pi}$. When $r = 0$, the new realization $\boldsymbol{\pi}^{new}$ coincides with $\boldsymbol{\pi}_0$, whereas, when $r = 1$, the new realization $\boldsymbol{\pi}^{new}$ is sampled from the prior distribution. The spatially correlated realization $\boldsymbol{\pi}_0$ as well as the new realization $\boldsymbol{\pi}^{new}$ can be generated using the sequential indicator simulation (SIS) method (Section 4.2) or any other geostatistical method, including multiple-point statistics algorithms (Section 4.5). The PPM approach can be integrated in an iterative stochastic optimization method for geophysical inverse problems, by generating, at each iteration, multiple realizations as a function of the parameter r and choosing the realization that best matches the measured data, according to an objective function.

The PPM approach for seismic facies classification problems can be summarized as follows:

1) We first generate an unconditional realization $\boldsymbol{\pi}_0$ of the facies, with prior probability $P(\boldsymbol{\pi})$ according to a spatial correlation model.
2) We simulate multiple realizations $\boldsymbol{\pi}^i(r)$ from the probability distribution $P^{new}(\boldsymbol{\pi})$ in Eq. (6.16), where the superscript i indicates the iteration number and it is initially set equal to 1.
3) We compute the data predictions $\mathbf{d}^i(r)$ using the geophysical forward operator, for example the rock physics and seismic models.
4) We solve a one-dimensional optimization on the perturbation parameter $r \in [0, 1]$ to select the model $\hat{\boldsymbol{\pi}}^i = \boldsymbol{\pi}^i(\hat{r})$ that best matches the data according to a predefined objective function, such as the L2-norm of the data residuals, $J^i(r) = \|\mathbf{d}^i(r) - \mathbf{d}\|^2$.
5) We calculate the indicator variable $I(\hat{\boldsymbol{\pi}}^i)$ of the selected facies realization $\hat{\boldsymbol{\pi}}^i$.
6) We compute a new probability perturbation $P^{new}(\boldsymbol{\pi})$ in Eq. (6.16), as a convex combination of the indicator variable $I(\hat{\boldsymbol{\pi}}^i)$ and the prior distribution $P(\boldsymbol{\pi})$.
7) We iteratively repeat steps 2–6 until convergence.

As in the gradual deformation approach, we reduce a high-dimensional optimization problem to a one-dimensional optimization. One of the advantages of this approach is the possibility of incorporating different types of conditioning data. If $\boldsymbol{\pi}$ is the model variable and $\mathbf{d}_1$ and $\mathbf{d}_2$ are two different conditioning datasets, then it is generally difficult to derive the analytical expression of the conditional probability $P(\boldsymbol{\pi}|\mathbf{d}_1, \mathbf{d}_2)$. Journel (2002) proposes an efficient method, often referred to as "Tau model" or "τ model" to integrate data from different sources. According to the τ model, if we assume that the prior distribution of the random variable $\boldsymbol{\pi}$ is $P(\boldsymbol{\pi})$ and that two conditional distributions $P(\boldsymbol{\pi}|\mathbf{d}_1)$ and $P(\boldsymbol{\pi}|\mathbf{d}_2)$

conditioned on the data $\boldsymbol{d}_1$ and $\boldsymbol{d}_2$ are available, then the conditional probability $P(\boldsymbol{\pi}|\boldsymbol{d}_1, \boldsymbol{d}_2)$ can be computed as:

$$P(\boldsymbol{\pi}|\boldsymbol{d}_1, \boldsymbol{d}_2) = \frac{1}{1+x}, \tag{6.17}$$

where:

$$\frac{x}{a} = \left(\frac{b}{a}\right)^{\tau_1}\left(\frac{c}{a}\right)^{\tau_2} \tag{6.18}$$

and

$$\begin{cases} a = \dfrac{1-P(\boldsymbol{\pi})}{P(\boldsymbol{\pi})} \\ b = \dfrac{1-P(\boldsymbol{\pi}|\boldsymbol{d}_1)}{P(\boldsymbol{\pi}|\boldsymbol{d}_1)} \\ c = \dfrac{1-P(\boldsymbol{\pi}|\boldsymbol{d}_2)}{P(\boldsymbol{\pi}|\boldsymbol{d}_2)}, \end{cases} \tag{6.19}$$

with τ_1 and τ_2 being constant parameters that account for the redundancy of the conditioning data $\boldsymbol{d}_1$ and $\boldsymbol{d}_2$. Assuming $\tau_1 = \tau_2 = 1$ is equivalent to assuming a form of conditional independence of the conditional probabilities $P(\boldsymbol{\pi}|\boldsymbol{d}_1)$ and $P(\boldsymbol{\pi}|\boldsymbol{d}_2)$ in terms of their ratios with respect to the prior probability $P(\boldsymbol{\pi})$. This assumption implies that the incremental contribution of the dataset $\boldsymbol{d}_1$ to the knowledge of $\boldsymbol{\pi}$ does not depend on the dataset $\boldsymbol{d}_2$, and vice versa, and it results in the following expression of the conditional probability $P(\boldsymbol{\pi}|\boldsymbol{d}_1, \boldsymbol{d}_2)$:

$$P(\boldsymbol{\pi}|\boldsymbol{d}_1, \boldsymbol{d}_2) = \frac{a}{a+bc}. \tag{6.20}$$

The τ model can be applied to hard and soft data (i.e. direct and indirect measurements) as well as data from different sources, such as seismic and electromagnetic data. It can be integrated into the PPM approach by combining Eq. (6.17) (or the special case in Eq. 6.20) with the probability in Eq. (6.16). An example of application of PPM, using the τ model, to seismic inversion problems for facies classification and simulation can be found in Grana et al. (2012b).

6.5 Other Stochastic Methods

As in the continuous case, several stochastic optimization algorithms have been proposed for facies classification and seismic inversion problems, including ensemble-based methods, simulated annealing, genetic algorithms (Sen and Stoffa 2013), as well as recent advances in machine learning methods, including convolutional neural networks, recurrent neural networks, and generative adversarial networks (Hall 2016; Wrona et al. 2018).

Stochastic optimization algorithms can be applied in conjunction with geostatistical simulations to predict and sample the probability distribution of facies and continuous reservoir properties conditioned on geophysical data. For example, pluri-Gaussian simulation

(Section 4.3) and sequential Gaussian mixture simulation (Section 3.5.2) are used in stochastic sampling and stochastic optimization algorithms such as simulated annealing and McMC algorithms.

Recent developments in machine learning provide an alternative to iterative stochastic methods. The term machine learning indicates a set of computational algorithms that learn from measured data to make predictions. Deep learning methods represent a subset of algorithms based on multilayered artificial neural networks and are now commonly applied in many domains of computer sciences. Several conventional algorithms are now classified as machine learning methods, such as gradient descent, genetic algorithms, simulated annealing, fuzzy decision tree, random forests, particle swarm optimization, and artificial neural networks. All these methods aim to find a model that minimizes an objective function, for example the data misfit in a geophysical inverse problem, through the back propagation of the error in the model space.

Convolutional and recurrent neural networks are specialized classes of deep neural networks for processing image or volumetric data and sequential data such as time signals. A convolutional neural network consists of an input and an output layer, as well as multiple hidden (convolutional) layers. A recurrent neural network consists of multiple layers in which the connections between nodes form a directed graph along a time sequence, allowing information cycles through a feedback loop that considers the current input as well as the information learned in the previous time steps. The feedback loops allow the neural network to capture long temporal dependencies in sequential data and allow an exhaustive analysis of time series data such as seismic reflection data. A comparison of several algorithms has been proposed in Wrona et al. (2018). Generally, the accuracy of the predictions obtained using deep learning algorithm is relatively high; however, the uncertainty quantification might be underestimated and strongly depends on the training dataset.

7

Integrated Methods

In the previous chapters, we provided a description of some of the most common tools and recent research developments for seismic reservoir characterization. The described workflow focuses on the predictions of reservoir properties based on seismic data or seismically derived attributes. In the proposed methods, we considered the uncertainty in the measured data owing to noise and limited resolution and the approximations in the physical relations. However, other sources of uncertainty affect the results of seismic inversion including the uncertainty associated with data processing (Claerbout 1976; Yilmaz 2001), geological structure (Thore et al. 2002; Mallet 2008; Caumon et al. 2009), anisotropy (Schoenberg and Sayers 1995; Tsvankin 2012; Mavko et al. 2020), and geomechanical processes (Sayers 2010; Zoback 2010; Tsvankin 2012). Different sources of data can be integrated into reservoir characterization studies, including electromagnetic and gravity data, to reduce the uncertainty in the model predictions (Hoversten et al. 2006; Constable 2010; Buland and Kolbjørnsen 2012; Gao et al. 2012; Dell'Aversana 2014). However, different geophysical datasets generally have different vertical and lateral resolution, and integrated modeling approaches should include the uncertainty in the upscaling or downscaling and domain changes. Geophysical surveys can also be acquired at different times to monitor the changes in the reservoir during injection and/or production. All the available data can be used, simultaneously or independently, to predict reservoir properties and reduce the prediction uncertainty. Geophysical model predictions are generally used to define the initial reservoir model for fluid flow studies that predict the fluid displacement and pressure changes owing to fluid injection and/or hydrocarbon production (Aziz 1979). Fluid flow simulations are commonly used in reservoir engineering and management to predict hydrocarbon production in oil and gas reservoirs and CO_2 plume location in carbon capture and storage studies. These models can be updated every time new data become available, including measurements from new boreholes or monitor geophysical surveys (Oliver et al. 2008; Evensen 2009). The updated model predictions are then used to optimize injection and production schedules and well placement. Decision-making processes in energy resources engineering are often based on subsurface numerical models and the uncertainty in the predictions. The value of information (VOI) of geophysical data depends on the particular decision, the preferences (utility function) of the decision maker, as well as the uncertainty in the available data and models (Caers 2011; Eidsvik et al. 2015).

Seismic Reservoir Modeling: Theory, Examples, and Algorithms, First Edition. Dario Grana, Tapan Mukerji, and Philippe Doyen.

All these topics widely overlap with seismic reservoir characterization methods in integrated studies. The goal of this chapter is to provide a general introduction and key references for each topic. We first discuss additional sources of uncertainty in seismic reservoir characterization, then we present the extension of inversion methodologies to time-lapse seismic data, the application to electromagnetic data, and the integration in seismic history matching studies, and we discuss the VOI of geophysical data in seismic reservoir modeling studies.

7.1 Sources of Uncertainty

Stochastic methods for reservoir characterization aim to predict reservoir properties conditioned on the available geophysical data. In the approaches proposed in Chapters 5 and 6, the data consist of partially stacked seismic amplitudes, often referred to as AVO data. In seismic and petrophysical inversion, we assume the magnitude of the errors associated with the measurements based on the signal-to-noise ratio and resolution of the data; however, we often ignore the uncertainty caused by the seismic processing. Several steps of the data processing, including velocity analysis, normal-moveout correction, multiple attenuation, stacking, and migration, might introduce additional errors in the measurements that are propagated from the raw measured data to the processed AVO data. These methods not only might increase the magnitude of the errors and the uncertainty in the data, but might also introduce systematic spatial correlations in the errors that are generally ignored in many seismic inversion methods. For further details on seismic processing we refer to Claerbout (1976), Yilmaz (2001), and Aki and Richards (2002).

One of the main limitations in reservoir modeling is associated with the isotropic assumption of petrophysical properties and the approximation in the geophysical relations and modeling methods that describe anisotropic properties, especially in fractured rocks. Anisotropy in rock properties can be caused by several geological factors, including but not limited to layering, alignment of minerals, orientation of fractures, and induced stresses, and it might create preferential fluid flow paths. In particular, in tight formations, such as shale, low-porosity sandstone, and carbonate, the identification of such fluid paths is a crucial factor for the management of energy resources. In these geological environments, the production of energy resources can be enhanced by exploiting preferential flow directions, owing to the preferential orientation of fractures caused by the stress field. Locating regions with high fracture density, determining the fracture orientation, and characterizing the connectivity of the pore space can improve the characterization of the reservoir, the prediction of rock heterogeneities, and the physical description of the fluid flow. Laboratory studies have proved the relation between anisotropic properties and elastic attributes at the core scale (Mavko et al. 2020); however, the development of anisotropic reservoir models at the field scale is still challenging.

In subsurface rocks, fractures generally show a preferential orientation in a certain direction, caused by the preferential orientation of the stress field in the subsurface. The resulting fracture geometry affects the elastic response of the rocks and results in the anisotropic behavior of seismic waves. In fractured rocks, seismic wave velocities of dry and saturated fractured rocks increase as a function of the confining pressure, whereas velocity changes in non-fractured rocks with stiff pores are generally negligible (Mavko et al. 2020). Even in rocks where the fracture porosity is isotropic at low-stress conditions, when uniaxial stress is applied, crack anisotropy is induced by preferentially closing cracks that are perpendicular or nearly perpendicular to the axis of compression (Nur 1971; Yin 1992; Zoback 2010). Therefore, seismic velocities generally vary with direction relative to the stress-induced crack alignment. The fluid effect and the fluid diffusivity are also affected by the anisotropic behavior of porous rocks. At low frequencies, wave-induced pore pressure gradients might have time to equilibrate throughout the fracture and matrix pore space, but physical models generally assume an isotropic behavior of the porous rocks (Mavko et al. 2020).

Cracks and fractures generally cause a decrease in the values of seismic velocities, and determine the stress dependence of velocities. Similarly, cracks and fractures increase the wave attenuation, leading to anisotropic effects on the seismic signature. The detection of systems of fractures is possible using various analyses to interpret azimuthal seismic data. In porous rocks affected by weak anisotropy, with a transverse isotropic symmetry, the elastic P-wave anisotropy can be quantified by three parameters, generally referred to as Thomsen's anisotropic parameters (Thomsen 1986) associated with the stiffness tensor of an anisotropic medium. Thomsen's parameters can then be correlated with fracture aperture, density, and orientation, using rock physics models (Hudson 1981). There are still several challenges in modeling anisotropic properties at the reservoir scale, and the application of these models at the field scale is limited by the discretization of the subsurface in grid cells that are orders of magnitude bigger than the average size of pores and fractures.

Geostatistical tools allow the investigation of the spatial uncertainty of reservoir properties assuming that the geological structure and the model grid are known. However, structural models can be very complex owing to the presence of multiple layers and faults. The derivation of the geometrical models that define structural horizons and faults is often based on the use of parametric surfaces and meshes, constrained by the available geophysical data and geological interpretation. Structural uncertainty mostly focuses on the uncertainty associated with the geometry and the topology of the structural models (Thore et al. 2002; Mallet 2008; Caumon et al. 2009).

The data most commonly used for structural modeling are seismic surveys and well log measurements. The structural model aims to define the boundaries of the geological layers. Horizons and faults are generally represented by surfaces approximated with planar polygons forming a two-dimensional mesh in the three-dimensional space. These surfaces are generally smooth to mimic the geological continuity but also represent discontinuities owing to variations in depositional events, erosion, reactive fluid transport, and tectonic activity such as faulting and folding. Structural modeling consists of several steps including interpretation of seismic data, construction of faults, connection of faults according to geological consistency rules, creation of horizon surfaces, and merging of faults and horizon surfaces (Caumon et al. 2009).

Assigning reservoir properties to each layer according to a spatial correlation model requires the construction of a geometrical grid, generally referred to as the stratigraphic grid, within the structural model. A stratigraphic grid is a Cartesian grid that is deformed to fit the layering of the structural model. Stratigraphic grids are generally irregular and they are defined by the coordinates of the corners of the hexahedral representing the deformed cell. Instead, seismic grids are regular Cartesian grids defined by inline, crossline, and travel time. Another difference between seismic and stratigraphic grids is that in seismic grids the properties (e.g. seismic amplitudes) are defined at the nodes of the grid, whereas in stratigraphic grids the variables (e.g. porosity) are defined within the cells. Indeed, stratigraphic grids are used to model reservoir properties defined in the geological layers, whereas seismic grids are used to model elastic contrasts at the interfaces.

Structural modeling is an iterative process that depends on the available measurements as well as the subjective interpretation of geophysical data and geological concepts. Modeling structural uncertainty is challenging owing to the various constraints of geological consistency and the difficulty of automating the model construction. For these reasons, in practical applications, the structural uncertainty is generally investigated by perturbing a reference model or a limited number of models. Sampling topological uncertainties and stochastic fault networks are still ongoing research topics (Aydin and Caers 2017).

7.2 Time-Lapse Seismic Inversion

Time-lapse seismic data, or 4D seismic data, refers to the acquisition of multiple seismic surveys at different times. For example, during hydrocarbon reservoir production or CO_2 injection, changes in dynamic reservoir properties, such as fluid saturations and pressure, can be monitored by repeatedly acquiring three-dimensional seismic data (Landrø 2001; Lumley 2001; Calvert 2005; Thore and Hubans 2012; Meadows and Cole 2013; Landrø 2015; Thore and Blanchard 2015; Maharramov et al. 2016).

The estimation of saturation and pressure from time-lapse seismic data requires a physical model relating the variations in reservoir properties to the changes in the seismic response. The saturation effect on seismic velocities can be generally described by Gassmann's equation (Section 2.6) combined with the mass balance equation for density (Section 2.1.4). For example, if water replaces hydrocarbon, the density and the P-wave velocity of the water-saturated rock typically increase, whereas the S-wave velocity decreases compared with the hydrocarbon-saturated rock. The pressure effect on seismic velocities (Section 2.7) has been described by several empirical equations (Eberhart-Phillips et al. 1989; MacBeth 2004; Sayers 2006). Generally, if effective pressure increases, both P-wave and S-wave velocities of the rock increase, whereas the pressure effect on density is often negligible unless the rock experiences a severe compaction. The increase in velocity is more significant at low effective pressure than high effective pressure. Indeed, many models are based on exponential relations that tend to an asymptotic value at high pressure. The seismic response, in terms of amplitude and travel time, can be modeled using the seismic convolutional model (Section 5.1).

Landrø (2001) presents a rock physics model approximation that expresses dynamic changes in density using a linear approximation in the changes in saturation (assuming the pressure effect on density to be negligible), and changes in velocities using an approximation that is linear in the saturation changes and quadratic in the pressure changes. According to Landrø's model, an approximation of the expressions for the relative changes of the elastic parameters owing to variations in water saturation and pore pressure, ΔS_w and ΔP_p, can be obtained as:

$$\begin{cases} \dfrac{\Delta V_P}{\overline{V_P}} \approx k_P \Delta S_w + l_P \Delta P_p + m_P \Delta {P_p}^2 \\ \dfrac{\Delta V_S}{\overline{V_S}} \approx k_S \Delta S_w + l_S \Delta P_p + m_S \Delta {P_p}^2 \\ \dfrac{\Delta \rho}{\overline{\rho}} \approx k_\rho \Delta S_w \end{cases}, \tag{7.1}$$

where k_P, l_P, m_P, k_S, l_S, m_S, and k_ρ are empirical parameters estimated from core sample measurements or rock physics model simulations. Landrø's model combines this formulation with a linearized approximation of Zoeppritz equations to derive a polynomial function that predicts the PP-reflectivity changes, $\Delta r_{PP}(\theta)$, between two repeated seismic surveys as a function of saturation and pressure variations and reflection angle θ as:

$$\begin{aligned} \Delta r_{PP}(\theta) \approx & \frac{1}{2}\left(k_\rho \Delta S_w + k_P \Delta S_w + l_P \Delta P_p + m_P \Delta {P_p}^2\right) \\ & + \frac{1}{2}\left(k_P \Delta S_w + l_P \Delta P_p + m_P \Delta {P_p}^2\right) \tan^2\theta - 4\frac{\overline{V_S}^2}{\overline{V_P}^2}\left(l_S \Delta P_p + m_S \Delta {P_p}^2\right) \sin^2\theta. \end{aligned} \tag{7.2}$$

This approximation assumes that the ratio $\overline{V_S}/\overline{V_P}$ is approximately constant between the two seismic surveys. This assumption might be valid for small changes in saturations; however, the $\overline{V_S}/\overline{V_P}$ ratio is generally different for hydrocarbon-saturated rocks compared with water-saturated rocks. The time-lapse inversion can then be solved using gradient-based methods or stochastic optimization algorithms. This approach has been successfully applied to hydrocarbon reservoir and CO_2 sequestration studies (Landrø et al. 2003; Veire et al. 2006; Trani et al. 2011; Grude et al. 2013; Bhakta and Landrø 2014).

Buland and El Ouair (2006) propose a Bayesian time-lapse inversion that extends the Bayesian linearized AVO approach (Section 5.3) to time-lapse seismic data. The Bayesian time-lapse inversion assumes that seismic data have been preliminary corrected for time shifts, owing to the changes in velocities between the two seismic surveys, to align the horizons of the base and monitor surveys, using, for example, a warping method (Hale 2009). The Bayesian time-lapse inversion in Buland and El Ouair (2006) applies the Bayesian linearized AVO approach to the difference of seismic amplitude of the base and monitor surveys and predicts the variations of the elastic properties.

We assume that two seismic surveys, $\boldsymbol{d}_1$ and $\boldsymbol{d}_2$, are available and were acquired according to the same geometry at two different times, t_1 and t_2, respectively. We also assume, for both seismic datasets, the same seismic modeling operator $\mathbf{G} = \mathbf{WAD}$ (Eqs. 5.10–5.15) based on the convolution of the wavelet and the linearized approximation of the Zoeppritz equations. Because the seismic operator is assumed to be linear, it can be applied to the seismic difference $\boldsymbol{\delta} = \boldsymbol{d}_2 - \boldsymbol{d}_1$. The convolutional model for the seismic difference can be then written as:

$$\boldsymbol{\delta} = \mathbf{G}\boldsymbol{\psi} + \boldsymbol{\varepsilon} = \mathbf{G}(\boldsymbol{m}_2 - \boldsymbol{m}_1) + \boldsymbol{\varepsilon}, \tag{7.3}$$

where the vector $\boldsymbol{\psi} = \boldsymbol{m}_2 - \boldsymbol{m}_1$ represents the change in the elastic properties along the seismic profile and $\boldsymbol{\varepsilon}$ represents the error term associated with the seismic difference $\boldsymbol{\delta}$. Since in the Bayesian linearized AVO approach (Section 5.3) the elastic model is expressed in terms of the logarithm of elastic properties, the model variable $\boldsymbol{\psi}$ is given by:

$$\boldsymbol{\psi} = \left[\ln\left(\frac{V_{P_2}}{V_{P_1}}\right), \ln\left(\frac{V_{S_2}}{V_{S_1}}\right), \ln\left(\frac{\rho_2}{\rho_1}\right) \right]^T, \tag{7.4}$$

where the subscripts 1 and 2 indicate the time of the base and monitor surveys, and where, for simplicity, we omitted the spatial coordinates. As in the Bayesian linearized AVO inversion, Buland and El Ouair (2006) show that if the prior distribution of the model variable $\boldsymbol{\psi}$ is Gaussian $\boldsymbol{\psi} \sim \mathcal{N}(\boldsymbol{\psi}; \boldsymbol{\mu}_\psi, \boldsymbol{\Sigma}_\psi)$ with prior mean $\boldsymbol{\mu}_\psi$ and prior covariance matrix $\boldsymbol{\Sigma}_\psi$ and the distribution of the data error $\boldsymbol{\varepsilon}$ is Gaussian $\boldsymbol{\varepsilon} \sim \mathcal{N}(\boldsymbol{\varepsilon}; \mathbf{0}, \boldsymbol{\Sigma}_\varepsilon)$ with $\mathbf{0}$ mean and covariance matrix $\boldsymbol{\Sigma}_\varepsilon$, and $\boldsymbol{\varepsilon}$ is independent of the model $\boldsymbol{\psi}$, then the posterior distribution of the elastic model $\boldsymbol{\psi}$ conditioned on the seismic difference $\boldsymbol{\delta}$ is also distributed according to a Gaussian distribution $\boldsymbol{\psi}|\boldsymbol{\delta} \sim \mathcal{N}\left(\boldsymbol{\psi}; \boldsymbol{\mu}_{\psi|\delta}, \boldsymbol{\Sigma}_{\psi|\delta}\right)$ with analytical expressions for the posterior mean $\boldsymbol{\mu}_{\psi|\delta}$ and posterior covariance matrix $\boldsymbol{\Sigma}_{\psi|\delta}$, analogous to the expressions in Eqs. (5.22) and (5.23). The variations of the elastic properties can then be estimated from the predicted mean or from the maximum a posteriori estimate. The estimation of the changes in saturation and pressure can be obtained by applying a Bayesian rock physics inversion using Gassmann's equation and a pressure model. Liu and Grana (2020) extend the ensemble smoother petrophysical inversion (Section 5.6.2) to time-lapse seismic data in a CO_2 sequestration study.

7.3 Electromagnetic Inversion

A potential alternative to seismic acquisition is given by controlled source electromagnetic (CSEM) data (Constable, 2010). CSEM data have been successfully utilized for hydrocarbon exploration in marine environments to map fluid saturations in the subsurface (Chen et al. 2007; Lien and Mannseth 2008; Key and Ovall 2011; Buland and Kolbjørnsen 2012). Electrical resistivity is particularly sensitive to fluid saturations and the value of CSEM data for the characterization of partially saturated porous rocks has been proven in several publications (Harris and MacGregor 2006; Hoversten et al. 2006; Constable 2010; MacGregor 2012). Because CSEM methods are sensitive to the entire interval from seafloor to target depth (MacGregor and Tomlinson 2014), accurate modeling and sensitivity analysis are required to ensure survey feasibility. Time-lapse CSEM data have been also used in reservoir monitoring studies (Orange et al. 2009; Shahin et al. 2012; Tveit et al. 2015; Rittgers et al. 2016). CSEM data have been applied to CO_2 sequestration in deep aquifers or depleted reservoirs for enhanced oil recovery studies (Bhuyian et al. 2012; Tveit et al. 2015) as well as for gas hydrate reservoirs (Weitemeyer et al. 2006).

The physical relations linking subsurface resistivity R and the associated electrical and magnetic fields, $\mathbf{E}$ and $\mathbf{H}$, are based on Faraday's and Ampère's laws, assuming that the displacement current is negligible compared with the conduction current, and they are generally expressed as a pair of coupled equations (Key and Ovall 2011) as a function of the electrical conductivity $\sigma = 1/R$:

$$\nabla \times \mathbf{E} - i\omega\mu\mathbf{H} = \mathbf{M}^{S} \tag{7.5}$$

$$\nabla \times \mathbf{H} - \sigma\mathbf{E} = \mathbf{J}^{S}, \tag{7.6}$$

where $\nabla \times \mathbf{E}$ and $\nabla \times \mathbf{H}$ are the curl of the electric and magnetic fields, μ and ω are constant values representing the magnetic permeability and the angular frequency, and $\mathbf{J}^{S}$ and $\mathbf{M}^{S}$ represent the electric and magnetic sources. The conductivity σ is in general complex given by $\sigma = \hat{\sigma} - i\omega\epsilon$, where $\hat{\sigma}$ is the real part of the electrical conductivity and ϵ is the dielectric permittivity; however, for geophysical applications, at typical CSEM frequencies (between 1 Hz and 10 Hz), $\hat{\sigma} \gg i\omega\epsilon$ and $\sigma \approx \hat{\sigma}$. On the other hand, at very high frequencies, appropriate for ground-penetrating radar applications (MHz or GHz), the subsurface property of interest is the dielectric permittivity. We generally assume that the electrical conductivity model is isotropic and we solve Eqs. (7.5) and (7.6) in the Fourier domain, assuming a time harmonic dependence of the form $e^{-i\omega t}$ (Key and Ovall 2011). By applying the Fourier transform, Eqs. (7.5) and (7.6) become a system of differential equations for the strike parallel components $\hat{E}_x$ and $\hat{H}_x$ of the electric and magnetic fields $\mathbf{E}$ and $\mathbf{H}$ in the wavenumber domain. The coupled differential equations are solved using a finite element approach, by discretizing the computational domain into an unstructured grid, and the results are transformed back in the spatial domain using the inverse Fourier transform (Key and Ovall 2011).

Bayesian approaches for the prediction of the posterior probability density function of resistivity and the prediction of the most likely resistivity model based on the measured CSEM data have been proposed in Trainor-Guitton and Hoversten (2011), Buland and Kolbjørnsen (2012), and Ray and Key (2012). The Bayesian CSEM inversion method can be combined with rock physics models linking saturation to resistivity to predict the posterior distribution of fluid saturations (Chen et al. 2007; Ayani et al. 2020; Tveit et al. 2020). Examples of rock physics models include Archie, Simandoux, and Poupon–Leveaux equations (Section 2.7).

Multi-physics methods for the joint inversion of seismic and CSEM data have been proposed in several studies. Hoversten et al. (2006) jointly invert marine seismic and CSEM data using a deterministic inversion method. Chen et al. (2007) use a Bayesian method for the joint inversion of seismic and CSEM data to invert for porosity and fluid saturations. Ayani et al. (2020) and Tveit et al. (2020) apply ensemble-based methods (Section 5.6.2) to time-lapse CSEM data for the prediction of the spatial distribution of saturation for CO_2 sequestration studies.

7.4 History Matching

Geophysical inverse problems can be coupled with fluid flow simulations to update the reservoir models over time and make predictions about the dynamic behavior of the reservoir system. Reservoir simulation is a subdiscipline of reservoir engineering that aims to develop mathematical–physical models to predict the flow of fluids, such as water, oil, and gas, through porous rocks. If the petrophysical properties, the temperature, and pressure conditions are known, then the spatial distribution of fluid saturations and pressures can be predicted by solving a system of partial differential equations (Aziz 1979).

The dynamic model that governs two-phase fluid flow in porous rocks is based on the constitutive equations of mass balance and Darcy's equation (Aziz 1979). In hydrocarbon reservoir studies, one of the common fluid flow models is the black oil simulator that includes a system of partial differential equations that describe the flux and concentration for each of the three phases, namely water, oil, and gas, and does not consider changes in the hydrocarbon composition, other than the dissolution of gas in oil (Aziz 1979). Recovery mechanisms that rely on mass transfer and thermodynamics processes require compositional balances in more advanced simulation models, namely compositional simulators.

The equations governing the black oil simulator can be derived from the fundamental laws of mass balance and Darcy's equation. The simulation assumes that the flow is isothermal, the gas can dissolve into oil but oil cannot be in the gas phase, the water component exists only in the water phase, and water is immiscible with both oil and gas. Based on these assumptions, if we indicate the fluid saturations with S_w, S_o, and S_g, and the fluid pressures with P_w, P_o, and P_g, where the subscripts w, o, and g stand for water, oil, and gas phase, respectively, then the resulting partial differential equations are:

$$\nabla \cdot \left[\frac{kk_{ro}}{B_o \mu_o} (\nabla P_o - \gamma_o \nabla d) \right] + q_o = \frac{\partial}{\partial t} \left(\frac{\phi S_o}{B_o} \right) \tag{7.7}$$

$$\nabla \cdot \left[\frac{kk_{rw}}{B_w \mu_w} (\nabla P_w - \gamma_w \nabla d) \right] + q_w = \frac{\partial}{\partial t} \left(\frac{\phi S_w}{B_w} \right) \tag{7.8}$$

$$\nabla \cdot \left[\frac{kk_{rg}}{B_g \mu_g} \left(\nabla P_g - \gamma_g \nabla d \right) + \frac{R_s kk_{ro}}{B_o \mu_o} (\nabla P_o - \gamma_o \nabla d) \right] + q_g = \frac{\partial}{\partial t} \left[\phi \left(\frac{R_s S_o}{B_o} + \frac{S_g}{B_g} \right) \right], \tag{7.9}$$

where k and k_r are the absolute and relative permeability of rocks, ϕ is the porosity, B is the formation volume factor, μ is the viscosity, γ is the specific gravity, q is the production rate, R_s is the gas–oil ratio, and d is the depth (Aziz 1979).

By assuming that $S_w + S_o + S_g = 1$, that the oil–water capillary pressure $P_{c_{o,w}}$ is a function of water saturation $P_{c_{o,w}} = P_w - P_o = f(S_w)$, and that the gas–oil capillary pressure $P_{c_{o,g}}$ is a function of gas saturation $P_{c_{o,g}} = P_o - P_g = h(S_g)$, then Eqs. (7.7)–(7.9) can be solved in a fully implicit manner for a reference pressure, generally P_o, and two saturations, typically S_w and S_g. Equations (7.7)–(7.9) are generally solved using finite difference or finite volume methods in a discrete reservoir grid, assuming that all the properties are constant within each grid cell (Aziz 1979). Fluid flow simulation models are commonly used to predict the production data, i.e. production rates and pressure conditions, at the borehole locations in hydrocarbon reservoir studies, and for the optimization of the well controls.

Seismic reservoir characterization models are commonly used to define the initial conditions, or static model, in fluid flow simulations. These models are often highly uncertain owing to the sparsity of direct measurements, the limited resolution of geophysical surveys, natural heterogeneity, and data errors. Such uncertainty affects the production predictions of the fluid flow simulation. Over time, predictions can be compared with the direct measurements in the borehole and the initial static model of reservoir properties can be updated to improve the accuracy and precision of future predictions. Borehole measurements generally include hydrocarbon production, water cut time, bottom hole pressure, and temperature. The process of updating the reservoir model is generally referred to as history matching (Oliver et al. 2008; Aanonsen et al. 2009; Evensen 2009). History matching is an inverse problem in which the model variables are updated every time new data are available and it can be solved using deterministic or stochastic methods. In geoscience applications, the physical operator is generally the dynamic simulation based on partial differential equations that model fluid flow in porous rocks (Eqs. 7.7–7.9). Mathematical methods, such as gradient-based and stochastic optimization algorithms, allow the derivation of potential solutions of the inverse problem given an initial model or set of models. However, the solution of the inverse problem is generally highly uncertain owing to the large number of variables and the limited number of measurements. In reservoir modeling problems, for example, the model parameters are the petrophysical properties at each grid cell in the reservoir model and the data are the production measurements at the well locations.

The integration of time-lapse seismic data in history matching, namely seismic history matching, can potentially reduce the uncertainty in the reservoir model and in its predictions (Aanonsen et al. 2003; Gosselin et al. 2003; Stephen et al. 2006; Roggero et al. 2007; Skjervheim et al. 2007; Aanonsen et al. 2009; Emerick and Reynolds 2012; Leeuwenburgh and Arts 2014; Liu and Grana 2020). Seismic history matching is a data assimilation problem, in which an initial model, generally obtained using reservoir characterization methods such as seismic and petrophysical inversion (Chapter 5), is updated every time new measurements, seismic or production data, are available. Traditional inverse methods, such as gradient-based approaches (gradient descent, Gauss–Newton, and Levenberg–Marquardt algorithms) provide a deterministic solution obtained by minimizing the objective function of the mismatch between model predictions and measurements. Because the forward model includes partial differential equations, the calculation and storage of the Jacobian and Hessian matrices of the predictions with respect to the model parameters are computationally demanding, in terms of time and memory. The large computational effort can be partially mitigated by applying adjoint methods to speed up the derivative computations. However, owing to the large number of variables, limited data availability, and measurement noise, the solution of the inverse problem is generally non-unique, and it is preferable to use probabilistic methods for the calculation of the posterior distributions of the model parameters conditioned on the data and the quantification of the uncertainty in the model predictions. The most common stochastic approach for history matching and seismic history matching is based on ensemble methods, such as the ensemble smoother (van Leeuwen and Evensen 1996), the ensemble Kalman filter (Evensen 2009), the ensemble randomized maximum likelihood method (Chen and Oliver 2012), and the ensemble smoother with multiple data assimilation (Emerick and Reynolds 2013); however, seismic history matching applications often require localization, regularization, or data re-parameterization approaches (Oliver and Chen 2011; Chen and Oliver 2017), to avoid severe underestimation of the posterior uncertainty owing to the large dimension of the data. Rock physics models can also be integrated in the seismic history matching workflow to predict the elastic response of the static reservoir model as well as the elastic changes owing to saturation and pressure variations (Liu and Grana 2020).

Seismic reservoir characterization methods are now commonly integrated in fluid flow simulation and optimization studies. In this context, the subsurface model is constantly updated as new data become available, and the model predictions and their uncertainties are included in the decision-making processes.

7.5 Value of Information

Several business decisions in reservoir management, such as drilling a new well or acquiring new data, are based on reservoir model predictions obtained from geophysical data and reservoir engineering models. Because we can only acquire a limited amount of measurements and several datasets only contain indirect measurements, the mathematical–physical models used to describe the subsurface reservoirs are generally uncertain (Section 7.1). The model uncertainty depends on a number of factors: sparse direct measurements (well log measurements), low resolution and low signal-to-noise ratio of indirect measurements (seismic and electromagnetic data), approximations in the physical models (approximations of complex physical equations), and model simplifications owing to computational costs (space and time discretizations).

To improve the reservoir description, multiple datasets can be integrated into the reservoir characterization study: well logs, seismic data, and electromagnetic data if available (Section 7.3). During production, time-lapse geophysical surveys (Section 7.2) and production data (Section 7.4) are used to monitor the reservoir conditions and update the initial model to improve the reservoir predictions. These datasets have different characteristics: well logs and production data are accurate measurements but they are only available at a few locations in the reservoir; seismic and electromagnetic data are measured in the entire field but they have lower resolution and signal-to-noise ratio than well logs. In general, a large number of measurements does not necessarily lead to a more accurate model. Indeed, especially when data are noisy and uncertain, multiple datasets can provide inconsistent information. Because the acquisition of geophysical data is expensive, it is necessary to study, prior to the acquisition, the added value of each dataset. For example, if rock physics models reveal that variations in saturation and pressure during production cause minimal changes in the elastic response (e.g. in a well-consolidated deep reservoir), time-lapse seismic data would not help to improve the reservoir model and reduce the uncertainty in the fluid flow simulations. Similarly, if the resolution of electromagnetic data is too low compared with the reservoir thickness, the integrated seismic and electromagnetic inversion results might be biased or more uncertain than the results obtained using seismic data only.

In order to quantify the added value of acquiring and integrating a new dataset, we can adopt a statistical concept, namely the VOI, commonly used in economics and finance. The VOI is a tool in decision theory to investigate the benefits, for example from an economical point of view, of acquiring additional information to make a decision. The VOI has been introduced in earth sciences to evaluate a set of decisions and determine the optimal one given the uncertainty in the data and in the model (Bickel et al. 2008; Eidsvik et al. 2008; Bratvold et al. 2009; Bhattacharjya et al. 2010; Caers 2011; Eidsvik et al. 2015). The VOI study allows the quantification of the advantages in acquiring new information before the data are purchased and links the prediction improvement to the monetary value (Caers 2011; Eidsvik et al. 2015).

In the VOI approach for reservoir management decisions, the decision-making process is generally formulated using decision trees or other probabilistic methods such as influence diagrams. First, we calculate the expected value of the decision under study, without the additional information of interest; then, we formulate the decision tree to include the additional information and calculate the expected value of the decision, assuming that the additional information is available. We can then compute the VOI by subtracting the expected value without information from the expected value with additional information. If the VOI is less than or equal to the acquisition value, then the additional information should not be purchased and the optimal decision is the same as the decision obtained without additional information. Eidsvik et al. (2015) show examples of uncertainty quantification and VOI applied to geophysical data in rock physics and reservoir characterization, using seismic and electromagnetic data.

8

Case Studies

In this chapter, we present the main results of four different applications: two hydrocarbon reservoir studies in the Norwegian Sea (Grana and Della Rossa 2010; Liu and Grana 2018), a carbonate field from offshore Brazil (de Figueiredo et al. 2018), and a CO_2 sequestration study from offshore Norway (Liu and Grana 2020).

8.1 Hydrocarbon Reservoir Studies

We illustrate three different applications of seismic reservoir characterization studies in hydrocarbon reservoirs. In the first application, we apply a Bayesian linearized approach; in the second example, we present an ensemble-based method; and in the third application, we propose a Markov chain Monte Carlo algorithm. The first two examples are from conventional clastic reservoirs, whereas the third study is a carbonate field.

8.1.1 Bayesian Linearized Inversion

We applied the Bayesian inversion methods described in Sections 5.3–5.5 to an oil field in the Norwegian Sea, where two wells and four seismic angle gathers are available. The complete study was published by Grana and Della Rossa (2010). The clastic reservoir is part of a fluvio-deltaic environment of the Middle–Late Triassic age characterized by a complex heterogeneous sand distribution. The basin has experienced significant subsidence and uplift events of approximately 700 – 900 m, with the most recent one having occurred in the late Pliocene. A complex fault system, identified from seismic data, divides the fluvio-deltaic system in several segments. The main reservoir is partially filled with oil with an average irreducible water saturation of approximately 0.10.

A three-dimensional seismic survey and a complete well log dataset from two wells are available for the seismic reservoir characterization study. The seismic data have been preprocessed to compute a set of four partially stacked seismic volumes corresponding to reflection angles of 8°, 20°, 32°, and 44°. The well log data include P-wave and S-wave velocity, density, gamma ray, neutron porosity, and resistivity, as well as the petrophysical curves obtained from quantitative log interpretation. The petrophysical properties of interest are porosity, clay volume, and water saturation. A log-facies classification is also available and is based on three litho-fluid facies: oil sand, water sand, and shale.

Seismic Reservoir Modeling: Theory, Examples, and Algorithms, First Edition. Dario Grana, Tapan Mukerji, and Philippe Doyen.

A rock physics model was calibrated using the well log data from one of the two boreholes (Grana and Della Rossa 2010). The model includes the Voigt–Reuss–Hill average to compute the effective elastic properties of the solid phase, the Reuss average to calculate the effective elastic properties of the fluid phase, the stiff sand model to predict the dry-rock elastic properties, and Gassmann's equations to estimate the saturated-rock elastic properties as well as the mass balance equation for density (Chapter 2). The elastic parameterization is expressed in terms of P- and S-impedance. A preliminary fluid substitution is performed on the log data to calibrate the stiff sand model parameters. The solid phase includes two mineral components, quartz and illite-rich clay. Of the input parameters required for the stiff sand model, the critical porosity is 0.4, the coordination number is 7, and the estimated effective pressure is 70 MPa. Figure 8.1 shows the calibration of the rock physics model in the porosity and P-wave velocity domain for water-saturated rocks. The rock physics model is combined with a seismic convolutional model of the Aki–Richards linearized approximation of Zoeppritz equations for vertical weak contrasts, expressed in terms of P- and S-impedance. The seismic wavelets are independently estimated for each angle stack, using a statistical approach applied to the calibration well dataset and the collocated seismic trace.

The prior distribution of porosity, clay volume, and water saturation is a trivariate Gaussian mixture with three components corresponding to the three litho-fluid facies (Figure 8.1). The weights of the mixture are assumed to represent the proportions of the litho-fluid facies. The prior parameters of the Gaussian mixture model (means, covariance matrices, and weights) are estimated from log measurements from the available well datasets and nearby fields. A normal score transformation is applied to water saturation, for each litho-fluid facies, whereas truncations are applied to porosity and clay volume. The inversion is conducted in the transformed domain, and the inversion results are then back-transformed into the original domain. The variograms of the petrophysical properties are estimated from the experimental variograms in each litho-fluid facies using an exponential function and they are included in the prior model (Figure 8.1).

The Bayesian inversion methodology implemented in Grana and Della Rossa (2010) is based on the methods described in Sections 5.3, 5.4.3, and 5.5. The authors apply the method to the whole seismic volume to obtain the posterior probability distributions of petrophysical variables and litho-fluid facies conditioned on partially stacked seismic data. The predictions are computed in the time domain of seismic data and transformed in the depth domain using a time-to-depth function calibrated using seismic velocities. The method is validated at the Well A location using synthetic seismograms with a signal-to-noise ratio of 5. The posterior distributions and the corresponding predictions are compared with the petrophysical curves of porosity, clay volume, and water saturation, and to the log-facies profile (Figure 8.2). The posterior distributions of the continuous properties in Figure 8.2 (top plots) capture the multimodality of the petrophysical properties. The posterior uncertainty is relatively large, especially within the sequence of alternating thin layers of sand and shale in the central part of the interval and for water saturation. The median provides a smooth prediction, whereas the maximum a posteriori estimate of the posterior distribution better describes the multimodal behavior of the properties. Figure 8.2 (bottom plots) shows the results of the seismic facies classification. The posterior distribution of litho-fluid facies shows high probability values for the oil sand class at the top of the reservoir, corresponding to the main high-porosity sand layer. The most likely facies model,

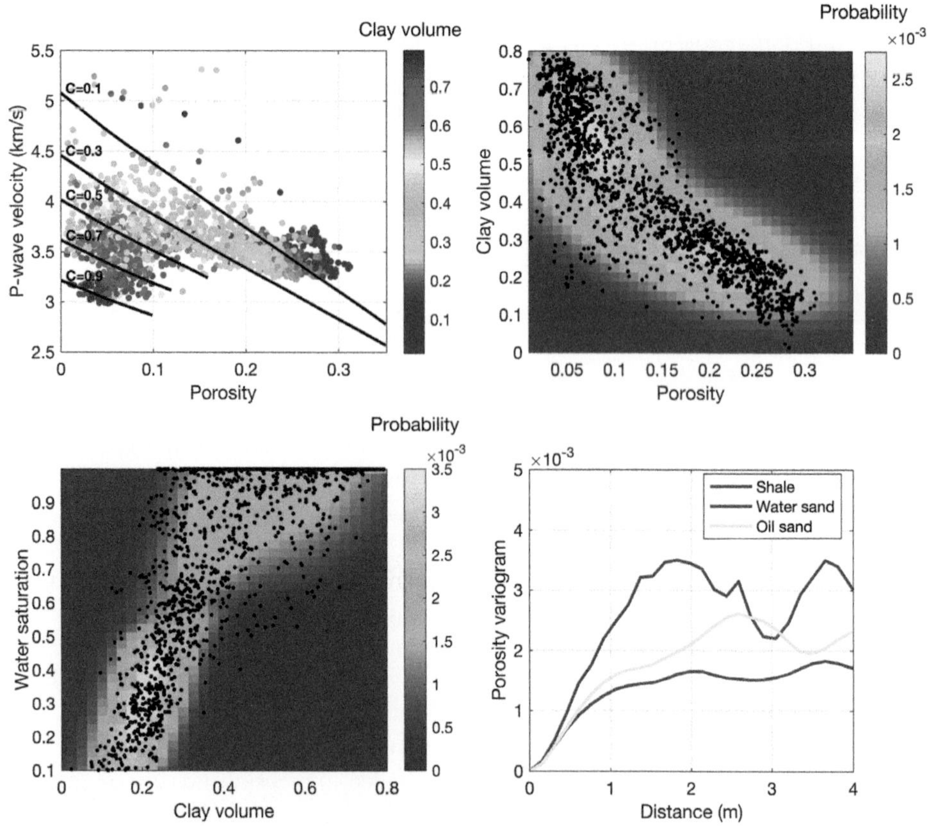

Figure 8.1 Rock physics relations and prior model: rock physics calibration in the porosity and P-wave velocity domain, color coded by clay volume, for water-saturated samples (the colored dots represent the well log data after fluid substitution, the black curves represent the rock physics model for different values of the clay volume); projections of the prior Gaussian mixture distribution in the porosity–clay volume and clay volume–water saturation domains (the black dots represent the well log data, the background color represents the prior probability); and experimental variograms of porosity in each litho-fluid facies (shale in green, water sand in brown, and oil sand in yellow). Modified after: Grana, D. and Della Rossa, E., 2010. Probabilistic petrophysical-properties estimation integrating statistical rock physics with seismic inversion. Geophysics, 75(3), pp.O21-O37

i.e. the maximum a posteriori of the facies probability, fails to capture the thin layers below seismic resolution. Grana and Della Rossa (2010) propose to integrate a stationary first-order Markov chain to sample from the posterior probability and obtain high-detailed facies profiles. The correlation coefficient of the median of the probability distribution of petrophysical properties and the actual measurements is 0.58 for porosity, 0.55 for clay volume, and 0.64 for water saturation, owing to the different frequencies of well log and seismic data. The correlation coefficient of the median and the borehole measurements upscaled to the seismic resolution is 0.89 for porosity, 0.81 for clay volume, and 0.84 for water saturation. The fraction of samples correctly classified in the actual facies is 0.70 for oil sand, 0.33 for water sand, and 0.69 for shale, owing to the resolution of seismic data. Grana and Della Rossa (2010) present a sensitivity analysis to investigate the effect of the resolution of seismic data on the predictions, and propose a comparison with a Bayesian inversion approach

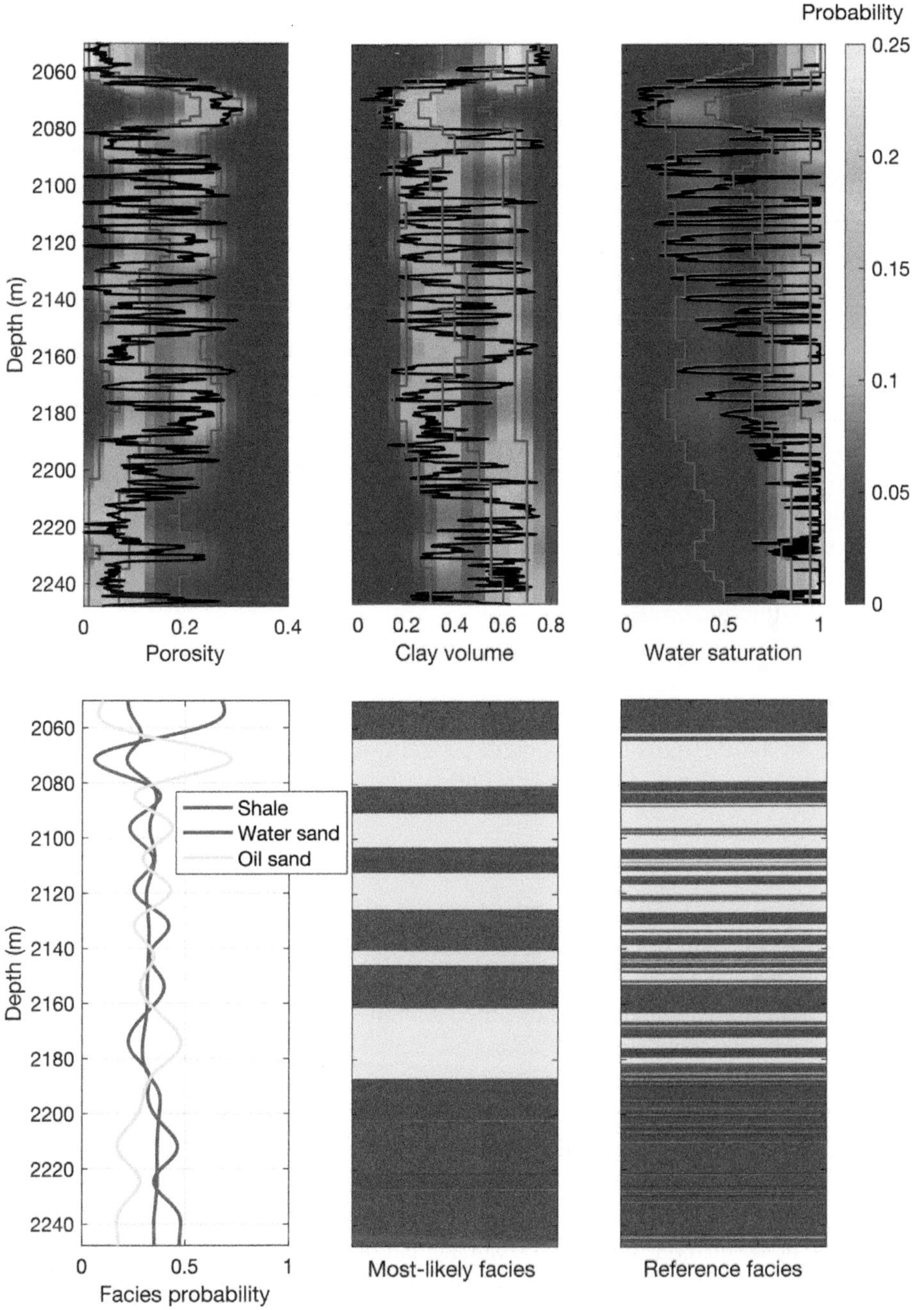

Figure 8.2 Posterior distributions of petrophysical properties and litho-fluid facies at the calibration well location (well A): (top) probability distributions of porosity, clay volume, and water saturation (the black curves represent the actual petrophysical curves, the red curves represent the median and the 0.90 confidence interval, and the background color represents the posterior probability); (bottom) probability distributions of litho-fluid facies, most likely facies, and actual facies classification (shale in green, water sand in brown, and oil sand in yellow). Modified after: Grana, D. and Della Rossa, E., 2010. Probabilistic petrophysical-properties estimation integrating statistical rock physics with seismic inversion. Geophysics, 75(3), pp.O21-O37.

based on non-parametric distributions showing consistent results with the Gaussian mixture case.

The inversion is applied trace by trace to the entire seismic volume. Figure 8.3 shows the near- and far-angle stacks along a seismic line interpolated between two wells. The signal-to-noise ratio of the measured data is estimated to be approximately 3 at the top of the interval and close to 1.5 at the bottom. Figure 8.3 also shows the predicted model from the posterior distribution of porosity, clay volume, and water saturation. In the upper part of the interval, we can identify the overlying clay and the oil-saturated high-porosity sand reservoir. In the lower part, the thin layers observed in the well logs are not detected and the uncertainty associated with the inverted properties increases, owing to the low quality of the seismic data.

Figure 8.4 illustrates the predictions of the Bayesian petrophysical inversion in the three-dimensional space, conditioned on a seismic sub-volume containing 10000 traces in a time window corresponding to a depth interval of approximately 250 m. The predictions of porosity and oil sand facies probability are shown along a crossline through Well B and an inline through Well A. In the bottom plot of Figure 8.4, we impose the three-dimensional isosurface of oil sand probability with values greater than 0.7 on top of the seismic lines. The isosurface delineates the main reservoir layer with average porosity greater than 0.20 and partial oil saturation up to 0.8.

The Bayesian linearized approach is based on analytical solutions, it is simple to implement and computationally fast. Several publications implemented the Bayesian linearized approach in applications to oil and gas reservoirs as well as CO_2 sequestration studies (Grana et al. 2017b), gas hydrate reservoirs (Dubreuil-Boisclair et al. 2012), and near-surface geophysics applications (Flinchum et al. 2018).

8.1.2 Ensemble Smoother Inversion

We applied the seismic and petrophysical inversion based on the ensemble smoother method described in Section 5.6.2 to the Norne oil field, located in the Norwegian Sea. The dataset includes prestack seismic data and well log measurements from one borehole. The complete study was presented by Liu and Grana (2018). The oil-saturated clastic reservoir has an average thickness of 200 m and porosity between 0.25 and 0.30. The seismic survey was acquired in 2001 and consists of three partial angle stacks, namely near, mid, and far stacks, corresponding to reflection angles of 10°, 23°, and 35°, respectively. The well log data include both sonic and petrophysical logs. The petrophysical curves obtained from the quantitative log interpretation, namely porosity, clay volume, and water saturation, are shown in Figure 8.5 (blue curves). Because of the high consolidation of the sandstone in the formation of interest, Liu and Grana (2018) adopt the stiff sand model (Section 2.4) for the rock physics analysis. The seismic forward model is a convolution of the angle-dependent wavelets and the Zoeppritz equations for reflected P-wave amplitudes (Liu and Grana 2018).

The inversion is based on the ensemble smoother with multiple data assimilation (ES-MDA) method (Section 5.6.2) with four iterations. A logit transformation is applied to the petrophysical properties, and the inverse transformation is applied to the inversion results to back-transform the predictions in the petrophysical model space. A data dimensionality reduction approach based on the singular value decomposition (SVD) is applied to reduce the dimension of the data space. The method is first validated at the well location.

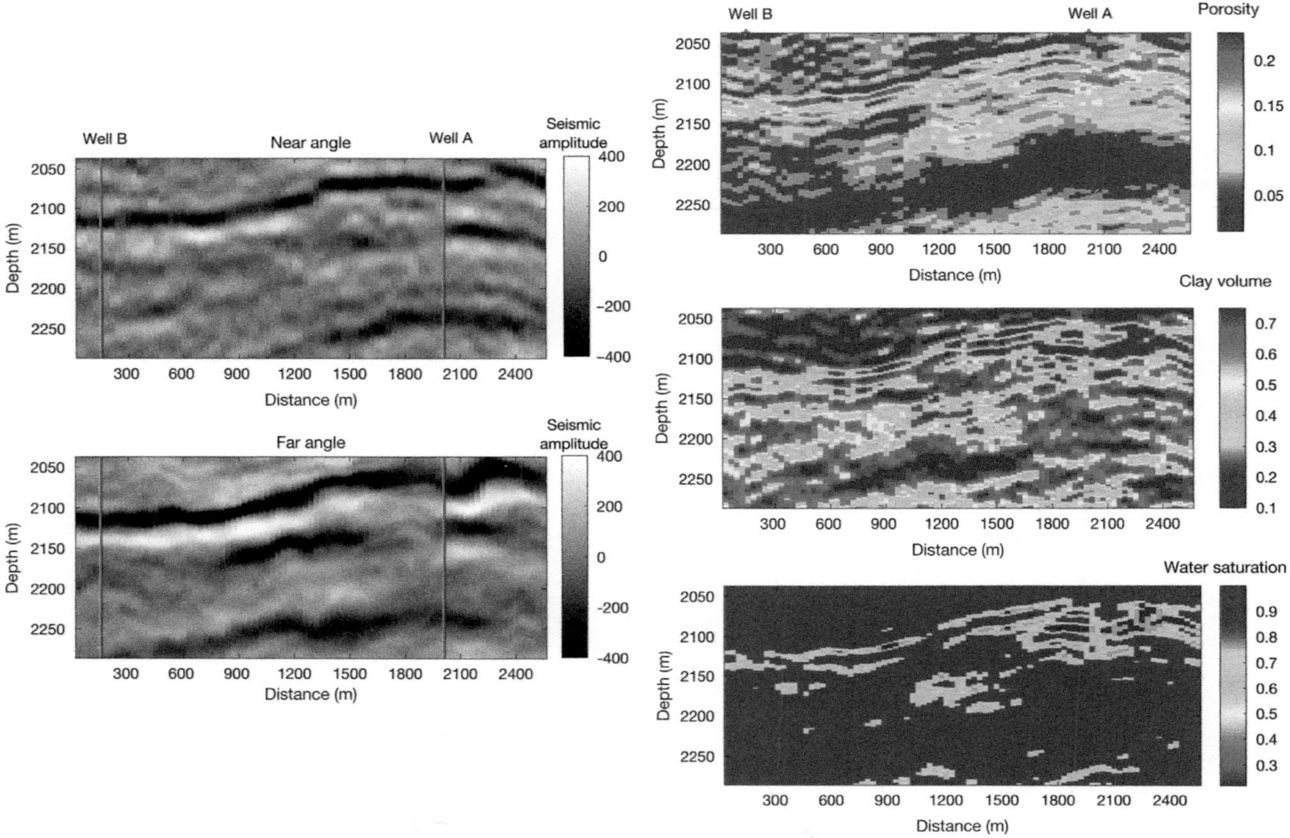

Figure 8.3 Partially stacked seismic data and corresponding petrophysical property predictions: (left) near- and far-angle stacks along a seismic line interpolated between two wells (the red lines show the locations of the wells); (right) predicted models of porosity, clay volume, and water saturation. Modified after: Grana, D. and Della Rossa, E., 2010. Probabilistic petrophysical-properties estimation integrating statistical rock physics with seismic inversion. Geophysics, 75(3), pp.O21-O37.

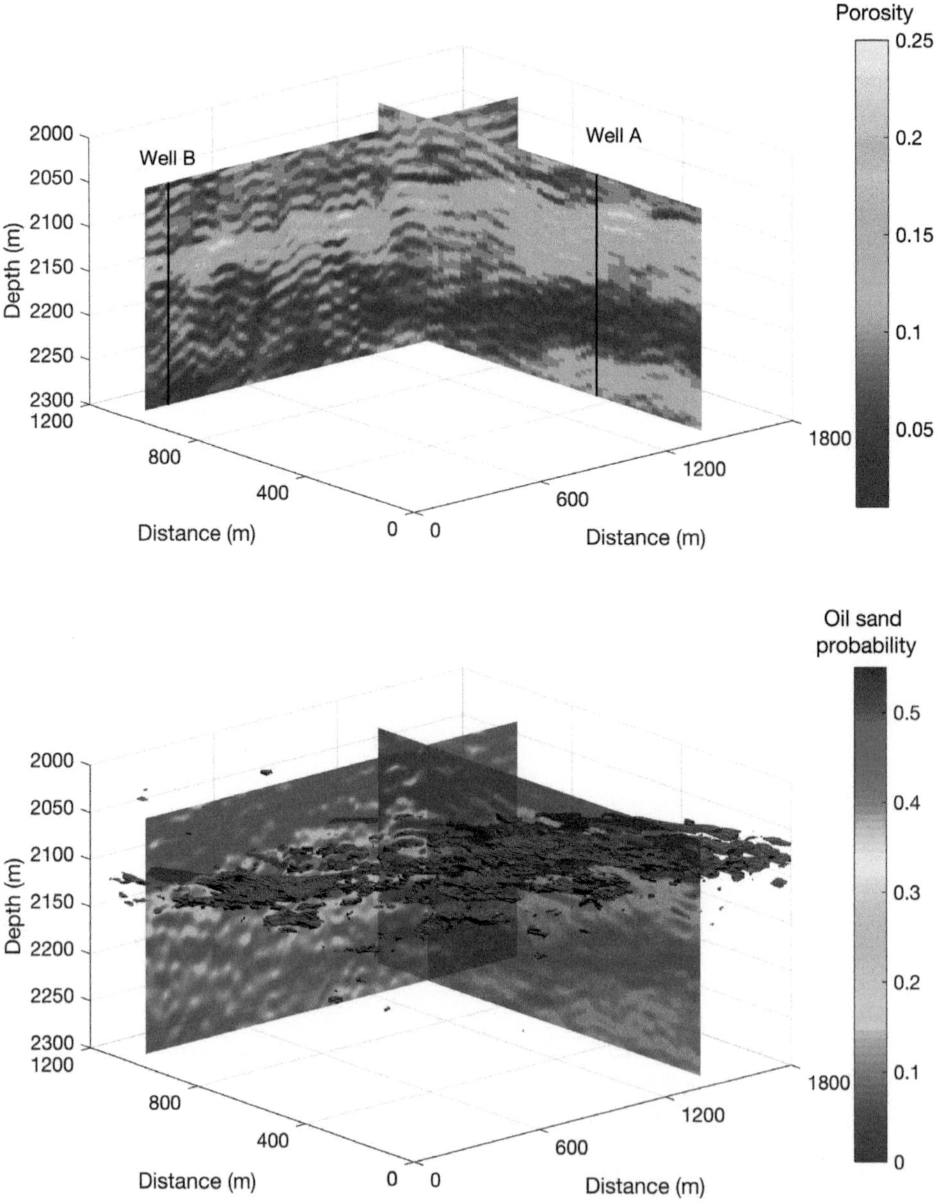

Figure 8.4 Predictions of porosity and oil sand facies probability in the three-dimensional space: (top) predicted porosity along two seismic lines extracted from the seismic volume and (bottom) corresponding oil sand facies probability. The red regions represent the isosurface of 0.7 probability of oil sand litho-fluid facies. Modified after: Grana, D. and Della Rossa, E., 2010. Probabilistic petrophysical-properties estimation integrating statistical rock physics with seismic inversion. Geophysics, 75(3), pp.O21-O37.

The prior ensemble includes 1000 initial profiles of porosity, clay volume, and water saturation, at the well location, generated from the prior distribution and the vertical covariance functions estimated from the well log data. For each model in the prior ensemble, the elastic response is computed using the stiff sand model and the corresponding synthetic seismic

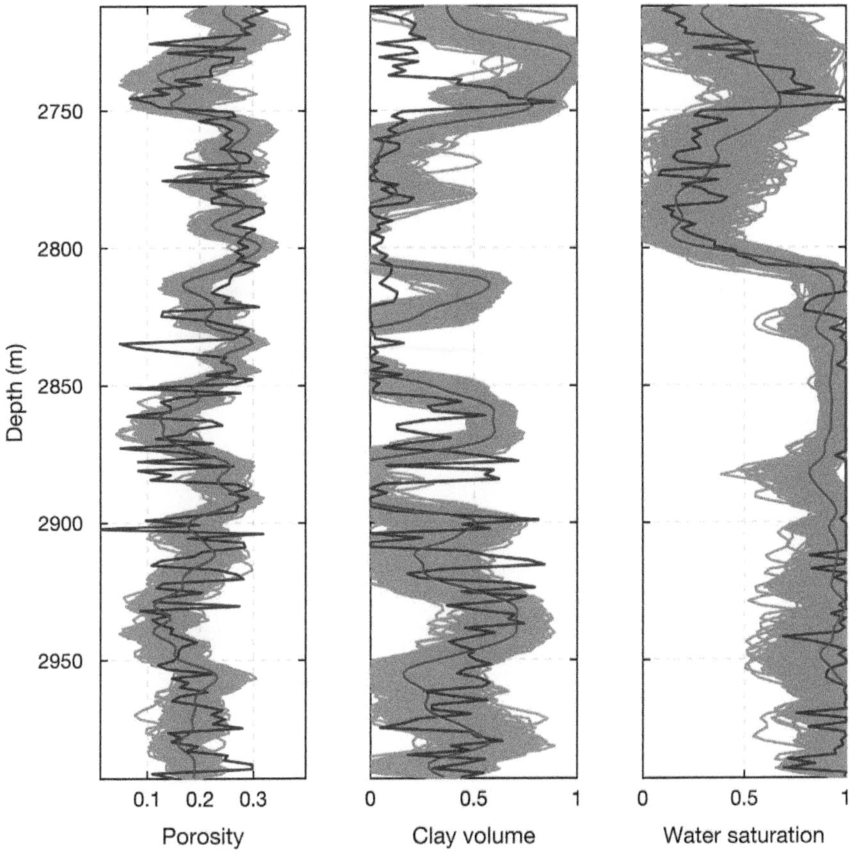

Figure 8.5 Posterior ensemble of petrophysical properties: porosity, clay volume, and water saturation. The gray curves represent 1000 posterior realizations, the red curves represent the posterior mean, and the blue curves represent the actual petrophysical curves. Modified after: Liu, M. and Grana, D., 2018. Stochastic nonlinear inversion of seismic data for the estimation of petroelastic properties using the ensemble smoother and data reparameterization. Geophysics, 83(3), pp.M25-M39.

trace is generated by convolving the reflection coefficients calculated using Zoeppritz equations and the corresponding wavelets. The dimensionality reduction is applied to the computed seismic data using the SVD and the ES-MDA method is performed in the reduced data space with constant inflation factors at each iteration. Figure 8.5 shows the posterior realizations of the petrophysical properties and their posterior means. The associated model uncertainty can be quantified using the point-wise empirical variances calculated from the updated ensemble members. The predicted models accurately match the actual petrophysical curves. The variability in the updated realizations of water saturation shows a larger uncertainty in the predictions compared with the other petrophysical properties. The correlation coefficient of the updated ensemble mean of petrophysical properties and the actual measurements is 0.74 for porosity, 0.59 for clay volume, and 0.71 for water saturation, owing to the different frequencies of well log and seismic data. The coverage ratio of the 0.90 confidence interval of the updated ensemble is 0.83 for porosity, 0.81 for clay volume, and 0.91 for water saturation. The coverage ratio of the predictions represents the fraction of measurements within the corresponding confidence interval. In this

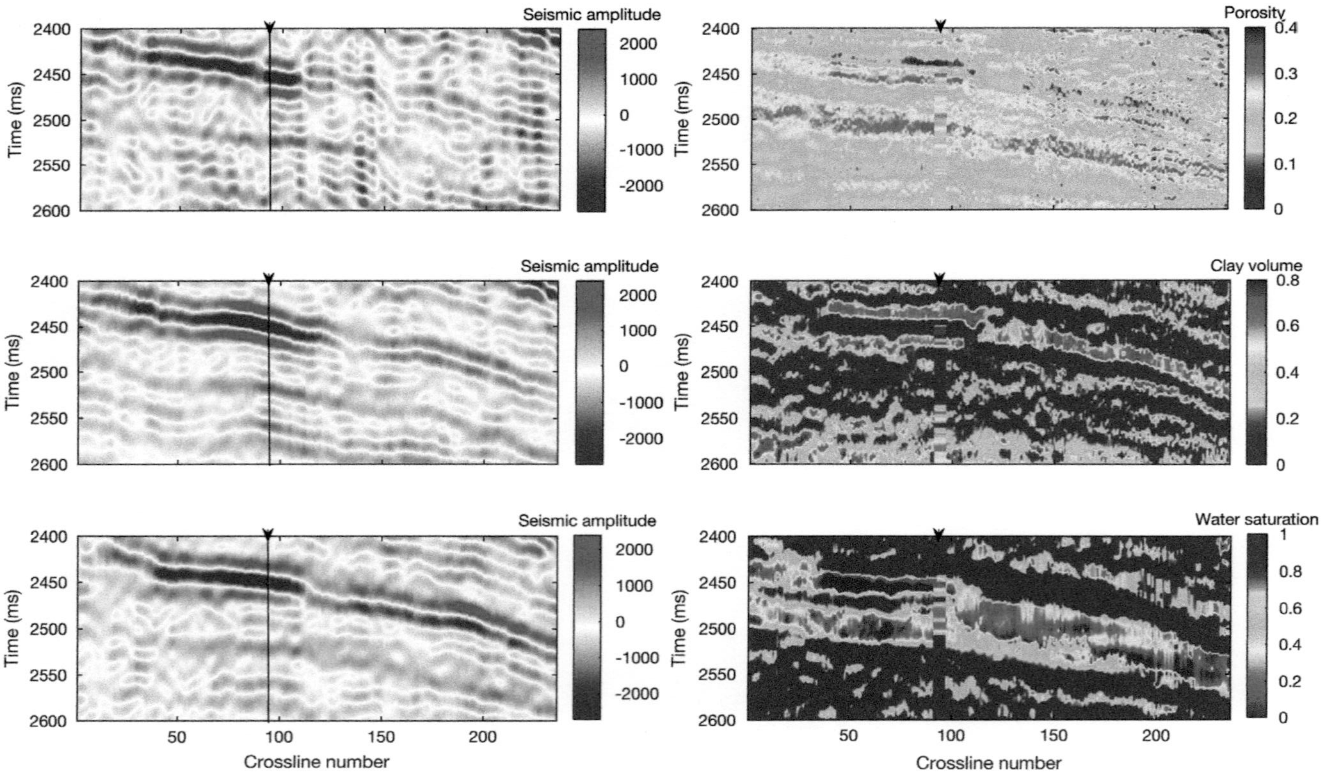

Figure 8.6 Partially stacked seismic data and corresponding petrophysical property predictions: (left) near-, mid-, and far-angle stacks along a seismic line crossing the calibration well (the black lines show the well location); (right) posterior ensemble mean of porosity, clay volume, and water saturation (the petrophysical curves filtered at the frequency bandwidth of the seismic data are plotted at the well location). Modified after: Liu, M. and Grana, D., 2018. Stochastic nonlinear inversion of seismic data for the estimation of petroelastic properties using the ensemble smoother and data reparameterization. Geophysics, 83(3), pp.M25-M39.

application, the uncertainty is slightly underestimated for porosity and clay volume, probably because of the noise in the actual petrophysical curves in the central part of the interval.

The inversion is then applied to a two-dimensional seismic line. Figure 8.6 shows the near-, mid-, and far-angle stacks along a seismic line passing across the well. The petrophysical models of the initial ensemble are generated using the fast Fourier transform – moving average (FFT-MA) method (Section 3.6) with a spatial correlation function estimated from the input seismic data and mean values inferred from the well logs. The models of the point-wise posterior ensemble mean of porosity, clay volume, and water saturation, are shown in Figure 8.6. The actual petrophysical curves filtered at the seismic frequency are superimposed at the well location and show an accurate match with the seismic inverted results. Liu and Grana (2018) propose a sensitivity analysis on the size of the ensemble and a comparison with the Bayesian linearized inversion. Liu and Grana (2020) apply the same approach to a CO_2 sequestration study including seismic and production data. In Liu and Grana (2020), the data re-parameterization is obtained by applying a machine learning technique, namely the deep convolutional autoencoder.

8.1.3 Multimodal Markov Chain Monte Carlo Inversion

We applied a joint seismic and petrophysical inversion based on the Markov chain Monte Carlo (McMC) algorithm described in Section 5.6.1 to an oil-saturated carbonate field, located offshore Brazil. The dataset includes three-dimensional partially stacked seismic data as well as borehole logs and core samples from three wells. The complete case study was proposed by de Figueiredo et al. (2018). The reservoir model is defined in a stratigraphic grid vertically limited by two horizons interpreted from seismic data. The seismic data have been processed to obtain three partial angle stacks, corresponding to 8°, 18°, and 28°, with average frequencies of the amplitude spectrum of 29, 24, and 22 Hz, respectively. The corresponding wavelets are obtained by fitting the wavelet amplitude and phase spectrum of the synthetic seismograms to the measured seismic data. The available well log data include P-wave velocity, S-wave velocity, density as well as porosity and water saturation computed from density and resistivity.

Three litho-fluid facies have been identified based on core samples: oil-saturated high-porosity carbonate (HPC), partially oil-saturated mid-porosity carbonate (MPC), and low-porosity carbonate (LPC). The average porosity is 0.13 in the HPC facies and 0.06 in the MPC, whereas the LPC has porosity close to 0. The rock physics model adopted in de Figueiredo et al. (2018) is based on an empirical rock physics relation for the dry-rock elastic properties combined with Gassmann's equations for the fluid effect.

The McMC method proposed in de Figueiredo et al. (2018) aims to simultaneously predict the petrophysical and elastic properties and the facies classification from seismic data. The method is based on two steps: first, an approximate posterior distribution of the model variables is computed using a Bayesian linearized approach at each location of the reservoir model, then a Gibbs sampling algorithm is applied to sample posterior realizations using the approximate distribution as the proposal distribution. The Gibbs algorithm computes the posterior distribution by repeatedly sampling the conditional distributions of each variable given all the other variables. The algorithm then samples multiple realizations according to the selected spatial correlation model using geostatistical algorithms. Discrete properties are simulated using the truncated Gaussian simulation method (Section 4.3),

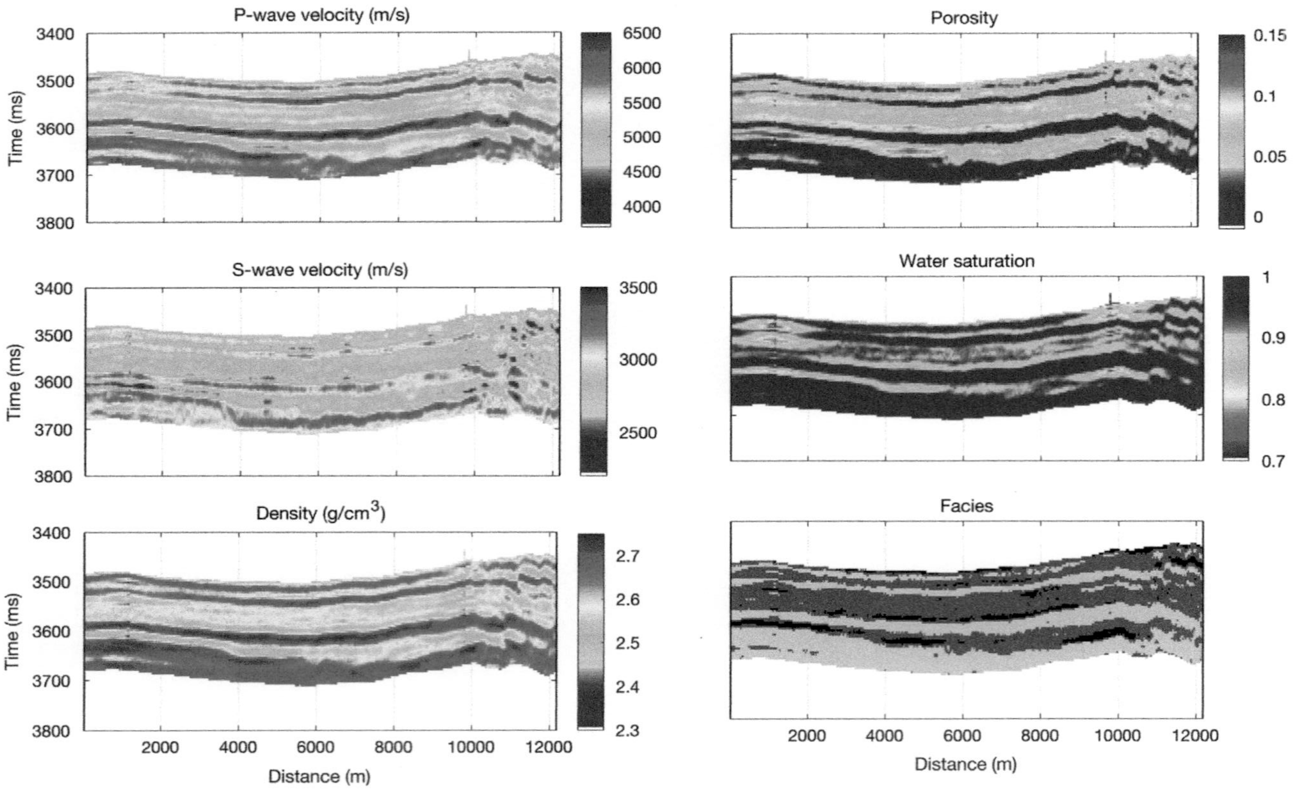

Figure 8.7 Predicted reservoir model of petro-elastic properties and facies along an arbitrary two-dimensional seismic line: the results show the posterior mean of P-wave and S-wave velocity, density, porosity, and water saturation, and the most likely facies model (HPC in black, MPC in brown, LPC in tan). Modified after: De Figueiredo, L.P., Grana, D., Bordignon, F.L., Santos, M., Roisenberg, M. and Rodrigues, B.B., 2018. Joint Bayesian inversion based on rock-physics prior modeling for the estimation of spatially correlated reservoir properties. Geophysics, 83(5), pp.M49-M61.

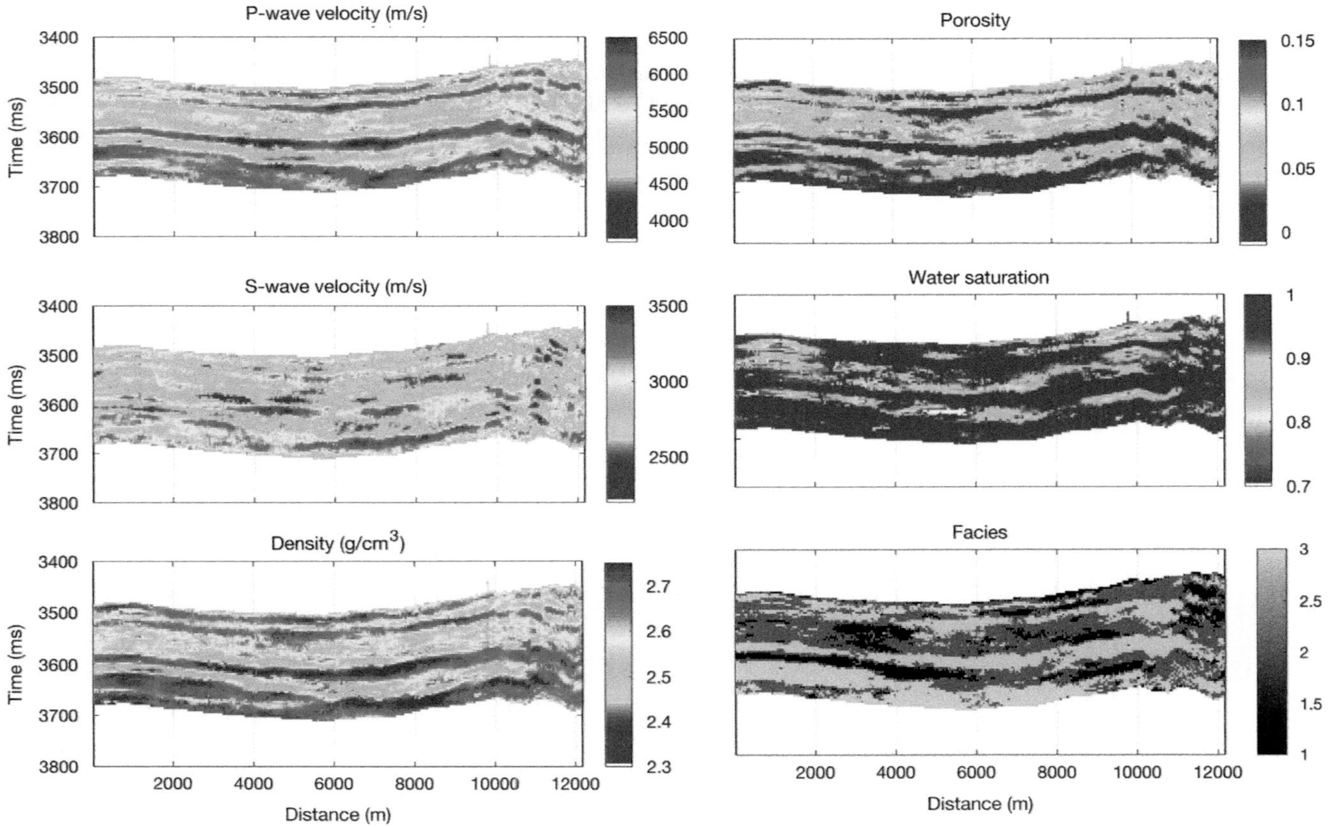

Figure 8.8 Stochastic realization of the reservoir model of petro-elastic properties and facies along an arbitrary two-dimensional seismic line: the results show one random realization of P-wave and S-wave velocity, density, porosity, and water saturation, and facies (HPC in black, MPC in brown, LPC in tan). Modified after: De Figueiredo, L.P., Grana, D., Bordignon, F.L., Santos, M., Roisenberg, M. and Rodrigues, B.B., 2018. Joint Bayesian inversion based on rock-physics prior modeling for the estimation of spatially correlated reservoir properties. Geophysics, 83(5), pp.M49-M61.

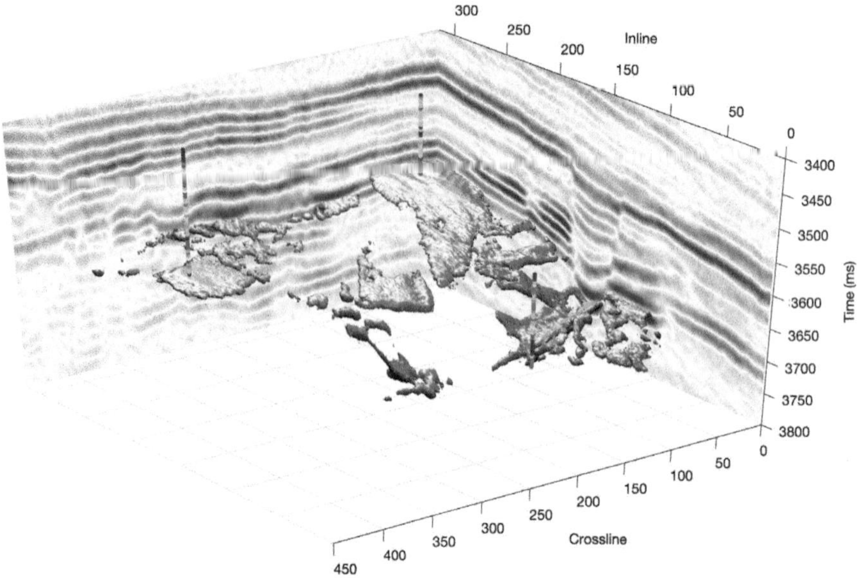

Figure 8.9 Isosurface of high-porosity carbonate probability higher than 0.65. The seismic data are shown in the background. Modified after: De Figueiredo, L.P., Grana, D., Bordignon, F.L., Santos, M., Roisenberg, M. and Rodrigues, B.B., 2018. Joint Bayesian inversion based on rock-physics prior modeling for the estimation of spatially correlated reservoir properties. Geophysics, 83(5), pp. M49-M61.

and continuous properties using the FFT-MA method (Section 3.6). The spatial correlation model includes three-dimensional spherical variograms with correlation lengths estimated from well log data for the vertical component and seismic data for the lateral component.

Figure 8.7 shows the mean of the model properties and the most likely facies model based on the Gibbs sampling realizations, whereas Figure 8.8 shows one random realization of each property. Both sets of solutions are equally consistent with the input seismic data but the simulations exhibit a higher frequency content controlled by the spherical covariance model.

In Figure 8.9, we display the isosurface of the HPC facies probability based on the Gibbs sampling realizations of the facies. The isosurface defines the regions with HPC facies probability higher than 0.65 in the three-dimensional reservoir model. The proposed approach takes approximately 10 hours, on a standard workstation, to calculate the posterior distribution and to sample 150 models (including 15 realizations of the burn-in period), for a three-dimensional model with 141750 seismic traces of 312 samples. De Figueiredo et al. (2018) apply the inversion using different initial models to validate the convergence of the algorithm and propose a comparison of the marginal distributions of the elastic properties obtained using the McMC method and those obtained using the Bayesian linearized approach, showing that the proposed McMC method better captures the multimodal behavior of the data.

8.2 CO_2 Sequestration Study

Several methodologies for seismic and petrophysical inversion have been developed for seismic reservoir characterization problems in hydrocarbon reservoirs. However, these methods can be applied to other subsurface modeling problems, including CO_2 sequestration, groundwater aquifers, geothermal energy, gas hydrate reservoirs, and near-surface geophysics studies.

Carbon dioxide sequestration in deep saline aquifers and depleted reservoirs can potentially reduce the CO_2 concentration in the atmosphere owing to anthropogenic activities. CO_2 sequestration requires accurate modeling methods for designing and monitoring carbon capture and storage to reduce the risks of leakage. Modeling approaches for assessing the storage capacity and for monitoring the CO_2 migration during and after injection are generally based on geophysical and fluid flow simulation methods (Davis et al. 2019; Ringrose 2020). The storage capacity, the CO_2 plume location, and the pressure extent are often uncertain owing to the heterogeneity of the petrophysical and geological properties in the subsurface. Predictive models and uncertainty quantification methods based on geophysical data have been proposed to estimate the spatial distribution of rock and fluid properties and the storage capacity in CO_2 studies in depleted reservoirs and deep saline aquifers (Davis et al. 2019; Ringrose 2020). Reservoir geophysics characterization and monitoring methods includes seismic and electromagnetic inversion and rock physics modeling.

In this section, we illustrate the application of a stochastic modeling approach to a CO_2 sequestration study to predict the spatial distribution of CO_2 saturation and its uncertainty during injection. The case study is based on the Johansen formation model (Eigestad et al. 2009; Bergmo et al. 2011). The Johansen formation is a deep saline aquifer located below the Troll hydrocarbon field, offshore Norway, and has been identified as a potential CO_2 storage unit, for the favorable geological features of the formation and of the overlying sealing, including pore volume, storage capacity, permeability, and pressure conditions. The structural model of the formation and the reservoir model of porosity and permeability have been presented by Bergmo et al. (2011). The model includes a major fault interpreted from seismic data. Porosity and permeability are relatively high in the upper layers of the model and tend to decrease in the lower layers. The CO_2 injection is simulated using the MRST-co2lab (Lie 2019) with a constant injection rate for a period of 100 years and the migration is simulated for 400 years after stopping CO_2 injection.

Figure 8.10 shows the reservoir model of porosity and permeability. Synthetic time-lapse seismic data have been generated in a sub-volume of the original model, centered around the injection well, every 5 years during the injection phase and every 20 years during the following migration phase. In this example, we predict the CO_2 saturation from seismic data, 10 years after stopping CO_2 injection. The base and monitor seismic surveys before and after injection are shown in Figure 8.10. The signal-to-noise ratio of the data is assumed to be 10, with uncorrelated data errors.

We apply an inversion method based on the ensemble smoother (Section 5.6.2) where the geophysical forward model includes a rock physics model to predict the elastic response of the fluid-saturated rocks based on porosity and fluid saturations and a seismic convolutional model (Liu and Grana 2020). The rock physics model is based on the stiff sand model (Section 2.4) and Gassmann's equations (Section 2.6), using the fluid parameters of brine and CO_2 at pressure and temperature conditions of the aquifer. The initial ensemble of models is generated in two steps: first, porosity and permeability realizations are simulated using the FFT-MA method (Section 3.6), then the CO_2 saturation realizations are generated using dynamic simulations of CO_2 injection and selecting a saturation model at a random time according to a uniform distribution of the simulation time. This approach guarantees a large variability in the initial ensemble and, at the same time, preserves a realistic shape of the

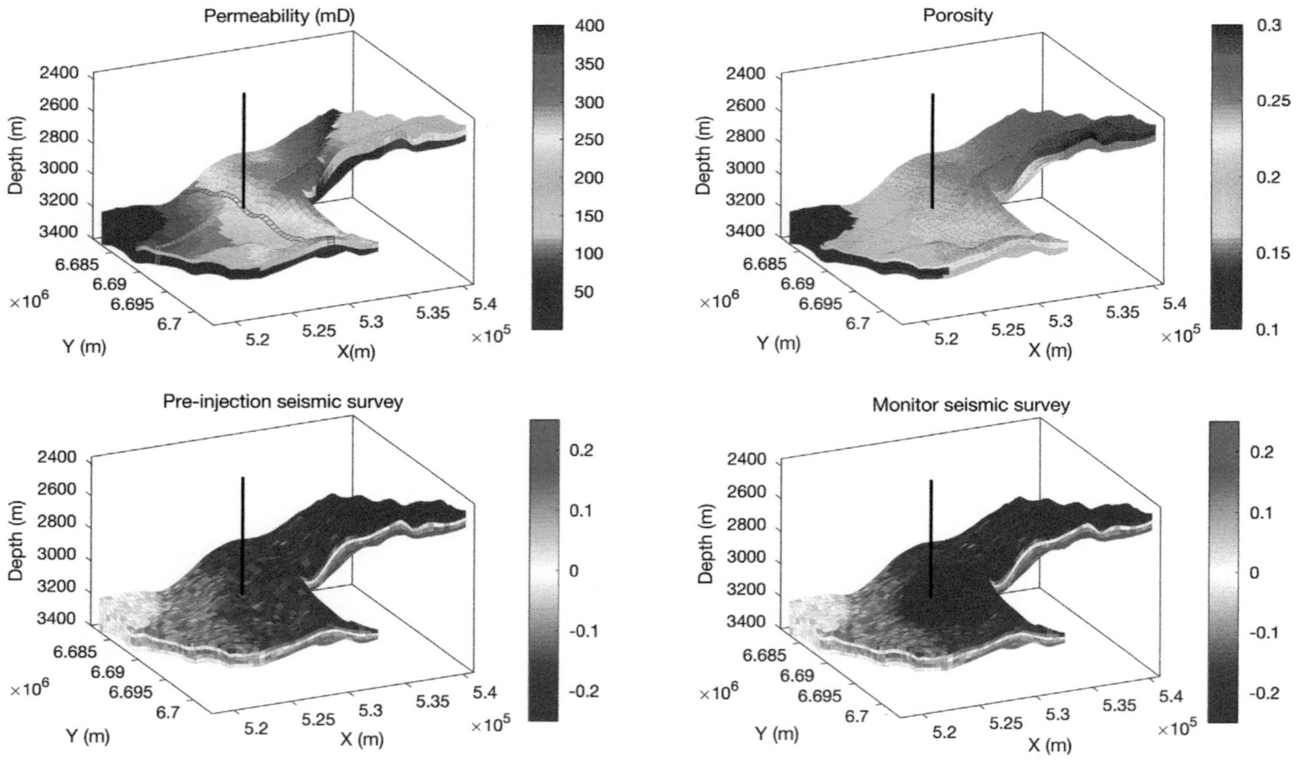

Figure 8.10 Structural and reservoir model of the Johansen formation and simulated geophysical data: (top) reservoir model of permeability and porosity; (bottom) synthetic time-lapse seismic data in a sub-volume centered around the injection well. The black line represents the location of the injection well.

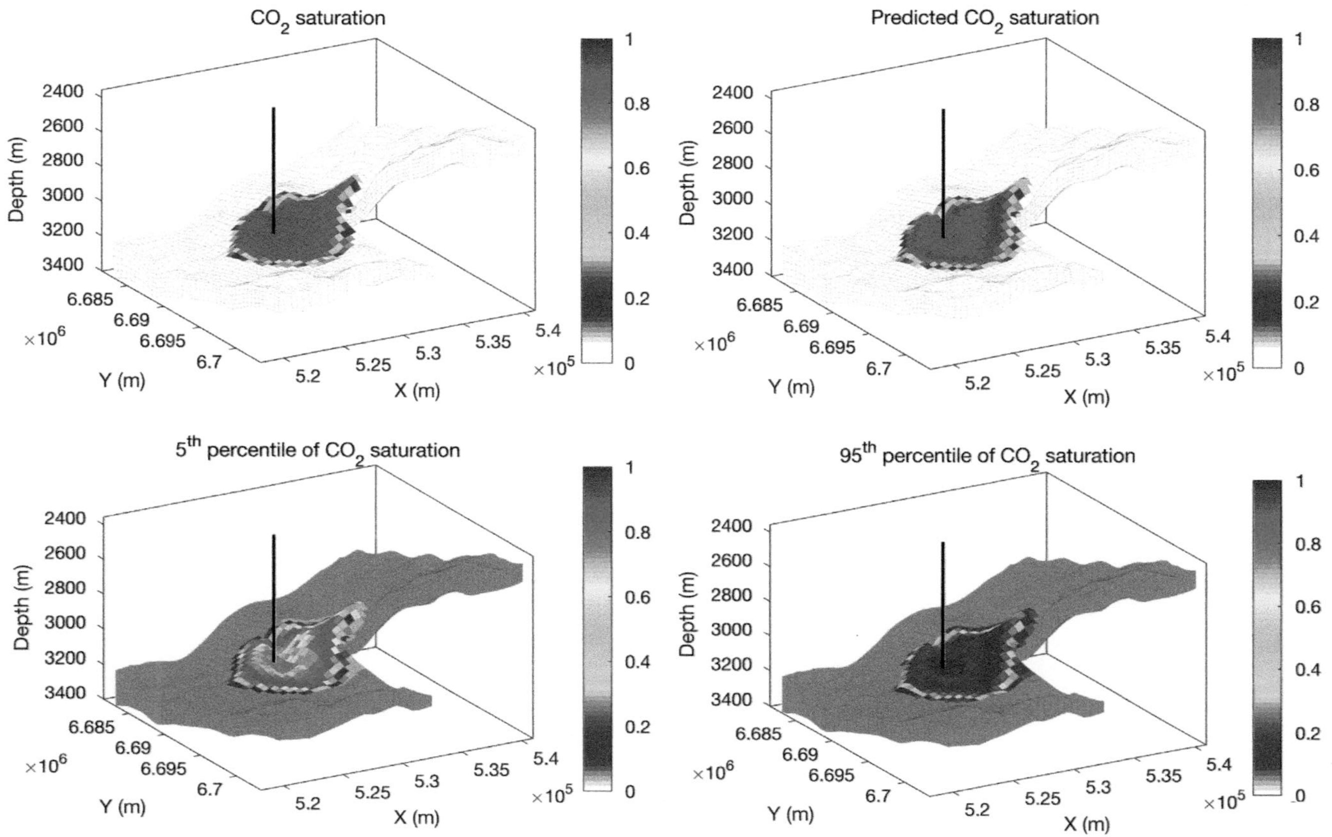

Figure 8.11 Inversion results for CO_2 saturation: (top) reference model of CO_2 saturation and posterior mean of CO_2 saturation predicted from time-lapse geophysical data; (bottom) 5th and 95th percentiles of CO_2 saturation.

CO_2 spatial distribution. We then compute the posterior probability distribution of the CO_2 saturation distribution conditioned on seismic data, using the ensemble smoother. A logit transformation is applied to the saturation models and a dimensionality reduction using the deep convolutional autoencoder is applied to the data (Liu and Grana 2020). The posterior distribution of the CO_2 saturation is shown in Figure 8.11, where the predicted mean is compared with the actual CO_2 model. The posterior uncertainty is represented by the 0.90 confidence interval and it is visualized using the 5th and 95th percentile maps of CO_2 saturation. Overall, the predictions are accurate and match the actual model. The correlation coefficient of the maximum a posteriori of CO_2 saturation and the actual model is 0.92. The confidence intervals are relatively narrow, except close to the transition from CO_2 to water, where the uncertainty is generally higher. The coverage ratio of the 0.90 confidence interval is 0.87. If we assume that porosity is unknown, and jointly estimate porosity and water saturation conditioned on the base and monitor surveys, the correlation coefficient of the predictions of CO_2 saturation and the actual model decreases to 0.81, whereas the coverage ratio of the 0.90 confidence interval becomes 0.86. These results also depend on the high resolution and signal-to-noise of the data as well as the perfect repeatability of the time-lapse seismic surveys.

The complete case studies for seismic and CSEM inversion are published in Liu and Grana (2020) and Ayani et al. (2020). The proposed methodology can also be applied to integrated studies of enhanced oil recovery (EOR) and carbon sequestration, also known as carbon capture, utilization, and storage projects, which include CO_2 injection and storage as part of EOR schemes. However, these studies generally require compositional models for the fluid flow simulation.

Appendix: MATLAB Codes

In this Appendix, we provide a detailed description of the MATLAB codes for rock physics, geostatistics, inversion, and facies modeling. The complete MATLAB package Seismic Reservoir Modeling (SeReM) and the Python version (SeReMpy) can be downloaded from the following website: https://seismicreservoirmodeling.github.io/SeReM/.

The SeReM package includes five folders:

- *Data*: this folder contains six datasets used for the examples and the elevation dataset from Yellowstone National Park shown in Figure 3.10.
- *RockPhysics*: this folder contains functions for several rock physics models described in Chapter 2.
- *Geostats*: this folder contains functions for kriging and geostatistical simulations of random variables described in Chapters 3 and 4.
- *Inversion*: this folder contains functions for seismic and rock physics inversion using the analytical and numerical solutions described in Chapter 5; the functions are subdivided into three subfolders: *Seismic*, *Petrophysical*, and *EnsembleSmoother*.
- *Facies*: this folder contains functions for facies classification and simulation described in Chapter 6.

The SeReM package includes multiple scripts with several examples. The following sections describe the MATLAB codes in detail.

A.1 Rock Physics Modeling

The folder *RockPhysics* includes the following rock physics models:

- *DensityModel*: this function implements the linear porosity-density relation to compute density.
- *MatrixFluidModel*: this function implements Voigt–Reuss–Hill averages to compute the elastic moduli and density of the solid and fluid phases.
- *GassmannModel*: this function implements Gassmann's equations to compute the elastic moduli of the fluid-saturated rock.
- *VelocityDefinitions*: this function implements the definitions of P-wave and S-wave velocity.

Seismic Reservoir Modeling: Theory, Examples, and Algorithms, First Edition. Dario Grana, Tapan Mukerji, and Philippe Doyen.

- *LinearizedRockPhysicsModel*: this function implements a linear rock physics model based on a multilinear regression to compute P-wave and S-wave velocity and density.
- *WyllieModel*: this function implements Wyllie's equation to compute P-wave velocity.
- *RaymerModel*: this function implements Raymer's equation to compute P-wave velocity.
- *SoftsandModel*: this function implements Dvorkin's soft sand model to compute P-wave and S-wave velocity.
- *StiffsandModel*: this function implements Dvorkin's stiff sand model to compute P-wave and S-wave velocity.
- *SphericalInclusionModel*: this function implements the inclusion model for spherical pores to compute P-wave and S-wave velocity.
- *BerrymanInclusionModel*: this function implements Berryman's inclusion model for prolate and oblate pores to compute P-wave and S-wave velocity.

The script *RockPhysicsModelDriver* illustrates the application of different rock physics models to compute P-wave and S-wave velocity and density from a well log of porosity, provided in the workspace *data1*. We assume that the mineral is quartz and the fluid is water.

We first load the data and define the elastic parameters for quartz and water.

```
addpath(genpath('../SeReM/'))
load Data/data1
Kmat=36;
Gmat=45;
Rhomat=2.65;
Kfl=2.25;
Gfl=0;
Rhofl=1;
```

In the first example, we compute the elastic properties using empirical models, namely Wyllie and Raymer's equations (Section 2.1). We first compute the velocities of the solid and fluid phases by definition.

```
Vpmat = VelocityDefinitions(Kmat, Gmat, Rhomat);
Vpfl = VelocityDefinitions(Kfl, Gfl, Rhofl);
```

We then apply Wyllie and Raymer's equations to compute P-wave velocity.

```
VpW = WyllieModel(Phi, Vpmat, Vpfl);
VpR = RaymerModel(Phi, Vpmat, Vpfl);
```

In the second example, we compute the elastic properties using granular media models, namely soft and stiff sand models (Section 2.4). We first define the parameters of the granular media model: critical porosity, coordination number, and pressure. We then compute the density using the linear porosity-density model, and the P-wave and S-wave velocity using the soft and stiff sand models.

```
criticalporo=0.4;
coordnumber=7;
pressure=0.02;
Rho = DensityModel(Phi, Rhomat, Rhofl);
```

```
  [VpSoft, VsSoft] = SoftsandModel(Phi, Rho, Kmat, Gmat, Kfl,
criticalporo, coordnumber, pressure);
  [VpStiff, VsStiff] = StiffsandModel(Phi, Rho, Kmat, Gmat, Kfl,
criticalporo, coordnumber, pressure);
```

In the third example, we compute the elastic properties using inclusion models for spherical and elliptical pores (Section 2.5). We first define the aspect ratio of the elliptical pores and then compute the density using the linear porosity-density model, and the P-wave and S-wave velocity using the inclusion models.

```
  Ar=0.2;
  Rho = DensityModel(Phi, Rhomat, Rhofl);
  [VpSph, VsSph] = SphericalInclusionModel(Phi, Rho, Kmat, Gmat,
Kfl);
  [VpEll, VsEll] = BerrymanInclusionModel(Phi, Rho, Kmat, Gmat,
Kfl, Ar);
```

The script provides different options to plot the results in the form of well log predictions or rock physics templates. The script can be modified to account for multiple mineral and fluid components using the function *MatrixFluidModel*, which returns the elastic moduli and density of the effective solid and fluid phases based on the mineral volumes and fluid saturations (Section 2.2).

A.2 Geostatistical Modeling

The folder *Geostatistics* includes the following functions for the interpolation and simulation of discrete and continuous random variables:

- *ExpCov*: this function implements the exponential spatial covariance model.
- *GauCov*: this function implements the Gaussian spatial covariance model.
- *SphCov*: this function implements the spherical spatial covariance model.
- *SpatialCovariance1D*: this function implements the 1D spatial covariance function according to one of the available models (exponential, Gaussian, and spherical).
- *RadialCorrLength*: this function implements the radial correlation length for two-dimensional spatial covariance functions.
- *SpatialCovariance2D*: this function implements the two-dimensional spatial covariance function according to one of the available models (exponential, Gaussian, and spherical).
- *SimpleKriging*: this function implements the simple kriging interpolation at a given location based on a set of measurements.
- *OrdinaryKriging*: this function implements the ordinary kriging interpolation at a given location based on a set of measurements.
- *IndicatorKriging*: this function implements the indicator kriging interpolation at a given location based on a set of measurements.
- *GaussianSimulation*: this function implements a Gaussian simulation to generate a sample of a random variable at a given location based on a set of measurements.

- *SeqGaussianSimulation*: this function implements the sequential Gaussian simulation method to generate spatially correlated realizations of a continuous random variable based on a set of measurements.
- *SeqIndicatorSimulation*: this function implements the sequential indicator simulation method to generate spatially correlated realizations of a discrete random variable based on a set of measurements.
- *CorrelatedSimulation*: this function implements a sampling approach to simulate spatially correlated stochastic realizations of multiple random variables.
- *MarkovChainSimulation*: this function implements a sampling approach to simulate multiple one-dimensional realizations of a discrete random variable based on a stationary first-order Markov chain.

The script *GeostatsContinuousDriver* illustrates two examples for the interpolation and the geostatistical simulation of a continuous random variable based on a set of measurements at different locations.

In the first example, we assume that four measurements are available as shown in Example 3.2 (Figure 3.9). We first define the coordinates and the values of the measurements as well as the coordinates of the location of the interpolation.

```
dcoords = [ 5 18; 15 13; 11 4; 1 9 ];
dvalues = [ 3.1 3.9 4.1 3.2 ]';
xcoord = [ 10 10 ];
```

We then define the prior mean, prior variance, and the parameters of the spatial covariance function.

```
xmean = 3.5;
xvar = 0.1;
l = 9;
type = 'exp';
```

We then compute the simple kriging estimate and the ordinary kriging estimate (Section 3.4) and draw multiple samples from a Gaussian distribution with the kriging parameters.

```
 [xsk, ~] = SimpleKriging(xcoord, dcoords, dvalues, xmean, xvar,
l, type);
 [xok, ~] = OrdinaryKriging(xcoord, dcoords, dvalues, xvar, l,
type);
 krig = 0;
 nsim = 100;
 gsim = zeros(nsim,1);
 for i=1:nsim
   gsim(i) = GaussianSimulation(xcoords, dcoords, dvalues,
xmean, xvar, l, type, krig);
 end
```

In the second example, we use a sparse subset of measurements from the Yellowstone dataset as shown in Example 3.3 (Figure 3.10). The coordinates of the simulation grid are stored in the workspace *ElevationData*, whereas the coordinates and the elevation values of the measurements of the sparse dataset are stored in *data6*. The workspace *ElevationData* also includes the entire dataset and a subset of 100 measurements.

```
load Data/ElevationData.mat
load Data/data6.mat
dcoords = [ dx dy ];
nd = size(dcoords,1);
xcoords = [ X(:) Y(:) ];
n = size(xcoords,1);
```

The prior mean, prior variance, and the parameters of the spatial covariance function have been estimated from the true dataset available in the workspace *ElevationData*.

```
xmean = 2476;
xvar = 8721;
l = 12.5;
type = 'exp';
```

We then compute the simple kriging estimate and the ordinary kriging estimate (Section 3.4) and generate multiple spatially correlated realizations using the sequential Gaussian simulation method (Section 3.5).

```
for i=1:n
  [xsk(i), ~] = SimpleKriging(xcoords(i,:), dcoords, dz, xmean,
xvar, l, type);
   [xok(i), ~] = OrdinaryKriging(xcoords(i,:), dcoords, dz,
xvar, l, type);
end
xsk = reshape(xsk,size(X));
xok = reshape(xok,size(X));
krig = 1;
nsim = 3;
sgsim = zeros(size(X,1),size(X,2),nsim);
for i=1:nsim
  sim = SeqGaussianSimulation(xcoords, dcoords, dz, xmean,
xvar, l, type, krig);
  sgsim(:,:,i) = reshape(sim,size(X,1),size(X,2));
end
```

The script *GeostatsDiscreteDriver* illustrates two examples for the interpolation and the geostatistical simulation of a discrete random variable and an example of simulation using Markov chain models. The first two examples are based on the same datasets of the continuous case. The binary variables are defined by applying a threshold to the continuous measurements.

In the first example, the discrete variable represents two geological facies, namely sand and shale, as shown in Example 4.1.

```
dcoords = [ 5 18; 15 13; 11 4; 1 9 ];
fvalues = [ 1 2 2 1 ]';
xcoords = [ 10 10 ];
nf = 2;
pprior = [ 0.5 0.5];
l = 9;
type = 'exp';
```

We compute the indicator kriging probability and the maximum a posteriori (Section 4.1), and we draw multiple samples from the indicator kriging probability using the function *RandDisc* in the *Facies* folder (Section A.4).

```
[ikp, ikmap] = IndicatorKriging(xcoords, dcoords, fvalues, nf,
pprior, l, type);
nsim = 1000;
isim = zeros(nsim,1);
for i=1:nsim
   isim(i) = RandDisc(ikp);
end
```

In the second example, the discrete variable represents valleys and peaks as shown in Example 4.2.

```
load Data/ElevationData.mat
load Data/data6.mat
dcoords = [ dx dy ];
nd = size(dcoords,1);
zmean = 2476;
df = ones(size(dz));
df(dz>zmean) = 2;
xcoords = [ X(:) Y(:) ];
n = size(xcoords,1);
pprior = [ 0.5 0.5 ];
l = 12.5;
type = 'exp';
```

We then compute the indicator kriging probability and the maximum a posteriori (Section 4.1) and we generate multiple spatially correlated realizations using the sequential indicator simulation method (Section 4.2).

```
ikp = zeros(n,nf );
for i=1:n
   [ikp(i,:), ikmap(i)] = IndicatorKriging(xcoords(i,:),
dcoords, df, nf, pprior, l, type);
end
ikp = reshape(ikp,size(X,1),size(X,2),nf );
ikmap = reshape(ikmap,size(X,1),size(X,2));
nsim = 3;
```

```
  sisim = zeros(size(X,1),size(X,2),nsim);
  for i=1:nsim
    sim = SeqIndicatorSimulation(xcoords, dcoords, df, nf,
pprior, 1, type);
    sisim(:,:,i) = reshape(sim,size(X,1),size(X,2));
  end
```

In the third example, we illustrate the simulation of multiple vertical profiles of facies based on a stationary first-order Markov chain (Section 4.4). We first define the initial parameters of the simulation, including the number of realizations, the number of samples, and the transition matrix. The driver illustrates three cases with different transition matrices as in Example 4.6.

```
nsim = 3;
ns = 100;
T1 = [ 0.5 0.5; 0.5 0.5];
T2 = [ 0.9 0.1; 0.1 0.9];
T3 = [ 0.1 0.9; 0.1 0.9];
```

We then generate one-dimensional stochastic realizations of the facies vertical profile using the function *MarkovChainSimulation.*

```
fsim1 = MarkovChainSimulation(T1, ns, nsim);
fsim2 = MarkovChainSimulation(T2, ns, nsim);
fsim3 = MarkovChainSimulation(T3, ns, nsim);
```

For more efficient implementations of sequential simulation methods, including anisotropic searching neighborhoods and extension to three-dimensional applications, we recommend the open-source software SGeMS (Remy et al. 2009), the library GSLib (Deutsch and Journel 1998), or the packages Gstat (Pebesma 2004) and mGstat (Hansen 2004).

A.3 Inverse Modeling

The folder *Inversion* includes functions for seismic and petrophysical inversion. We first introduce the seismic forward model, then we present the Bayesian linearized AVO inversion and the Bayesian rock physics inversion, and we illustrate the stochastic inversion based on the ensemble smoother multiple data assimilation (ES-MDA) method.

The folder *Inversion* includes the following functions subdivided into three subfolders, namely *Seismic*, *Petrophysical*, and *EnsembleSmoother*:

- *RickerWavelet*: this function computes a Ricker wavelet with given dominant frequency.
- *AkiRichardsCoefficientsMatrix*: this function computes the Aki–Richards coefficient matrix.
- *DifferentialMatrix*: this function computes the differential matrix for discrete differentiation.
- *WaveletMatrix*: this function computes the wavelet matrix for discrete convolution.

- *SeismicModel*: this function computes synthetic seismic data according to a linearized seismic model based on the convolution of a wavelet and the linearized approximation of Zoeppritz equations.
- *SeismicInversion*: this function computes the posterior distribution of elastic properties according to the Bayesian linearized AVO inversion.
- *RockPhysicsLinGaussInversion*: this function computes the posterior distribution of petrophysical properties conditioned on elastic properties assuming a Gaussian distribution and a linear rock physics model.
- *RockPhysicsLinGaussMixInversion*: this function computes the posterior distribution of petrophysical properties conditioned on elastic properties assuming a Gaussian mixture distribution and a linear rock physics model.
- *RockPhysicsGaussInversion*: this function computes the posterior distribution of petrophysical properties conditioned on elastic properties assuming a Gaussian distribution estimated from a training dataset.
- *RockPhysicsGaussMixInversion*: this function computes the posterior distribution of petrophysical properties conditioned on elastic properties assuming a Gaussian mixture distribution estimated from a training dataset.
- *RockPhysicsKDEInversion*: this function computes the posterior distribution of petrophysical properties conditioned on elastic properties assuming a non-parametric distribution estimated from a training dataset using kernel density estimation.
- *EpanechnikovKernel*: this function computes the Epanechnikov kernel used in kernel density estimation.
- *EnsembleSmootherMDA*: this function computes the updated model realizations of the model variables conditioned on seismic data using the ES-MDA method.
- *LogitBounded*: this function computes the logit transformation for bounded properties.
- *InvLogitBounded*: this function computes the inverse logit transformation for bounded properties.

A.3.1 Seismic Inversion

The script *SeismicModelDriver* illustrates the implementation of the seismic forward model based on the convolution of a wavelet and the linearized approximation of Zoeppritz equations (Section 5.1). The model is applied to a set of well logs, including P-wave and S-wave velocity and density, saved in the workspace *data2*.

We first load the data from the workspace and define the number of variables of the parameterization (three in this example) and the reflection angles of the seismograms.

```
load Data/data2.mat
nv = 3;
theta = [15, 30, 45];
```

Because the data are in depth domain, we compute the corresponding two-way travel time and interpolate the well logs on a time vector with constant sampling rate. The number of samples of the seismogram, for each angle, is the number of samples of the velocity log in time domain minus 1.

```
dt = 0.001;
t0 = 1.8;
TimeLog = [t0; t0 + 2*cumsum(diff(Depth)./(Vp(2:end,:)))];
Time = (TimeLog(1):dt:TimeLog(end))';
Vp = interp1(TimeLog, Vp, Time);
Vs = interp1(TimeLog, Vs, Time);
Rho = interp1(TimeLog, Rho, Time);
nd = length(Vp)-1;
```

We define a Ricker wavelet with a given dominant frequency.

```
freq = 45;
ntw = 64;
[wavelet, tw] = RickerWavelet(freq, dt, ntw);
```

We then compute the seismic data using the function *SeismicModel*.

```
[Seis, TimeSeis] = SeismicModel (Vp, Vs, Rho, Time, theta,
wavelet);
Snear = Seis(1:nd);
Smid = Seis(nd+1:2*nd);
Sfar = Seis(2*nd+1:end);
```

The script *SeismicInversionDriver* illustrates the implementation of the Bayesian seismic inversion (Section 5.3) based on the linearized AVO approximation implemented in *SeismicModel*. The inversion is applied to a set of three seismograms, including near-, mid- and far-angle stacks, saved in the workspace *data3*. The model properties of interest are P-wave and S-wave velocity and density.

We first load the data from the workspace and define the number of variables of the parameterization (three in this example) and the reflection angles of the seismograms. The number of samples of the model, for each elastic variable, is the number of samples of the seismogram plus 1. We also define the parameters of the distribution of the data error.

```
load Data/data3.mat
nm = size(Snear,1)+1;
nv = 3;
theta = [15, 30, 45];
ntheta = length(theta);
dt = TimeSeis(2)-TimeSeis(1);
varerr = 10^-4;
sigmaerr = varerr*eye(ntheta*(nm-1));
```

We define a Ricker wavelet with a given dominant frequency. In practical applications, the wavelet should be estimated from the measured data.

```
freq = 45;
ntw = 64;
[wavelet, tw] = RickerWavelet(freq, dt, ntw);
```

We introduce a prior model by filtering the well log data at the seismic resolution and we define the time covariance matrix as the Kronecker product of the stationary covariance matrix of the data and the time-dependent correlation matrix.

```
nfilt = 3;
cutofffr = 0.04;
[b, a] = butter(nfilt, cutofffr);
Vpprior = filtfilt(b, a, Vp);
Vsprior = filtfilt(b, a, Vs);
Rhoprior = filtfilt(b, a, Rho);
corrlength = 5*dt;
trow = repmat(0:dt:(nm-1)*dt,nm,1);
tcol = repmat((0:dt:(nm-1)*dt)',1,nm);
tdis = abs(trow-tcol);
sigmatime = exp(-(tdis./corrlength).^2);
sigma0 = cov([log(Vp),log(Vs),log(Rho)]);
sigmaprior = kron(sigma0, sigmatime);
```

We then compute the statistical parameters of the posterior Gaussian distribution of elastic properties conditioned on seismic data, using the function *SeismicInversion*. The function returns the maximum a posteriori and the lower and upper bounds of the 0.95 confidence interval.

```
Seis = [Snear; Smid; Sfar];
[mmap, mlp, mup, Time] = SeismicInversion(Seis, TimeSeis, Vpprior,
Vsprior, Rhoprior, sigmaprior, sigmaerr, wavelet, theta, nv);
Vpmap = mmap(1:nm);
Vsmap = mmap(nm+1:2*nm);
Rhomap = mmap(2*nm+1:end);
Vplp = mlp(1:nm);
Vslp = mlp(nm+1:2*nm);
Rholp = mlp(2*nm+1:end);
Vpup = mup(1:nm);
Vsup = mup(nm+1:2*nm);
Rhoup = mup(2*nm+1:end);
```

A.3.2 Petrophysical Inversion

The script *RockPhysicsInversionDriver* illustrates the implementation of the Bayesian rock physics inversion methods described in Section 5.4, for linear and non-linear forward models assuming Gaussian, Gaussian mixture, and non-parametric distributions of the model variables. The inversion is applied to a set of well logs, including P-wave and S-wave velocity and density, saved in the workspace *data4*. The model properties of interest are porosity, clay volume, and water saturation. The well logs of the petrophysical properties are included in the workspace for the validation of the results.

We first load the data from the workspace including the measured well logs and a training dataset. The training dataset includes rock physics model predictions obtained by applying a non-linear rock physics model to the well logs of petrophysical properties. The training dataset also includes a facies classification, for the application of the Gaussian mixture model.

```
load Data/data4.mat
mtrain = [Phi Clay Sw];
nv = size(mtrain,2);
dtrain = [Vprpm Vsrpm Rhorpm];
nd = size(dtrain,2);
dcond = [Vp Vs Rho];
ns = size(dcond,1);
nf = max(unique(Facies));
```

We define a discretized domain for the evaluation of the posterior distribution using a multidimensional grid.

```
phidomain = (0:0.005:0.4);
cdomain = (0:0.01:0.8);
swdomain = (0:0.01:1);
[P,V,S] = ndgrid(phidomain, cdomain, swdomain);
mdomain = [P(:) V(:) S(:)];
```

We define a linearized rock physics model by estimating the coefficients of the multilinear regression from the training dataset. The coefficients are stored in the matrix associated with the rock physics model. We also define the covariance matrix of the data errors.

```
R = zeros(nd,nv+1);
X = [mtrain ones(size(Phi))];
R(1,:) = regress(Vprpm,X);
R(2,:) = regress(Vsrpm,X);
R(3,:) = regress(Rhorpm,X);
sigmaerr = 10^-2*eye(nd,nd);
```

In the first example, we assume a Gaussian distribution of the model variables and a linearized rock physics model. The parameters of the Gaussian distribution are estimated from the actual well logs.

```
mum = mean(mtrain);
sm = cov(mtrain);
G = R(:,1:nv);
datacond = dcond-R(:,end)';
```

We then apply the Bayesian linearized rock physics inversion described in Section 5.4.1 to obtain the posterior distribution of the petrophysical properties using the function *RockPhysicsLinGaussInversion*. The function returns the posterior mean, the posterior covariance matrix, and the full posterior distribution evaluated on the multidimensional grid for each measured data. The marginal distributions are obtained by numerically integrating the posterior distribution.

```
  [mupost, sigmapost, Ppost] = RockPhysicsLinGaussInversion(mum,
sm, G, mdomain, datacond, sigmaerr);
  Phipost = mupost(:,1);
  Cpost = mupost(:,2);
  Swpost = mupost(:,3);
```

In the second example, we assume a Gaussian mixture distribution of the model variables and a linearized rock physics model. The parameters of the Gaussian mixture distribution are estimated from the actual well logs and the corresponding facies classification.

```
  mum = zeros(nf,nv);
  sm = zeros(nv,nv,nf );
  pf = zeros(nf,1);
  for k=1:nf
    pf(k) = sum(Facies==k)/ns;
    mum(k,:) = mean(mtrain(Facies==k,:));
    sm(:,:,k) = cov(mtrain(Facies==k,:));
  end
```

We then apply the Bayesian linearized rock physics inversion described in Section 5.4.2 to obtain the posterior distribution of the petrophysical properties using the function *RockPhysicsLinGaussMixInversion*. The function returns the full posterior distribution evaluated on the multidimensional grid for each measurement. The marginal distributions are obtained by numerically integrating the posterior distribution. The maximum a posteriori is also numerically evaluated.

```
  [~, ~, ~, Ppost] = RockPhysicsLinGaussMixInversion(pf, mum, sm,
G, mdomain, datacond, sigmaerr);
  for i=1:ns
    Ppostjoint = reshape(Ppost(i,:),length(phidomain),
  length(cdomain),length(swdomain));
    Ppostphi(i,:)=sum(squeeze(sum(squeeze(Ppostjoint),3)),2);
    Ppostclay(i,:)=sum(squeeze(sum(squeeze(Ppostjoint),3)),1);
    Ppostsw(i,:)=sum(squeeze(sum(squeeze(Ppostjoint),2)),1);
    Ppostphi(i,:)=Ppostphi(i,:)/sum(Ppostphi(i,:));
    Ppostclay(i,:)=Ppostclay(i,:)/sum(Ppostclay(i,:));
    Ppostsw(i,:)=Ppostsw(i,:)/sum(Ppostsw(i,:));
    [~,Phimapind]=max(Ppostphi(i,:));
    [~,Cmapind]=max(Ppostclay(i,:));
    [~,Swmapind]=max(Ppostsw(i,:));
    Phimap(i)=phidomain(Phimapind);
    Cmap(i)=cdomain(Cmapind);
    Swmap(i)=swdomain(Swmapind);
  end
```

In the third example, we estimate the parameters of the Gaussian mixture joint distribution of the model and data variables from the training dataset and we apply the Bayesian rock physics inversion described in Section 5.4.3 to obtain the posterior distribution of the petrophysical properties conditioned on the elastic data using the function *RockPhysicsGaussMixInversion*.

```
[~, ~, ~, Ppost] = RockPhysicsGaussMixInversion(Facies, mtrain,
dtrain, mdomain, dcond, sigmaerr);
```

The same approach can be applied assuming a Gaussian joint distribution of the model and data variables by applying the function *RockPhysicsGaussInversion*. Both functions return the full posterior distribution evaluated on the multidimensional grid for each data measurement. The marginal distributions and the maximum a posteriori are numerically computed as in the previous example.

In the fourth example, we numerically estimate the joint non-parametric distribution of the model and data variables from the training dataset using kernel density estimation. We first define a discretized domain for the evaluation of the joint and conditional distributions using a multidimensional grid.

```
ndiscr = 25;
phidomain = linspace(0, 0.4, ndiscr)';
cdomain = linspace(0, 0.8, ndiscr)';
swdomain = linspace(0, 1, ndiscr)';
mdomain = [phidomain cdomain swdomain];
vpdomain = linspace(min(Vp), max(Vp),ndiscr)';
vsdomain = linspace(min(Vs), max(Vs),ndiscr)';
rhodomain = linspace(min(Rho), max(Rho),ndiscr)';
ddomain =[vpdomain vsdomain rhodomain];
```

We define the kernel bandwidths as fractions of the ranges of the properties.

```
h = 5;
hm(1) = (max(phidomain)-min(phidomain))/h;
hm(2) = (max(cdomain)-min(cdomain))/h;
hm(3) = (max(swdomain)-min(swdomain))/h;
hd(1) = (max(vpdomain)-min(vpdomain))/h;
hd(2) = (max(vsdomain)-min(vsdomain))/h;
hd(3) = (max(rhodomain)-min(rhodomain))/h;
```

We then apply the Bayesian rock physics inversion described in Section 5.4.4 to obtain the posterior distribution of the petrophysical properties conditioned on the elastic data using the function *RockPhysicsKDEInversion*.

```
Ppost = RockPhysicsKDEInversion(mtrain, dtrain, mdomain,
ddomain, dcond, hm, hd);
```

The marginal distributions and the maximum a posteriori are then numerically computed as in the previous example.

A.3.3 Ensemble Smoother Inversion

The script *ESSeisInversionDriver* illustrates the implementation of the seismic inversion method based on the ES-MDA (Section 5.6.2) to predict the posterior distribution of elastic properties conditioned on seismic data. The inversion is applied to a set of three seismograms, including near-, mid- and far-angle stacks, saved in the workspace *data3*. The model properties of interest are P-wave and S-wave velocities and density.

We first load the data from the workspace and define the number of variables of the parameterization and the reflection angles of the seismograms. The number of samples of the model, for each elastic variable, is the number of samples of the seismogram plus 1. We also define the parameters of the distribution of the data error.

```
load Data/data3.mat
nm = size(Snear,1)+1;
nd = size(Snear,1);
nv = 3;
theta = [15, 30, 45];
ntheta = length(theta);
dt = TimeSeis(2)-TimeSeis(1);
varerr = 10^-4;
sigmaerr = varerr*eye(ntheta*(nm-1));
```

As for the Bayesian linearized AVO inversion (Section A.3.1), we define the wavelet, the prior model, and the spatial correlation function.

```
freq = 45;
ntw = 64;
[wavelet, tw] = RickerWavelet(freq, dt, ntw);
nfilt = 3;
cutofffr = 0.04;
[b, a] = butter(nfilt, cutofffr);
Vpprior = filtfilt(b, a, Vp);
Vsprior = filtfilt(b, a, Vs);
Rhoprior = filtfilt(b, a, Rho);
mprior = [Vpprior Vsprior Rhoprior];
corrlength = 5*dt;
trow = repmat(0:dt:(nm-1)*dt,nm,1);
tcol = repmat((0:dt:(nm-1)*dt)',1,nm);
tdis = trow-tcol;
sigmatime = exp(-(tdis./corrlength).^2);
sigma0 = cov([Vp,Vs,Rho]);
```

We generate an ensemble of prior realizations of the model variables using the function *CorrelatedSimulation* available in the *Geostats* folder and compute the seismic response of each realization.

```
nsim = 500;
for i=1:nsim
  msim = CorrelatedSimulation(mprior, sigma0, sigmatime);
  Vpsim(:,i) = msim(:, 1);
  Vssim(:,i) = msim(:, 2);
```

```
    Rhosim(:,i) = msim(:, 3);
    [SeisPred(:,i), TimeSeis] = SeismicModel (Vpsim(:,i), Vssim
(:,i), Rhosim(:,i), Time, theta, wavelet);
  end
```

We then apply the ES-MDA to obtain the ensemble of updated realizations of the model variables and their posterior mean.

```
  niter = 4;
  alpha = 1/niter;
  PriorModels = [Vpsim; Vssim; Rhosim];
  SeisData = [Snear; Smid; Sfar];
  PostModels = PriorModels;
  for j=1:niter
    [PostModels, KalmanGain] = EnsembleSmootherMDA(PostModels,
SeisData, SeisPred, alpha, sigmaerr);
    Vppost = PostModels(1:nm,:);
    Vspost = PostModels(nm+1:2*nm,:);
    Rhopost = PostModels(2*nm+1:end,:);
    for i=1:nsim
      [SeisPred(:,i), TimeSeis] = SeismicModel (Vppost(:,i),
Vspost(:,i), Rhopost(:,i), Time, theta, wavelet);
    end
  end
  mpost = mean(PostModels, 2);
  Vpmean = mpost(1:nm);
  Vsmean = mpost(nm+1:2*nm);
  Rhomean = mpost(2*nm+1:end);
```

The script *ESPetroInversionDriver* illustrates the implementation of the petrophysical inversion method based on the ES-MDA (Section 5.6.2) to predict the posterior distribution of petrophysical properties conditioned on seismic data. The inversion is applied to a set of three seismograms, including near-, mid- and far-angle stacks, saved in the workspace *data5*. The model properties of interest are porosity, clay volume, and water saturation.

The problem setting and the parameter initialization are the same as in the seismic case; however, this approach includes a rock physics model. For simplicity, we adopt a linearized rock physics formulation, but any model in the *RockPhysics* folder could be used. We generate an ensemble of prior realizations of the model variables using the function *CorrelatedSimulation*. For simplicity, we apply truncations to avoid unphysical values; however, a logit transformation is generally recommended.

```
  for i=1:nsim
    msim = CorrelatedSimulation(mprior, sigma0, sigmatime);
    Phisim(:,i) = msim(:,1);
    Claysim(:,i) = msim(:,2);
    Swsim(:,i) = msim(:,3);
  end
```

```
  Phisim(Phisim0.4)=0.4;
  Claysim(Claysim0.8)=0.8;
  Swsim(Swsim1)=1;
  [Vpsim, Vssim, Rhosim] = LinearizedRockPhysicsModel(Phisim,
Claysim, Swsim, R);
  for i=1:nsim
    [SeisPred(:,i), TimeSeis] = SeismicModel (Vpsim(:,i), Vssim
(:,i), Rhosim(:,i), Time, theta, wavelet);
  end
```

We then apply the ES-MDA to obtain the ensemble of updated realizations of the model variables and their posterior mean.

```
  niter = 4;
  alpha = 1/niter;
  PriorModels = [Phisim; Claysim; Swsim];
  SeisData = [Snear; Smid; Sfar];
  PostModels = PriorModels;
  for j=1:niter
    [PostModels, KalmanGain] = EnsembleSmootherMDA(PostModels,
SeisData, SeisPred, alpha, sigmaerr);
    Phipost = PostModels(1:nm,:);
    Claypost = PostModels(nm+1:2*nm,:);
    Swpost = PostModels(2*nm+1:end,:);
    Phipost(Phipost0.4)=0.4;
    Claypost(Claypost0.8)=0.8;
    Swpost(Swpost1)=1;
    [Vppost, Vspost, Rhopost] = LinearizedRockPhysicsModel
(Phipost, Claypost, Swpost, R);
    for i=1:nsim
      [SeisPred(:,i), TimeSeis] = SeismicModel (Vppost(:,i),
Vspost(:,i), Rhopost(:,i), Time, theta, wavelet);
    end
  end
  mpost = mean(PostModels, 2);
  Phimean = mpost(1:nm);
  Claymean = mpost(nm+1:2*nm);
  Swmean = mpost(2*nm+1:end);
```

A.4 Facies Modeling

The folder *Facies* includes the following functions for facies classification and simulation:

- *BayesGaussFaciesClass*: this function computes the Bayesian facies classification assuming a multivariate Gaussian distribution of the continuous properties.
- *BayesKDEFaciesClass*: this function computes the Bayesian facies classification assuming a multivariate non-parametric distribution of the continuous properties.

- *RandDisc*: this function simulates samples of a discrete random variable with given probability mass function.
- *ConfusionMatrix*: this function computes the classification confusion matrix.

The script *FaciesClassificationDriver* illustrates the Bayesian facies classification (Section 6.1) of a set of well logs, including P-wave velocity and density. In the first example, we assume a Gaussian distribution of the model properties in each facies, whereas in the second example, we adopt a non-parametric distribution estimated using kernel density estimation.

We first load the dataset from the workspace *data4* and define the domain of the variables for the evaluation of the likelihood function.

```
data = [Vp Rho];
ns = size(data,1);
nv = size(data,2);
v = (3:0.005:5);
r = (2:0.01:2.8);
[V, R] = meshgrid(v,r);
domain = [V(:) R(:)];
```

In the first example, we assume that there are two possible facies, namely sand and shale (facies 1 and 2), and that the model variables (P-wave velocity and density) are distributed according to a bivariate Gaussian distribution in each facies.

```
nf = max(unique(Facies));
fp = zeros(nf,1);
mup = zeros(nf,nv);
sp = zeros(nv,nv,nf);
for k=1:nf
  fp(k) = sum(Facies==k)/ns;
  mup(k,:) = mean(data(Facies==k,:));
  sp(:,:,k) = cov(data(Facies==k,:));
end
```

The likelihood function is evaluated for plotting purposes only. The numerical evaluation of the joint distribution is not necessary because the solution of the Bayesian classification can be analytically computed (Section 6.1). We then apply the Bayesian facies classification to obtain the posterior distribution of the facies using the function *BayesGaussFaciesClass*. The function returns the facies probability and the maximum a posteriori for each data measurement.

```
for k=1:nf
  lf = mvnpdf(domain,mup(k,:),sp(:,:,k));
  lf = lf/sum(lf);
  GaussLikeFun(:,:,k) = reshape(lf,length(r),length(v));
end
[fmap, fpost] = BayesGaussFaciesClass(data, fp, mup, sp);
```

By comparing the predictions to the actual classification provided in the workspace, we can perform the statistical contingency analysis and compute the confusion matrix, the reconstruction and recognition rates, and the estimation index as in Example 6.4.

```
confmat = ConfusionMatrix(Facies, fmap, nf );
reconstrate = confmat./repmat(sum(confmat,2),1,nf );
recognrate = confmat./repmat(sum(confmat,1),nf,1);
estimindex = reconstrate-recognrate;
```

In the second example, we use the same dataset as in the previous case, and we assume that the model variables (P-wave velocity and density) are distributed according to a bivariate non-parametric distribution in each facies. We apply the Bayesian facies classification to obtain the posterior distribution of the facies using the function *BayesKDEFaciesClass*, where we estimate the joint distribution of the data variables using kernel density estimation. As in the Gaussian case, the function returns the facies probability and the maximum a posteriori for each data measurement.

```
[fpred, fpost] = BayesKDEFaciesClass(data, dtrain, ftrain, fp, domain);
confmat = ConfusionMatrix(Facies, fmap, nf );
```

The driver illustrates the application to a dataset with two facies only; however, the code can be applied to any finite number of facies.

References

Aanonsen, S.I., Aavatsmark, I., Barkve, T., Cominelli, A., Gonard, R., Gosselin, O., Kolasinski, M. and Reme, H. (2003). Effect of scale dependent data correlations in an integrated history matching loop combining production data and 4D seismic data. *SPE Reservoir Simulation Symposium*, Houston, Texas (February 3–5). Society of Petroleum Engineers.

Aanonsen, S.I., Nævdal, G., Oliver, D.S., Reynolds, A.C. and Vallès, B. (2009). The ensemble Kalman filter in reservoir engineering – a review. *SPE Journal* **14** (03): 393–412.

Aki, K. and Richards, P.G. (2002). *Quantitative Seismology*. W. H. Freeman & Co.

Alabert, F. (1987). The practice of fast conditional simulations through the LU decomposition of the covariance matrix. *Mathematical Geology* **19** (5): 369–386.

Archie, G.E. (1942). The electrical resistivity log as an aid in determining some reservoir characteristics. *Transactions of AIME* **146** (01): 54–62.

Armstrong, M., Galli, A., Beucher, H., Loc'h, G., Renard, D., Doligez, B., Eschard, R. and Geffroy, F. (2011). *Plurigaussian Simulations in Geosciences*. Springer Science & Business Media.

Arpat, G.B. and Caers, J. (2007). Conditional simulation with patterns. *Mathematical Geology* **39** (2): 177–203.

Aster, R.C., Borchers, B., and Thurber, C.H. (2018). *Parameter Estimation and Inverse Problems*. Elsevier.

Avseth, P., Mukerji, T., and Mavko, G. (2010). *Quantitative Seismic Interpretation: Applying Rock Physics Tools to Reduce Interpretation Risk*. Cambridge University Press.

Ayani, M., Liu, M., and Grana, D. (2020). Stochastic inversion method of time-lapse controlled source electromagnetic data for CO_2 plume monitoring. *International Journal of Greenhouse Gas Control* **100**: 103098.

Aydin, O. and Caers, J.K. (2017). Quantifying structural uncertainty on fault networks using a marked point process within a Bayesian framework. *Tectonophysics* **712**: 101–124.

Azevedo, L. and Soares, A. (2017). *Geostatistical Methods for Reservoir Geophysics*. Springer.

Aziz, K. (1979). *Petroleum Reservoir Simulation*. Applied Science Publishers.

Bachrach, R. (2006). Joint estimation of porosity and saturation using stochastic rock-physics modeling. *Geophysics* **71** (5): O53–O63.

Batzle, M. and Wang, Z. (1992). Seismic properties of pore fluids. *Geophysics* **57** (11): 1396–1408.

Bergmo, P.E.S., Grimstad, A.A., and Lindeberg, E. (2011). Simultaneous CO_2 injection and water production to optimise aquifer storage capacity. *International Journal of Greenhouse Gas Control* **5** (3): 555–564.

Seismic Reservoir Modeling: Theory, Examples, and Algorithms, First Edition. Dario Grana, Tapan Mukerji, and Philippe Doyen.

Berryman, J.G. (1995). Mixture theories for rock properties. In: *Rock Physics and Phase Relations: A Handbook of Physical Constants*, vol. **3** (ed. T.J. Ahrens), 205–228. Wiley.

Bhakta, T. and Landrø, M. (2014). Estimation of pressure-saturation changes for unconsolidated reservoir rocks with high V_P/V_S ratio. *Geophysics* **79** (5): M35–M54.

Bhattacharjya, D., Eidsvik, J., and Mukerji, T. (2010). The value of information in spatial decision making. *Mathematical Geosciences* **42** (2): 141–163.

Bhuyian, A.H., Landrø, M., and Johansen, S.E. (2012). 3D CSEM modeling and time-lapse sensitivity analysis for subsurface CO_2 storage. *Geophysics* **77** (5): E343–E355.

Bickel, J.E., Gibson, R.L., McVay, D.A., Pickering, S. and Waggoner, J.R. (2008). Quantifying the reliability and value of 3D land seismic. *SPE Reservoir Evaluation & Engineering* **11** (05): 832–841.

Boggs, S. (2001). *Principles of Sedimentology and Stratigraphy*. Prentice-Hall.

Bornard, R., Allo, F., Coléou, T., Freudenreich, Y., Caldwell, D.H. and Hamman, J.G. (2005). Petrophysical seismic inversion to determine more accurate and precise reservoir properties. *SPE Europec/EAGE Annual Conference*, Madrid, Spain (June 13–16). Society of Petroleum Engineers.

Bortoli, L.J., Alabert, F., Haas, A., and Journel, A. (1993). Constraining stochastic images to seismic data. In: *Geostatistics Tróia '92* (ed. A. Soares), 325–337. Springer.

Bosch, M. (1999). Lithologic tomography: from plural geophysical data to lithology estimation. *Journal of Geophysical Research* **104**: 749–766.

Bosch, M., Carvajal, C., Rodrigues, J., Torres, A., Aldana, M. and Sierra, J. (2009). Petrophysical seismic inversion conditioned to well-log data: methods and application to a gas reservoir. *Geophysics* **74** (2): O1–O15.

Bosch, M., Mukerji, T., and González, E.F. (2010). Seismic inversion for reservoir properties combining statistical rock physics and geostatistics: a review. *Geophysics* **75** (5): 75A165–75A176.

Bourbié, T., Coussy, O., Zinszner, B., and Junger, M.C. (1992). *Acoustics of Porous Media*. Institut français du pétrole publications.

Bowman, A.W. and Azzalini, A. (1997). *Applied Smoothing Techniques for Data Analysis: The Kernel Approach with S-Plus Illustrations*. Oxford University Press.

Bratvold, R.B., Bickel, J.E., and Lohne, H.P. (2009). Value of information in the oil and gas industry: past, present, and future. *SPE Reservoir Evaluation & Engineering* **12** (04): 630–638.

Brie, A., Pampuri, F., Marsala, A.F., and Meazza, O. (1995). Shear sonic interpretation in gas-bearing sands. *SPE Annual Technical Conference and Exhibition*, Dallas, Texas (October 22–25). Society of Petroleum Engineers.

Brown, R.J. and Korringa, J. (1975). On the dependence of the elastic properties of a porous rock on the compressibility of the pore fluid. *Geophysics* **40** (4): 608–616.

Buland, A. and El Ouair, Y. (2006). Bayesian time-lapse inversion. *Geophysics* **71** (3): R43–R48.

Buland, A. and Kolbjørnsen, O. (2012). Bayesian inversion of CSEM and magnetotelluric data. *Geophysics* **77** (1): E33–E42.

Buland, A. and Omre, H. (2003). Bayesian linearized AVO inversion. *Geophysics* **68** (1): 185–198.

Buland, A., Kolbjørnsen, O., and Omre, H. (2003). Rapid spatially coupled AVO inversion in the Fourier domain. *Geophysics* **68** (3): 824–836.

Buland, A., Kolbjørnsen, O., Hauge, R., Skjæveland, Ø. and Duffaut, K. (2008). Bayesian lithology and fluid prediction from seismic prestack data. *Geophysics* **73** (3): C13–C21.

Caers, J. (2011). *Modeling Uncertainty in the Earth Sciences*. Wiley.

Caers, J. and Hoffman, T. (2006). The probability perturbation method: a new look at Bayesian inverse modeling. *Mathematical Geology* **38** (1): 81–100.

Calvert, R. (2005). *Insights and Methods for 4D Reservoir Monitoring and Characterization*. Society of Exploration Geophysicists and European Association of Geoscientists and Engineers.

Carman, P.C. (1937). Fluid flow through granular beds. *Transactions of the Institution of Chemical Engineers* **15**: 150–166.

Caumon, G., Collon-Drouaillet, P.L.C.D., De Veslud, C.L.C., Viseur, S. and Sausse, J. (2009). Surface-based 3D modeling of geological structures. *Mathematical Geosciences* **41** (8): 927–945.

Chen, J., Hoversten, G.M., Vasco, D., Rubin, Y. and Hou, Z. (2007). A Bayesian model for gas saturation estimation using marine seismic AVA and CSEM data. *Geophysics* **72** (2): WA85–WA95.

Chen, Y. and Oliver, D.S. (2012). Ensemble randomized maximum likelihood method as an iterative ensemble smoother. *Mathematical Geosciences* **44** (1): 1–26.

Chen, Y. and Oliver, D.S. (2017). Localization and regularization for iterative ensemble smoothers. *Computational Geosciences* **21** (1): 13–30.

Chilès, J.P. and Delfiner, P. (2009). *Geostatistics: Modeling Spatial Uncertainty*. Wiley.

Christensen, N.I. and Wang, H.F. (1985). The influence of pore pressure and confining pressure on dynamic elastic properties of Berea sandstone. *Geophysics* **50** (2): 207–213.

Claerbout, J.F. (1976). *Fundamentals of Geophysical Data Processing*, vol. **274**. McGraw-Hill.

Coléou, T., Allo, F., Bornard, R., Hamman, J. and Caldwell, D. (2005). Petrophysical seismic inversion. *SEG Technical Program Expanded Abstracts*: 1355–1358.

Connolly, P.A. and Hughes, M.J. (2016). Stochastic inversion by matching to large numbers of pseudo-wells. *Geophysics* **81** (2): M7–M22.

Connolly, P.A. and Kemper, M. (2007). Statistical uncertainty of seismic net pay estimations. *The Leading Edge* **26** (10): 1284–1289.

Constable, S. (2010). Ten years of marine CSEM for hydrocarbon exploration. *Geophysics* **75** (5): 75A67–75A81.

Daly, C. (2005). Higher order models using entropy, Markov random fields and sequential simulation. In: *Geostatistics Banff 2004* (eds. O. Leuangthong and C.V. Deutsch), 215–224. Springer.

Davis, M.W. (1987). Production of conditional simulations via the LU triangular decomposition of the covariance matrix. *Mathematical Geology* **19** (2): 91–98.

Davis, T.L., Landrø, M., and Wilson, M. (eds.) (2019). *Geophysics and Geosequestration*. Cambridge University Press.

Dell'Aversana, P. (2014). *Integrated Geophysical Models*. European Association of Geoscientists and Engineers.

Deutsch, C.V. (1992). Annealing techniques applied to reservoir modeling and the integration of geological and engineering (well test) data. PhD dissertation. Stanford University.

Deutsch, C.V. (2002). *Geostatistical Reservoir Modelling*. Oxford University Press.

Deutsch, C.V. and Journel, A.G. (1998). *GSLIB: Geostatistical Software Library and User's Guide*. Oxford University Press.

Doyen, P.M. (1988). Porosity from seismic data: a geostatistical approach. *Geophysics* **53** (10): 1263–1275.

Doyen, P.M. (2007). *Seismic Reservoir Characterization: An Earth Modelling Perspective*. European Association of Geoscientists and Engineers.

Doyen, P.M. and Den Boer, L.D. (1996). Bayesian sequential Gaussian simulation of lithology with non-linear data. US Patent 5,539,704.

Dubreuil-Boisclair, C., Gloaguen, E., Bellefleur, G., and Marcotte, D. (2012). Non-Gaussian gas hydrate grade simulation at the Mallik site, Mackenzie Delta, Canada. *Marine and Petroleum Geology* **35** (1): 20–27.

Dubrule, O. (2003). *Geostatistics for Seismic Data Integration in Earth Models.* Society of Exploration Geophysicists and European Association of Geoscientists and Engineers.

Duijndam, A.J.W. (1988a). Bayesian estimation in seismic inversion, part I: principles. *Geophysical Prospecting* **36**: 878–898.

Duijndam, A.J.W. (1988b). Bayesian estimation in seismic inversion, part II: uncertainty analysis. *Geophysical Prospecting* **36**: 899–918.

Dvorkin, J. (2008). Yet another V_S equation. *Geophysics* **73** (2): E35–E39.

Dvorkin, J. and Nur, A. (1996). Elasticity of high-porosity sandstones: theory for two North Sea data sets. *Geophysics* **61** (5): 1363–1370.

Dvorkin, J., Gutierrez, M.A., and Grana, D. (2014). *Seismic Reflections of Rock Properties.* Cambridge University Press.

Eberhart-Phillips, D., Han, D.H., and Zoback, M.D. (1989). Empirical relationships among seismic velocity, effective pressure, porosity, and clay content in sandstone. *Geophysics* **54** (1): 82–89.

Eidsvik, J., Avseth, P., Omre, H., Mukerji, T. and Mavko, G. (2004a). Stochastic reservoir characterization using prestack seismic data. *Geophysics* **69** (4): 978–993.

Eidsvik, J., Mukerji, T., and Switzer, P. (2004b). Estimation of geological attributes from a well log: an application of hidden Markov chains. *Mathematical Geology* **36** (3): 379–397.

Eidsvik, J., Bhattacharjya, D., and Mukerji, T. (2008). Value of information of seismic amplitude and CSEM resistivity. *Geophysics* **73** (4): R59–R69.

Eidsvik, J., Mukerji, T., and Bhattacharjya, D. (2015). *Value of Information in the Earth Sciences: Integrating Spatial Modeling and Decision Analysis.* Cambridge University Press.

Eigestad, G.T., Dahle, H.K., Hellevang, B., Riis, F., Johansen, W.T. and Øian, E. (2009). Geological modeling and simulation of CO_2 injection in the Johansen formation. *Computational Geosciences* **13** (4): 435.

Elfeki, A. and Dekking, M. (2001). A Markov chain model for subsurface characterization: theory and applications. *Mathematical Geology* **33** (5): 569–589.

Emerick, A.A. and Reynolds, A.C. (2012). History matching time-lapse seismic data using the ensemble Kalman filter with multiple data assimilations. *Computational Geosciences* **16** (3): 639–659.

Emerick, A.A. and Reynolds, A.C. (2013). Ensemble smoother with multiple data assimilation. *Computers & Geosciences* **55**: 3–15.

Evensen, G. (2009). *Data Assimilation: The Ensemble Kalman Filter.* Springer Science & Business Media.

Fernández Martínez, J.L., Mukerji, T., Garcia Gonzalo, E., and Suman, A. (2012). Reservoir characterization and inversion uncertainty via a family of particle swarm optimizers. *Geophysics* **77** (1): M1–M16.

de Figueiredo, L.P., Grana, D., Bordignon, F.L., Santos, M., Roisenberg, M. and Rodrigues, B.B. (2018). Joint Bayesian inversion based on rock-physics prior modeling for the estimation of spatially correlated reservoir properties. *Geophysics* **83** (5): M49–M61. https://doi.org/10.1190/geo2017-0463.1.

de Figueiredo, L.P., Grana, D., Roisenberg, M., and Rodrigues, B.B. (2019a). Gaussian mixture Markov chain Monte Carlo method for linear seismic inversion. *Geophysics* **84** (3): R463–R476.

de Figueiredo, L.P., Grana, D., Roisenberg, M., and Rodrigues, B.B. (2019b). Multimodal Markov chain Monte Carlo method for nonlinear petrophysical seismic inversion. *Geophysics* **84** (5): M1–M13.

Fjeldstad, T. and Grana, D. (2018). Joint probabilistic petrophysics-seismic inversion based on Gaussian mixture and Markov chain prior models. *Geophysics* **83** (1): R31–R42.

Flinchum, B.A., Holbrook, W.S., Grana, D., Parsekian, A.D., Carr, B.J., Hayes, J.L. and Jiao, J. (2018). Estimating the water holding capacity of the critical zone using near-surface geophysics. *Hydrological Processes* **32** (22): 3308–3326.

Frankel, A. and Clayton, R.W. (1986). Finite difference simulations of seismic scattering: implications for the propagation of short-period seismic waves in the crust and models of crustal heterogeneity. *Journal of Geophysical Research. Solid Earth* **91** (B6): 6465–6489.

Gal, D., Dvorkin, J., and Nur, A. (1998). A physical model for porosity reduction in sandstones. *Geophysics* **63** (2): 454–459.

Gao, G., Abubakar, A., and Habashy, T.M. (2012). Joint petrophysical inversion of electromagnetic and full-waveform seismic data. *Geophysics* **77** (3): WA3–WA18.

Gassmann, F. (1951). Über die elastizität poröser medien. *Quarterly publication of the Natural Research Society in Zurich* **96**: 1–23.

Gineste, M., Eidsvik, J., and Zheng, Y. (2020). Ensemble-based seismic inversion for a stratified medium. *Geophysics* **85** (1): R29–R39.

González, E.F., Mukerji, T., and Mavko, G. (2008). Seismic inversion combining rock physics and multiple-point geostatistics. *Geophysics* **73** (1): R11–R21.

Goodfellow, I., Bengio, Y., and Courville, A. (2016). *Deep Learning*. MIT Press.

Goovaerts, P. (1997). *Geostatistics for Natural Resources Evaluation*. Oxford University Press.

Gosselin, O., Aanonsen, S.I., Aavatsmark, I., Cominelli, A., Gonard, R., Kolasinski, M., Ferdinandi, F., Kovacic, L. and Neylon, K. (2003). History matching using time-lapse seismic (HUTS). *SPE Annual Technical Conference and Exhibition*, Denver, Colorado (October 5–8). Society of Petroleum Engineers.

Grana, D. (2016). Bayesian linearized rock-physics inversion. *Geophysics* **81** (6): D625–D641.

Grana, D. (2018). Joint facies and reservoir properties inversion. *Geophysics* **83** (3): M15–M24.

Grana, D. and Della Rossa, E. (2010). Probabilistic petrophysical-properties estimation integrating statistical rock physics with seismic inversion. *Geophysics* **75** (3): O21–O37. https://doi.org/10.1190/1.3386676.

Grana, D., Mukerji, T., Dovera, L., and Della Rossa, E. (2012a). Sequential simulations of mixed discrete-continuous properties: sequential Gaussian mixture simulation. In: *Geostatistics Oslo 2012* (eds. P. Abrahamsen, R. Hauge and O. Kolbjørnsen), 239–250. Springer.

Grana, D., Mukerji, T., Dvorkin, J., and Mavko, G. (2012b). Stochastic inversion of facies from seismic data based on sequential simulations and probability perturbation method. *Geophysics* **77** (4): M53–M72.

Grana, D., Fjeldstad, T., and Omre, H. (2017a). Bayesian Gaussian mixture linear inversion for geophysical inverse problems. *Mathematical Geosciences* **49** (4): 493–515.

Grana, D., Verma, S., Pafeng, J., Lang, X., Sharma, H., Wu, W., McLaughlin, F., Campbell, E., Ng, K., Alvarado, V. and Mallick, S. (2017b). A rock physics and seismic reservoir characterization study of the Rock Springs uplift, a carbon dioxide sequestration site in Southwestern Wyoming. *International Journal of Greenhouse Gas Control* **63**: 296–309.

Greenberg, M.L. and Castagna, J.P. (1992). Shear-wave velocity estimation in porous rocks: theoretical formulation, preliminary verification and applications. *Geophysical Prospecting* **40** (2): 195–209.

Grude, S., Landrø, M., and Osdal, B. (2013). Time-lapse pressure–saturation discrimination for CO2 storage at the Snøhvit field. *International Journal of Greenhouse Gas Control* **19**: 369–378.

Guardiano, F.B. and Srivastava, R.M. (1993). Multivariate geostatistics: beyond bivariate moments. In: *Geostatistics Tróia '92* (ed. A. Soares), 133–144. Springer.

Gunning, J. and Glinsky, M.E. (2004). Delivery: an open-source model-based Bayesian seismic inversion program. *Computers & Geosciences* **30** (6): 619–636.

Gunning, J. and Glinsky, M.E. (2007). Detection of reservoir quality using Bayesian seismic inversion. *Geophysics* **72** (3): R37–R49.

Gutierrez, M.A., Braunsdor, N.R., and Couzens, B.A. (2006). Calibration and ranking of pore-pressure prediction models. *The Leading Edge* **25** (12): 1516–1523.

Haas, A. and Dubrule, O. (1994). Geostatistical inversion – a sequential method of stochastic reservoir modelling constrained by seismic data. *First Break* **12** (11): 561–569.

Hale, D. (2009). A method for estimating apparent displacement vectors from time-lapse seismic images. *Geophysics* **74** (5): V99–V107.

Hall, B. (2016). Facies classification using machine learning. *The Leading Edge* **35** (10): 906–909.

Han, D.H. (1986). Effects of porosity and clay content on acoustic properties of sandstones and unconsolidated sediments. PhD dissertation. Stanford University.

Hansen, T.M. (2004). Mgstat: a geostatistical MATLAB toolbox. http://mgstat.sourceforge.net.

Hansen, T.M., Journel, A.G., Tarantola, A., and Mosegaard, K. (2006). Linear inverse Gaussian theory and geostatistics. *Geophysics* **71** (6): R101–R111.

Harris, P.E. and MacGregor, L.M. (2006). Determination of reservoir properties from the integration of CSEM, seismic, and well-log data. *First Break* **24** (11): 15–21.

Hastie, T., Tibshirani, R., and Friedman, J. (2009). *The Elements of Statistical Learning: Data Mining, Inference, and Prediction*. Springer Science & Business Media.

Honarkhah, M. and Caers, J. (2010). Stochastic simulation of patterns using distance-based pattern modeling. *Mathematical Geosciences* **42** (5): 487–517.

Hoversten, G.M., Cassassuce, F., Gasperikova, E., Newman, G.A., Chen, J., Rubin, Y., Hou, Z. and Vasco, D. (2006). Direct reservoir parameter estimation using joint inversion of marine seismic AVA and CSEM data. *Geophysics* **71** (3): C1–C13.

Hu, L.Y. (2000). Gradual deformation and iterative calibration of Gaussian-related stochastic models. *Mathematical Geology* **32** (1): 87–108.

Hu, L.Y. (2008). Extended probability perturbation method for calibrating stochastic reservoir models. *Mathematical Geosciences* **40** (8): 875–885.

Hudson, J.A. (1981). Wave speeds and attenuation of elastic waves in material containing cracks. *Geophysical Journal International* **64** (1): 133–150.

Ikelle, L.T., Yung, S.K., and Daube, F. (1993). 2-D random media with ellipsoidal autocorrelation functions. *Geophysics* **58** (9): 1359–1372.

Isaaks, E.H. and Srivastava, R.M. (1989). *An Introduction to Applied Geostatistics*. Oxford University Press.

Jeong, C., Mukerji, T., and Mariethoz, G. (2017). A fast approximation for seismic inverse modeling: adaptive spatial resampling. *Mathematical Geoscience* **49**: 845–869.

Journel, A.G. (2002). Combining knowledge from diverse sources: an alternative to traditional data independence hypotheses. *Mathematical Geology* **34** (5): 573–596.

Journel, A.G. and Gomez-Hernandez, J.J. (1993). Stochastic imaging of the Wilmington clastic sequence. *SPE Formation Evaluation* **8** (01): 33–40.

Journel, A.G. and Huijbregts, C.J. (1978). *Mining Geostatistics*. Academic press.

Kemper, M. and Gunning, J. (2014). Joint impedance and facies inversion–seismic inversion redefined. *First Break* **32** (9): 89–95.

Kennett, B.L.N. (1984). Guided wave propagation in laterally varying media – I. theoretical development. *Geophysical Journal International* **79** (1): 235–255.

Key, K. and Ovall, J. (2011). A parallel goal-oriented adaptive finite element method for 2.5-D electromagnetic modelling. *Geophysical Journal International* **186** (1): 137–154.

Kitanidis, P.K. (1997). *Introduction to Geostatistics: Applications in Hydrogeology*. Cambridge University Press.

Kozeny, J. (1927). Über kapillare Leitung der Wasser im Boden. *Royal Academy of Science, Vienna, Proceedings Class I* **136**: 271–306.

Krief, M., Garat, J., Stellingwerff, J. and Ventre, J. (1990). A petrophysical interpretation using the velocities of P and S waves (full-waveform sonic). *The Log Analyst* **31** (06): 355–369.

Krumbein, W.C. and Dacey, M.F. (1969). Markov chains and embedded Markov chains in geology. *Journal of the International Association for Mathematical Geology* **1** (1): 79–96.

Kuster, G.T. and Toksöz, M.N. (1974). Velocity and attenuation of seismic waves in two-phase media: part I. theoretical formulations. *Geophysics* **39** (5): 587–606.

Landrø, M. (2001). Discrimination between pressure and fluid saturation changes from time-lapse seismic data. *Geophysics* **66** (3): 836–844.

Landrø, M. (2015). 4D seismic. In: *Petroleum Geoscience* (ed. K. Bjørlykke), 489–514. Springer.

Landrø, M., Veire, H.H., Duffaut, K., and Najjar, N. (2003). Discrimination between pressure and fluid saturation changes from marine multicomponent time-lapse seismic data. *Geophysics* **68** (5): 1592–1599.

Lang, X. and Grana, D. (2018). Bayesian linearized petrophysical AVO inversion. *Geophysics* **83** (3): M1–M13.

Lantuéjoul, C. (2013). *Geostatistical Simulation: Models and Algorithms*. Springer Science & Business Media.

Larsen, A.L., Ulvmoen, M., Omre, H., and Buland, A. (2006). Bayesian lithology/fluid prediction and simulation on the basis of a Markov-chain prior model. *Geophysics* **71** (5): R69–R78.

Le Ravalec, M. (2005). *Inverse Stochastic Modeling of Flow in Porous Media*. Editions Technip.

Le Ravalec, M., Noetinger, B., and Hu, L.Y. (2000). The FFT moving average (FFT-MA) generator: an efficient numerical method for generating and conditioning Gaussian simulations. *Mathematical Geology* **32** (6): 701–723.

Leeuwenburgh, O. and Arts, R. (2014). Distance parameterization for efficient seismic history matching with the ensemble Kalman filter. *Computational Geosciences* **18** (3–4): 535–548.

Lie, K.A. (2019). *An Introduction to Reservoir Simulation Using MATLAB/GNU Octave: User Guide for the MATLAB Reservoir Simulation Toolbox (MRST)*. Cambridge University Press.

Lien, M. and Mannseth, T. (2008). Sensitivity study of marine CSEM data for reservoir production monitoring. *Geophysics* **73** (4): F151–F163.

Lindberg, D.V. and Omre, H. (2014). Blind categorical deconvolution in two-level hidden Markov models. *IEEE Transactions on Geoscience and Remote Sensing* **52** (11): 7435–7447.

Liu, M. and Grana, D. (2018). Stochastic nonlinear inversion of seismic data for the estimation of petroelastic properties using the ensemble smoother and data reparameterization. *Geophysics* **83** (3): M25–M39. https://doi.org/10.1190/geo2017-0713.1.

Liu, M. and Grana, D. (2020). Petrophysical characterization of deep saline aquifers for CO_2 storage using ensemble smoother and deep convolutional autoencoder. *Advances in Water Resources* **142**: 103634.

Lumley, D.E. (2001). Time-lapse seismic reservoir monitoring. *Geophysics* **66** (1): 50–53.

MacBeth, C. (2004). A classification for the pressure-sensitivity properties of a sandstone rock frame. *Geophysics* **69** (2): 497–510.

MacGregor, L. (2012). Integrating seismic, CSEM and well log data for reservoir characterization. *The Leading Edge* **31** (3): 258–265.

MacGregor, L. and Tomlinson, J. (2014). Marine controlled-source electromagnetic methods in the hydrocarbon industry: a tutorial on method and practice. *Interpretation* **2** (3): SH13–SH32.

Maharramov, M., Biondi, B.L., and Meadows, M.A. (2016). Time-lapse inverse theory with applications. *Geophysics* **81** (6): R485–R501.

Mallet, J.L. (2008). *Numerical Earth Models*. European Association of Geoscientists and Engineers.

Mallick, S. (1995). Model-based inversion of amplitude-variations-with-offset data using a genetic algorithm. *Geophysics* **60** (4): 939–954.

Mallick, S. (1999). Some practical aspects of prestack waveform inversion using a genetic algorithm: an example from the East Texas woodbine gas sand. *Geophysics* **64** (2): 326–336.

Mariethoz, G. and Caers, J. (2014). *Multiple-Point Geostatistics: Stochastic Modeling with Training Images*. Wiley.

Mariethoz, G., Renard, P., and Straubhaar, J. (2010). The direct sampling method to perform multiple-point geostatistical simulations. *Water Resources Research* **46** (11).

Matheron, G. (1969). *Le krigeage universel*. École nationale supérieure des mines de Paris.

Matheron, G. (1970). *La théorie des variables régionalisées et ses applications*. Masson.

Matheron, G. (1973). The intrinsic random functions and their applications. *Advances in Applied Probability* **5** (3): 439–468.

Mavko, G. and Nur, A. (1997). The effect of a percolation threshold in the Kozeny-Carman relation. *Geophysics* **62** (5): 1480–1482.

Mavko, G., Chan, C., and Mukerji, T. (1995). Fluid substitution: estimating changes in V_p without knowing V_s. *Geophysics* **60** (6): 1750–1755.

Mavko, G., Mukerji, T., and Dvorkin, J. (2020). *The Rock Physics Handbook*. Cambridge University Press.

Mazzotti, A. and Zamboni, E. (2003). Petrophysical inversion of AVA data. *Geophysical Prospecting* **51** (6): 517–530.

Meadows, M.A. and Cole, S.P. (2013). 4D seismic modeling and CO_2 pressure-saturation inversion at the Weyburn field, Saskatchewan. *International Journal of Greenhouse Gas Control* **16**: S103–S117.

Menke, W. (2018). *Geophysical Data Analysis: Discrete Inverse Theory*. Academic Press.

Mosegaard, K. (1998). Resolution analysis of general inverse problems through inverse Monte Carlo sampling. *Inverse Problems* **14** (3): 405.

Mosegaard, K. and Tarantola, A. (1995). Monte Carlo sampling of solutions to inverse problems. *Journal of Geophysical Research. Solid Earth* **100** (B7): 12431–12447.

Mukerji, T., Mavko, G., Mujica, D., and Lucet, N. (1995). Scale-dependent seismic velocity in heterogeneous media. *Geophysics* **60** (4): 1222–1233.

Mukerji, T., Jørstad, A., Avseth, P., Mavko, G. and Granli, J.R. (2001). Mapping lithofacies and pore-fluid probabilities in a North Sea reservoir: seismic inversions and statistical rock physics. *Geophysics* **66** (4): 988–1001.

Nur, A. (1971). Effects of stress on velocity anisotropy in rocks with cracks. *Journal of Geophysical Research* **76** (8): 2022–2034.

Nur, A., Marion, D., and Yin, H. (1991). Wave velocities in sediments. In: *Shear Waves in Marine Sediments* (eds. J.M. Hovem, M.D. Richardson and R.D. Stoll), 131–140. Springer.

Okabe, H. and Blunt, M.J. (2005). Pore space reconstruction using multiple-point statistics. *Journal of Petroleum Science and Engineering* **46** (1–2): 121–137.

Oliver, D.S. (1995). Moving averages for Gaussian simulation in two and three dimensions. *Mathematical Geology* **27** (8): 939–960.

Oliver, D.S. and Chen, Y. (2011). Recent progress on reservoir history matching: a review. *Computational Geosciences* **15** (1): 185–221.

Oliver, D.S., Reynolds, A.C., and Liu, N. (2008). *Inverse Theory for Petroleum Reservoir Characterization and History Matching*. Cambridge University Press.

Omre, H. (1987). Bayesian kriging – merging observations and qualified guesses in kriging. *Mathematical Geology* **19** (1): 25–39.

Orange, A., Key, K., and Constable, S. (2009). The feasibility of reservoir monitoring using time-lapse marine CSEM. *Geophysics* **74** (2): F21–F29.

Papoulis, A. and Pillai, S.U. (2002). *Probability, Random Variables, and Stochastic Processes*. McGraw-Hill.

Pardo-Iguzquiza, E. and Chica-Olmo, M. (1993). The Fourier integral method: an efficient spectral method for simulation of random fields. *Mathematical Geology* **25** (2): 177–217.

Pebesma, E.J. (2004). Multivariable geostatistics in S: the gstat package. *Computers & Geosciences* **30** (7): 683–691.

Peredo, O. and Ortiz, J.M. (2011). Parallel implementation of simulated annealing to reproduce multiple-point statistics. *Computers & Geosciences* **37** (8): 1110–1121.

Poupon, A. and Leveaux, J. (1971). Evaluation of water saturations in shaly formations. *The Log Analyst* **12** (4): 1–6.

Pyrcz, M.J. and Deutsch, C.V. (2014). *Geostatistical Reservoir Modeling*. Oxford University Press.

Ray, A. and Key, K. (2012). Bayesian inversion of marine CSEM data with a trans-dimensional self-parametrizing algorithm. *Geophysical Journal International* **191** (3): 1135–1151.

Raymer, L.L., Hunt, E.R., and Gardner, J.S. (1980). An improved sonic transit time-to-porosity transform. *SPWLA 21st Annual Logging Symposium*, Lafayette, Louisiana (July 8–11). Houston, TX: Society of Petrophysicists and Well-Log Analysts.

Remy, N., Boucher, A., and Wu, J. (2009). *Applied Geostatistics with SGeMS: a User's Guide*. Cambridge University Press.

Rimstad, K. and Omre, H. (2010). Impact of rock-physics depth trends and Markov random fields on hierarchical Bayesian lithology/fluid prediction. *Geophysics* **75** (4): R93–R108.

Rimstad, K., Avseth, P., and Omre, H. (2012). Hierarchical Bayesian lithology/fluid prediction: a North Sea case study. *Geophysics* **77** (2): B69–B85.

Ringrose, P. (2020). *How to Store CO_2 Underground: Insights from Early-Mover CCS Projects*. Springer.

Rittgers, J.B., Revil, A., Mooney, M.A., Karaoulis, M., Wodajo, L. and Hickey, C.J. (2016). Time-lapse joint inversion of geophysical data with automatic joint constraints and dynamic attributes. *Geophysical Journal International* **207** (3): 1401–1419.

Roggero, F., Ding, D.Y., Berthet, P., Lerat, O., Cap, J. and Schreiber, P.E. (2007). Matching of production history and 4D seismic data – application to the Girassol Field, offshore Angola.

SPE Annual Technical Conference and Exhibition, Anaheim, California (November 11–14). Society of Petroleum Engineers.

Russell, B.H. (1988). *Introduction to Seismic Inversion Methods.* Society of Exploration Geophysicists.

Sambridge, M. and Mosegaard, K. (2002). Monte Carlo methods in geophysical inverse problems. *Reviews of Geophysics* **40** (3): 3–1.

Sayers, C.M. (2006). Sensitivity of time-lapse seismic to reservoir stress path. *Geophysical Prospecting* **54** (3): 369–380.

Sayers, C.M. (2010). *Geophysics under Stress: Geomechanical Applications of Seismic and Borehole Acoustic Waves.* Society of Exploration Geophysicists and European Association of Geoscientists and Engineers.

Scales, J.A. and Tenorio, L. (2001). Prior information and uncertainty in inverse problems. *Geophysics* **66** (2): 389–397.

Schoenberg, M. and Sayers, C.M. (1995). Seismic anisotropy of fractured rock. *Geophysics* **60** (1): 204–211.

Sen, M.K. and Stoffa, P.L. (1991). Nonlinear one-dimensional seismic waveform inversion using simulated annealing. *Geophysics* **56** (10): 1624–1638.

Sen, M.K. and Stoffa, P.L. (2013). *Global Optimization Methods in Geophysical Inversion.* Cambridge University Press.

Shahin, A., Key, K., Stoffa, P., and Tatham, R. (2012). Petro-electric modeling for CSEM reservoir characterization and monitoring. *Geophysics* **77** (1): E9–E20.

Sheriff, R.E. and Geldart, L.P. (1995). *Exploration Seismology.* Cambridge University Press.

Shuey, R.T. (1985). A simplification of the Zoeppritz equations. *Geophysics* **50** (4): 609–614.

Silverman, B.W. (1986). *Density Estimation for Statistics and Data Analysis.* CRC Press.

Simandoux, P. (1963). Dielectric measurements on porous media, application to the measurements of water saturation: study of behavior of argillaceous formations. *Revue de l'Institut Français du Pétrole* **18**: 93–215.

Simm, R. and Bacon, M. (2014). *Seismic Amplitude: An Interpreter's Handbook.* Cambridge University Press.

Skjervheim, J.A., Evensen, G., Aanonsen, S.I., Ruud, B.O. and Johansen, T.A. (2007). Incorporating 4D seismic data in reservoir simulation models using ensemble Kalman filter. *SPE Journal* **12** (3): 282–292.

Soares, A. (2001). Direct sequential simulation and cosimulation. *Mathematical Geology* **33** (8): 911–926.

Spikes, K., Mukerji, T., Dvorkin, J., and Mavko, G. (2007). Probabilistic seismic inversion based on rock-physics models. *Geophysics* **72** (5): R87–R97.

Srivastava, R.M. (1992). Reservoir characterization with probability field simulation. *SPE Annual Technical Conference and Exhibition*, Washington, DC (October 4–7). Society of Petroleum Engineers.

Stephen, K.D., Soldo, J., Macbeth, C., and Christie, M.A. (2006). Multiple model seismic and production history matching: a case study. *SPE Journal* **11** (04): 418–430.

Stien, M. and Kolbjørnsen, O. (2011). Facies modeling using a Markov mesh model specification. *Mathematical Geosciences* **43** (6): 611.

Stolt, R.H. and Weglein, A.B. (1985). Migration and inversion of seismic data. *Geophysics* **50** (12): 2458–2472.

Strebelle, S.B. (2002). Conditional simulation of complex geological structures using multiple-point statistics. *Mathematical Geology* **34** (1): 1–21.

Strebelle, S.B. and Journel, A.G. (2001). Reservoir modeling using multiple-point statistics. *SPE Annual Technical Conference and Exhibition*, New Orleans, Louisiana (September 30 – October 3). Society of Petroleum Engineers.

Tahmasebi, P., Hezarkhani, A., and Sahimi, M. (2012). Multiple-point geostatistical modeling based on the cross-correlation functions. *Computational Geosciences* **16** (3): 779–797.

Tahmasebi, P., Sahimi, M., and Caers, J. (2014). MS-CCSIM: accelerating pattern-based geostatistical simulation of categorical variables using a multi-scale search in Fourier space. *Computers & Geosciences* **67**: 75–88.

Tarantola, A. (2005). *Inverse Problem Theory and Methods for Model Parameter Estimation*. Society for Industrial and Applied Mathematics.

Tarantola, A. and Valette, B. (1982). Inverse problems = quest for information. *Journal of Geophysics* **50** (1): 159–170.

Te Wu, T. (1966). The effect of inclusion shape on the elastic moduli of a two-phase material. *International Journal of Solids and Structures* **2** (1): 1–8.

Terzaghi, K. (1943). *Theoretical Soil Mechanics*. Wiley.

Thomsen, L. (1986). Weak elastic anisotropy. *Geophysics* **51** (10): 1954–1966.

Thore, P. and Blanchard, T.D. (2015). 4D propagated layer-based inversion. *Geophysics* **80** (1): R15–R29.

Thore, P. and Hubans, C. (2012). 4D seismic-to-well tying, a key step towards 4D inversion. *Geophysics* **77** (6): R227–R238.

Thore, P., Shtuka, A., Lecour, M., Ait-Ettajer, T. and Cognot, R. (2002). Structural uncertainties: determination, management, and applications. *Geophysics* **67** (3): 840–852.

Thurin, J., Brossier, R., and Métivier, L. (2019). Ensemble-based uncertainty estimation in full waveform inversion. *Geophysical Journal International* **219** (3): 1613–1635.

Tjelmeland, H. and Besag, J. (1998). Markov random fields with higher-order interactions. *Scandinavian Journal of Statistics* **25** (3): 415–433.

Toftaker, H. and Tjelmeland, H. (2013). Construction of binary multi-grid Markov random field prior models from training images. *Mathematical Geosciences* **45** (4): 383–409.

Torres-Verdín, C., Victoria, M., Merletti, G. and Pendrel, J. (1999). Trace-based and geostatistical inversion of 3-D seismic data for thin-sand delineation: an application in San Jorge Basin, Argentina. *The Leading Edge* **18** (9): 1070–1077.

Trainor-Guitton, W. and Hoversten, G.M. (2011). Stochastic inversion for electromagnetic geophysics: practical challenges and improving convergence efficiency. *Geophysics* **76** (6): F373–F386.

Trani, M., Arts, R., Leeuwenburgh, O., and Brouwer, J. (2011). Estimation of changes in saturation and pressure from 4D seismic AVO and time-shift analysis. *Geophysics* **76** (2): C1–C17.

Tsvankin, I. (2012). *Seismic Signatures and Analysis of Reflection Data in Anisotropic Media*. Society of Exploration Geophysicists.

Tveit, S., Bakr, S.A., Lien, M., and Mannseth, T. (2015). Ensemble-based Bayesian inversion of CSEM data for subsurface structure identification. *Geophysical Journal International* **201** (3): 1849–1867.

Tveit, S., Mannseth, T., Park, J., Sauvin, G. and Agersborg, R. (2020). Combining CSEM or gravity inversion with seismic AVO inversion, with application to monitoring of large-scale CO_2 injection. *Computational Geosciences* **24**: 1201–1220.

Ulrych, T.J., Sacchi, M.D., and Woodbury, A. (2001). A Bayes tour of inversion: a tutorial. *Geophysics* **66** (1): 55–69.

Ulvmoen, M. and Omre, H. (2010). Improved resolution in Bayesian lithology/fluid inversion from prestack seismic data and well observations: part 1 – methodology. *Geophysics* **75** (2): R21–R35.

Ursin, B. and Stovas, A. (2006). Traveltime approximations for a layered transversely isotropic medium. *Geophysics* **71** (2): D23–D33.

Van Leeuwen, P.J. and Evensen, G. (1996). Data assimilation and inverse methods in terms of a probabilistic formulation. *Monthly Weather Review* **124** (12): 2898–2913.

Veire, H.H., Borgos, H.G., and Landrø, M. (2006). Stochastic inversion of pressure and saturation changes from time-lapse AVO data. *Geophysics* **71** (5): C81–C92.

Wackernagel, H. (2003). *Multivariate Geostatistics*. Springer.

Weitemeyer, K.A., Constable, S.C., Key, K.W. and Behrens, J.P. (2006). First results from a marine controlled-source electromagnetic survey to detect gas hydrates offshore Oregon. *Geophysical Research Letters* **33** (3): L03304.

Worthington, P.F. (1985). Evolution of shaley sand concepts in reservoir evaluation. *The Log Analyst* **26**: 23–40.

Wrona, T., Pan, I., Gawthorpe, R.L., and Fossen, H. (2018). Seismic facies analysis using machine learning. *Geophysics* **83** (5): O83–O95.

Wyllie, M.R.J., Gregory, A.R., and Gardner, L.W. (1956). Elastic wave velocities in heterogeneous and porous media. *Geophysics* **21** (1): 41–70.

Xu, W. and Journel, A.G. (1993). GTSIM: Gaussian truncated simulations of reservoir units in a W. Texas carbonate field. Society of Petroleum Engineers.

Xu, W., Tran, T.T., Srivastava, R.M. and Journel, A.G. (1992). Integrating seismic data in reservoir modeling: the collocated cokriging alternative. *SPE Annual Technical Conference and Exhibition*, Washington, DC (October 4–7). Society of Petroleum Engineers.

Yao, T. (1998). Conditional spectral simulation with phase identification. *Mathematical Geology* **30** (3): 285–308.

Yilmaz, Ö. (2001). *Seismic Data Analysis: Processing, Inversion, and Interpretation of Seismic Data*. Society of Exploration Geophysicists.

Yin, H. (1992), Acoustic velocity and attenuation of rocks: isotropy, intrinsic anisotropy, and stress-induced anisotropy. PhD dissertation. Stanford University.

Zhang, T., Switzer, P., and Journel, A. (2006). Filter-based classification of training image patterns for spatial simulation. *Mathematical Geology* **38** (1): 63–80.

Zimmerman, R.W. (1990). *Compressibility of Sandstones*. Elsevier.

Zoback, M.D. (2010). *Reservoir Geomechanics*. Cambridge University Press.

Zoeppritz, K. (1919). On the reflection and propagation of seismic waves. *Gottinger Nachrichten* **1** (5): 66–84.

Zunino, A., Lange, K., Melnikova, Y., Hansen, T.M. and Mosegaard, K. (2014). Reservoir modeling combining geostatistics with Markov chain Monte Carlo inversion. In: *Mathematics of Planet Earth* (eds. E. Pardo-Igúzquiza, C. Guardiola-Albert, J. Heredia, L. Moreno-Merino, J.J. Durán and J.A. Vargas-Guzmán), 683–687. Springer.

Index

Seismic Reservoir Modeling: Theory, Examples, and Algorithms, First Edition. Dario Grana, Tapan Mukerji, and Philippe Doyen.